Conversion Factors
From Imperial Units to SI - Units

Units	Imperial Units	SI - Units
MOMENTS	1 ft.kip	1.356 kN.m
	1 ft.kip	138.2 kg.m
UNIFORM LOADING	1 lb/in	0.1785 kg/m
	1 lb/ft	1.488 kg/m
	1 k/in	1.216 kN/m
	1 k/in	178.5 kg/m
	1 k/ft	14.59 kN/m
DENSITY	$1 \ lb/in^3$	$0.0277 \ kg/cm^3$
	$1 \ lb/ft^3$	$16.018 \ kg/m^3$
TEMPERATURE	T_F	$1.8T_C + 32$
	T_C	$\dfrac{T_F - 32}{1.8}$

PRESTRESSED STEEL BRIDGES
BRIDGES
THEORY AND DESIGN

PRESTRESSED STEEL BRIDGES THEORY AND DESIGN

M. S. Troitsky, D.Sc
Professor of Engineering
Concordia University
Montreal

BRIDGESERIES

VAN NOSTRAND REINHOLD COMPANY
New York

Printed in the United States of America

Van Nostrand Reinhold
115 Fifth Avenue
New York, New York 10003

Van Nostrand Reinhold International Company Limited
11 New Fetter Lane
London EC4P 4EE, England

Van Nostrand Reinhold
480 La Trobe Street
Melbourne, Victoria 3000, Australia

Nelson Canada
1120 Birchmount Road
Scarborough, Ontario M1K 5G4, Canada

16 15 14 13 12 11 10 9 8 7 6 5 4 3 2 1

Library of Congress Cataloging-in-Publication Data

Troitsky, M. S.
  Prestressed steel bridges : theory and design / by M. S. Troitsky.
    p. cm.
  Includes bibliographies and index.
  ISBN 0-442-31922-3
  1. Bridges, Iron and steel—Design and construction.
2. Prestressed steel construction. I. Title.
TG380.T76 1990
624'.25—dc20   89-16479
                                                   CIP

To the memory of my parents
Lydia and Serge

Contents

Foreword

During the nineteenth century timber, stone, and masonry were the common materials for bridge construction. Then iron, steel, concrete, and reinforced concrete emerged successively as favorite materials, culminating with the use of prestressed concrete in the twentieth century. For many years, in the long span bridge field, structural steel was the dominant material of choice, it might even be said that structural steel had a monopoly on long span bridges. However, after World War II, starting in about the 1950s in Europe and later in about the 1970s in the United States, two new concepts evolved. These were the emergence of the cable-stayed and the prestressed concrete segmental type of bridge construction.

Today's implementation of concrete cable-stayed bridges and prestressed concrete segmental bridges has not only penetrated the long span bridge market, but also threatens to replace structural steel as the dominant material of choice. The only totally structural steel cable-stayed bridge in the United States was the Luling Bridge in Louisiana, completed in 1983. Since then the so-called steel cable-stayed bridges have been, in reality, composite structures. Aside from the cable stays, structural steel is only found in the longitudinal edge girders and floor beams. The application of prestressing to steel bridges, as presented in this volume, will provide increased competitiveness for structural steel and, in all probability, emerge as a major technological advancement in the next decade.

In order for the reader, whether a seasoned veteran or a novice, to comprehend the theory and design of prestressed steel bridges requires a basic understanding of the concept of prestressing. The basic principle of prestressing is that of the introduction of internal stresses of such magnitude and distribution that the stresses resulting from given external loadings are counteracted to a desired degree. That is to say, the theory of prestressed design provides for the elimination of undesirable stresses in a load carrying structure or member by introducing into it artificial stresses which are directly opposed to those to which it will be subjected when the load is imposed. The key word is "principle." There are obvious differences in the application of the principle to various materials, however, the principle remains valid regardless of what material is being used. There is an apparent "mental block" tendency which considers prestressing as being applicable only to concrete. In fact, it has been and is being applied to structural steel. Prestressing is being used for timber structures, and in theory could be applied to other materials such as plastics. Prestressing is a basic principle of design for structures of all kinds and materials.

In the case of prestressed steel bridges the elimination of these undesirable stresses makes possible savings in the quantity of structural steel required to support a given load, as compared with conventional structural steel design. The savings in structural steel weight has been estimated to be in a range of approximately 10 to 30 percent, depending on the specifics of application in a particular design or project.

The concept of prestressing steel structures has been traced back to Squire Whipple who, in 1837, overcame the brittleness of cast iron tension members in trusses by prestressing them. During the late 1800s and early 1900s bridge designers, undoubtedly without the realization of the principles of prestressing, strengthened bridges by the use of the king and queen post trussed floor beams. Starting in about the mid 1930s, F. Dischinger in Germany, Gustave Magnel in Belgium and others in Europe began to study, investigate, test, and construct prestressed steel structures.

Magnel in an article titled ''Precompressed Structural Steel Construction'' published in the June 1950 issue of *L'Ossature Metalique* (Brussels) urged United States engineers to explore the use of prestressed steel. In another article titled ''Prestressed Steel Structures'' published in *The Structural Engineer* (November 1950) he stated:

> The specialists in structural steel have the great advantage of having at their disposal proper means of prestressing. The girder tested in our laboratory has been prestressed with the same sandwich cable, which we are using for prestressed concrete and with the same jacks. This fortunate fact will allow prestressed steel to make much quicker progress than has been possible in prestressed concrete.

Numerous articles and papers have appeared in the technical literature. Rudolph Szilard wrote ''Strengthening Steel Structures by Means of Prestressing'' (*The Engineering Journal* (Canada), October 1955) and ''Design of Prestressed Composite Steel Structures'' (*Journal of the Structural Division*, ASCE, November 1959). Subcommittee 3 on Prestressed Steel of Joint ASCE-AASHO Committee on Steel Flexural Members published a report on the ''Development and Use of Prestressed Steel Flexural Members'' in the ASCE *Journal of the Structural Division* (September 1968). Two articles by Homer M. Hadley in the ASCE *Civil Engineering Magazine* (May 1960 and May 1966) describe the design and construction of two prestressed steel bridges in the state of Washington.

The June 1960 issue of *Transportnoye Stroitelstvo* describes a prestressed steel bridge over the River Tom in western Siberia. This bridge consists of five spans, one of 240 ft., three of 358 ft. and one span of 240 ft., and was designed for vehicular traffic as well as for use as a temporary railway siding. The use of prestressed steel bridges of this magnitude had not previously been attempted in the Soviet Union. The reader will note from the examples and references in this book that the Soviets are undoubtedly the leaders in this technology.

The reader will also note that all of the developmental activity in prestressed steel parallels, at least in the time period under consideration, the development of prestressed concrete and may question why the prestressing of steel structures never reached a stage of exploitation similar to that of prestressed concrete. Certainly there is no technological reason why the concept of prestressing steel structures should not have advanced further than it has. Within the context of the historical time period, it must be remembered that structural steel was the dominant material and reinforced concrete was not very competitive, at least in the medium to long span range of bridges. From a commercial point of view, the use of prestressing for steel bridges would only reduce the amount of structural steel necessary, and thus be of no commercial advantage to the steel industry. The rapid technological advances of prestressed concrete, coupled with the concepts of segmental and cable-stayed bridge construction of the last two decades in the United States, have placed prestressed concrete in a very strong competitive position against structural steel, and recent trends may have made it the more dominant material. This course of events represents a classical example of a dominant competitor underestimating its competition.

The resurrection of the concept of prestressed steel bridges can reverse this trend, or at least make structural steel much more competitive than it has been in recent times. If this is true, then we must consider how the prestressing of steel can be used to best advantage. In the bridge construction field there are two broad areas where the prestressing of steel can be used to advantage: the design and construction of new longer span bridges, and the strengthening and rehabilitation of existing structures.

There is a definite trend in new structures to increasing span lengths. In the case of bridges crossing navigable waterways there are several valid reasons for spans longer than what would be required strictly for the navigation channel width. Removal of bridge piers from the waterway to the river bank, or at least to shallower water, reduces construction cost which would be greater when using deep water piers. Longer spans reduce the hazard of ship impact with the piers from errant vessels which could produce a potential failure with attendant loss of life or environmentally objectional cargo spills. Removal of the piers from the confines of the river eliminates the potential of pier scour. Longer achievable spans with prestressed steel bridges, consistent with geotechnical constraints, would require fewer piers and thus minimize impact upon environmentally or ecologically sensitive areas. Where pier locations are based upon other constraints it may be possible to reduce the span-to-depth ratio, by the use of prestressing, so that the length of the approach structure or approach embankment is reduced, thus providing economies.

The 1989 *Report of the Secretary of Transportation to the United States Congress on the Status of the Nation's Highways and Bridges* indicates that of the number of bridges inventoried and classified (577,710), approximately 26 percent (151,330) are or should be load posted. Bridges that require load posting fall into two groups. One group includes structurally deficient bridges that have deteriorated to such a condition that they cannot carry the load for which they were designed. The second group includes functionally obsolete bridges that are in good condition, but whose current state legal load exceeds the original design load and therefore, the bridges require posting.

Numerous states in the United States are confronted with the problem that many of the steel beam composite concrete deck bridges constructed between 1940 and 1960 are not in compliance with today's bridge standards. This condition results from an increase in legal load limitations, changes in specifications, and increased dead weight from resurfacing with concrete overlays. There are three possible solutions to this problem. The first solution is bridge replacement, which would be extremely expensive, not only because of the tangible cost of reconstruction, but also because of the intangible costs of inconvenience to the traveling public, longer time and additional distance traveled as a result of detours, and increased fuel consumption. The second solution is posting load restrictions, on higher type primary roads; trucks with loads exceeding the limits would be required to take alternative routes, with attendant intangible costs as indicated in the first solution. The third solution is to strengthen these existing bridges by prestressing. This can be an extremely cost-effective method.

The objective of this book is to summarize in one volume the current state-of-the-art of design and construction methods for all types of prestressed steel bridges, and thus to serve as a ready reference source for engineering faculties, practicing engineers, contractors, sub-contractors, and local, state, and federal bridge engineers. It presents guidelines for the structural analysis and design of prestressed steel bridges such as plate girders, trusses, arch bridges, cable-truss bridges, and guidelines for the rehabilitation and strengthening of existing bridges.

Chapter 1 is a quick review of the historical evolution of the subject to the current state-of-the-art. It offers the reader an appreciation of the way in which prestressing of steel bridges developed, the basic concepts of design and analysis as well as economies that are achievable by prestressing.

Chapter 2 discusses the methods of achieving prestress in a structure or in individual members, and provides examples of applications, while Chapter 3 discusses the all important subject of the prestressing tendons and anchorages through which the prestressing is induced into the structure.

Chapters 4 through 6 deal essentially with plate girder bridges, including composite girder bridges, and discusses such parameters as tendon geometry, optimum cross-sections, stress conditions, the effect of creep and shrinkage of the composite concrete deck, and continuous prestressed girders. Prestressing of steel box girders is considered in Chapter 7.

Although Chapters 4 through 7 deal with girder type bridges, prestressing can effectively be applied to other types of steel bridges. The analysis, design, and application of prestressing to trusses, tied arches, and cable-truss type bridges are treated in Chapters 8, 9 and 10, respectively.

The important topic of the application of prestressing to the strengthening and rehabilitation of existing steel bridges is presented in Chapter 11.

Perhaps the most important feature of this book, aside from the presentation of the theory and application of prestressing to steel bridges, is the authors effective utilization, throughout the book, of numerous illustrations, examples of practical applications and detailed design examples. In this way, the author has presented to the designer and engineer a series of guidelines for the successful application of the concept of prestressing to steel bridges.

There has been considerable change in bridge design and construction in the last decade, much of which has been evolving since the 1950s. As we progress into the next decade, there must be a conscious awareness of change. Research is being conducted to improve materials that may have a dramatic impact on the industry; and application of new systems to existing bridge types, along with new and improved types of bridge structures, is being attempted. Many of the improved materials and new concepts will reach practical application, others will be abandoned for technical or economic reasons. Improvements in materials, and new types of bridge structures that are currently unknown, are certain to evolve in the next decade. The next decade promises to be one of excitement and challenge.

Engineers must be constantly open to new concepts and ideas that will technically and economically improve the structures they build. However, they must also anticipate a new generation of problems that changes in methodology are certain to bring. In particular, they must avoid the trap of over-sophistication in design at the expense of simplicity in construction, and thus at the expense of economy.

As we strive for longer spans and improved means of constructing bridges, it would be well to remember the words of F. Stussi, an eminent Swiss engineer;

The problem of long spans has always fascinated the specialist as well as the layman. The realization of a bridge with a length of span hitherto unattained, not only requires great technical knowledge and capability, but also intuition and creative courage; it signifies a victory over the forces of nature and progress in the battle against human insufficiency.

This philosophy not only applies to the achievement of longer spans, but also, to the changing technology of the future. The next decade will continue to provide technological change and advancement, foremost among these technological advances will most assuredly be the concept of prestressed steel bridges. Professor Troitsky's Herculean effort in producing this book surely represents his contribution to the great technical knowledge, intuition, and creative courage that will be required to design and build the innovative and cost effective bridges of the future. It only remains for the rest of us to exploit this effort.

Walter Podolny, Jr.
Bridge Division
Office of Engineering
Federal Highway Administration

Preface

The prestressing of steel bridges is one of the best methods of providing economy in steel and reducing construction costs. The technical and economical usefulness of prestressing has been influenced widely by its application in prestressed concrete, but has seen relatively slow development in prestressed steel structures. The usefulness of prestressing in steel structures consists mainly in the economy of the material in the construction of new bridges as well as in the strengthening of old ones.

In contrast to the cross sections of concrete members, which cannot take tensile stresses, steel cross sections do not require special stress distribution. In addition, in steel structures the prestressed tendons do not cause substantial losses due to friction as do the tendons in prestressed concrete. However, the development of prestressed concrete structures indicates that, generally, similar principles of analysis may be applied to prestressed steel structures.

The modern development of prestressed steel structures started more than three decades ago, when Dischinger and Magnel initiated the application of prestressed steel structures. This new technique developed very quickly in highly industrialized countries such as the United States, the USSR, England, Belgium, and Germany, as well as in other parts of the world. As a result of this wide experience in the design of prestressed steel structures, they are no longer subject to tests and are equal in quality to conventional steel structures.

This book presents a guideline for the structural analysis and design of prestressed steel bridges, such as plate girders, box-type structures, arches, and cable trusses. In this text, methods of prestressing steel structures, including the widely used method of tendons, are discussed.

The successful development of prestressed steel bridges has created a need for comprehensive presentation of the theory and design of this structurally efficient and economical method. Until now, such information has been available mainly in technical journals, industrial bulletins, and various scattered publications, most of them foreign.

This text is an attempt to summarize experience with this method of structural design for constructing new prestressed steel bridges and strengthening old bridges and to produce an up-to-date practical reference book on the subject. It is also intended to provide designers of prestressed steel bridges and postgraduate students having a special interest in this subject with criteria and methods for the design of such structures.

The material contained in this book is presented in the following chapters:

Chapter 1   provides an introduction to the subject and brief historical development.

Chapter 2   discusses the methods of prestressing and application of prestressed steel structures.

Chapter 3   treats tendons and different types of anchorages.

Chapter 4   presents the analysis and design of prestressed steel plate girders using straight, parabolic, and polygonal types of tendons.

Chapter 5   discusses the analysis and design of prestressed composite steel girders having concrete slabs.

Chapter 6   treats continuous prestressed steel girders of large spans.

Chapter 7   is concerned with prestressed steel box-type girders.

Chapter 8   treats analysis and design of prestressed steel trusses.

Chapter 9   discusses prestressed tied arch steel bridges.

Chapter 10   provides analysis of prestressed cable truss bridges.

Chapter 11   treats rehabilitation of steel bridges.

It must be recognized that this text is by no means complete. However, it gives in a systematic form, as much as possible, the state of knowledge available at the time of the preparation of this book.

Since many variables affect the analysis and design presented in the book, the author would appreciate having called to his attention any errors that have escaped the author's editing efforts.

PRESTRESSED STEEL BRIDGES
THEORY AND DESIGN

Chapter 1
Prestressed Steel Bridge Systems

1.1 BASIC IDEA OF STEEL PRESTRESSING

Prestressing is the introduction and distribution of exactly defined stresses in member cross sections to increase the strength of the structure. This method may use the material more economically. Also, the prestressing of steel in tension results in compression, which provides the structure with added stiffening.

The tendency in prestressed concrete is to achieve, in each cross section under the influence of prestressing and loading, predominantly compressive stresses with limited or no tensile stress. However, at equal cross sections of prestressed steel structures under loading, there exist both tension and compression. Consequently, the acting cross-sectional area of steel structures is greater than that of concrete structures, so prestressed steel structures are more economical than prestressed concrete structures.

The concept of prestressing may be summarized as follows:

1. The controlled introduction of stressing in one part of the structure or the whole structure, which is under the influence of external loading, so that stresses opposite to those originating from its own weight and live loading are produced.
2. Regulation of the prestressing magnitude in the structure as requested by the design.

1.2 BRIEF HISTORICAL REVIEW

1.2.1 General Data

Historically, the principle of prestressing was employed long before the word was coined, and the principle is used today in some everyday objects. The oldest prestressed structures were Egyptian sailing ships, built in 2700 B.C. They were prestressed by a system of wire ropes—a main rope along the ship and a number of ropes in the transverse direction (Fig. 1.1).[1,2] The Romans countered the problem of arches tending to overthrow piers by putting a large weight on the pier in order to counteract the tensional stresses due to the arch thrust (Fig. 1.2).[3]

1

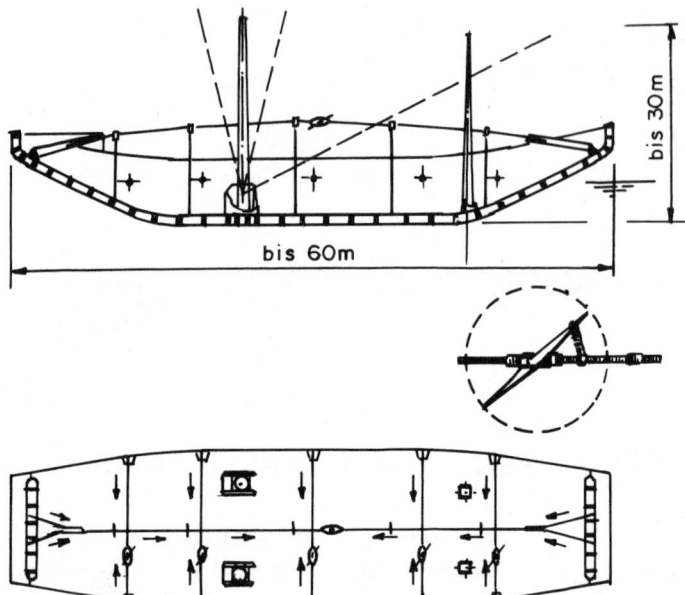

Fig. 1.1 Old prestressed Egyptian ship.

Wheels and timber barrels are both poststressed by drawn-out iron rings. Materials such as cast iron, which are strong in compression but weak in tension, require compressive prestressing to make them more effective. In the fifteenth century, Leonardo da Vinci, suggested that cast-iron cannons would burst less frequently when fired if the barrels were tightly wound with iron wire; centuries

Fig. 1.2 Arch of Constantine, Rome. (Courtesy of the Institution of Structural Engineers, London).

later, the idea was adopted in wire-wound guns. In 1861, A. V. Gadolin suggested winding artillery barrels with hot high-strength wire which, after cooling, would compress the barrel and therefore reduce tensile stresses in it after the charges exploded.[4]

1.2.2 Timber Bridges

In the United States in 1840, Howe was granted a patent on a timber truss in single or double systems, with timber diagonals and vertical iron ties prestressed by nuts, as shown in Fig. 1.3.[5] At that time, no theory of truss analysis existed, and in the original Howe truss all tensioned members were of the same cross section. Progress in analyzing Howe trusses was made by D. I. Jourawski who in 1844 proved that web members should be of changeable cross section when under the influence of changeable loading positions and their resulting forces.[6]

1.2.3 Metal Bridges

From 1847 to 1850, H. Rider designed prestressed trusses, as shown in Fig. 1.4, with the upper chord and verticals made from cast iron and the diagonals and bottom chord made from plates of wrought iron.[7,8] The prestressing was performed on the diagonals.

In Czechoslovakia in the nineteenth century, a prestressed truss system was introduced, applying the principle of the Howe truss, as shown in Fig. 1.5.[9] The upper chord and diagonals were made from cast iron, the bottom chord from plates of wrought iron, and the vertical bars from wrought iron. The poststressing was introduced in the vertical by nuts, which resulted in compression in the diagonals and tension in both chords. Because the tensile strength of cast iron is small, the compressive strength should never reach the zero value. From this condition, the necessary stresses in the verticals were determined. The bridges consisted of single, double, triple, or quadruple trusses, depending on the span. This system was built for the first time in 1858 as a bridge over the Tizera River. During the next ten years, 163 such bridges were built.

A project by B. Schnirch, which is older than the system mentioned above,

(a)

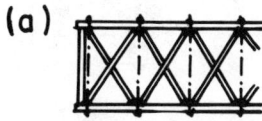

(b)

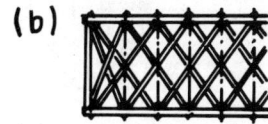

Fig. 1.3 Prestressed timber truss.

(a)

(b)

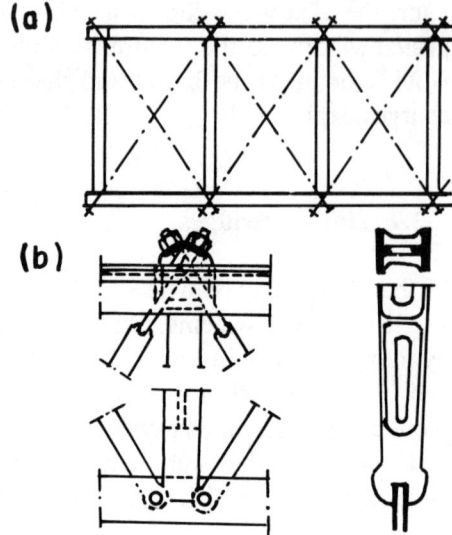

Fig. 1.4 Prestressed truss by H. Rider.

deserves mention.[10] It consists of a force-regulated bridge structure, representing a beam frame bridge, as illustrated in Fig. 1.6.

With the bottom chord under compression, the "Foepple truss" was "prestressed" (Fig. 1.7). It is interesting to note that the bearings are inclined since the surface of the abutment is also inclined.[11]

The "Bährecke truss," as seen in Fig. 1.8, has inclined-pendulum support instead of movable bearings.[12]

Robert Stephenson's great tubular Britannia Bridge, over the Menai Straits in England, completed in 1850, illustrates another application of prestressing.[13]

(a)

(b)   (c)

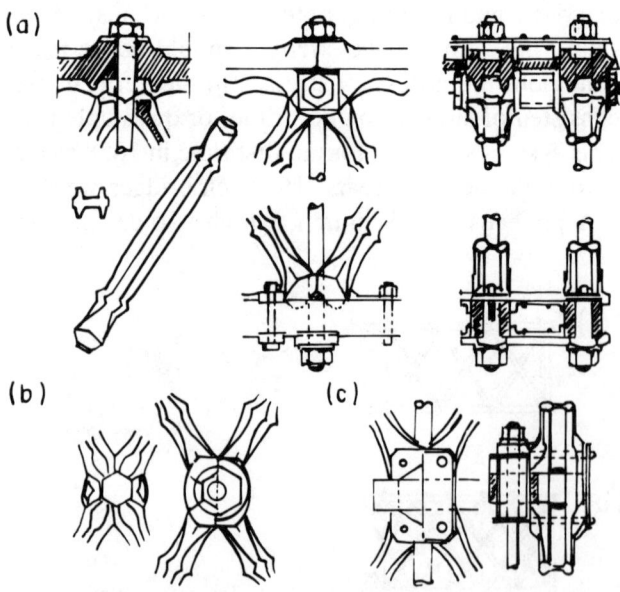

Fig. 1.5 Elements of a prestressed truss. (a) Upper and bottom chords; connections of diagonals and verticals. (b) Detail of diagonal intersection. (c) Intersection of diagonals at middle chord and transverse beam suspension.

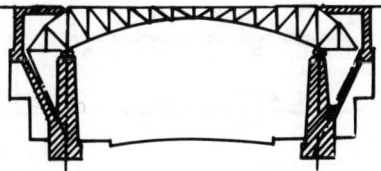

Fig. 1.6 Bridge by B. Schnirch project.

(a)

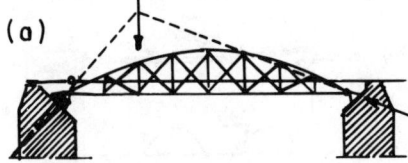

Fig. 1.7 Prestressed truss by Foepple.

Each railway track was carried inside a continuous wrought-iron box girder by two main spans of 460 ft each and two side spans of 230 ft each. In a bridge of this size, for reasons of economy in material, and when the bridge carried only its own weight, the maximum mid-span bending moment and the maximum bending moment over the piers should be as nearly as possible the same in magnitude, as this would keep the girder stresses to a minimum.

The spans were separately raised into position on the piers, and if they had simply been riveted together after erection, the bending moments at the piers would have been very small compared to those at the mid-span, as shown in Fig. 1.9.

Before any connections were made, girders B and D were jacked up above their final horizontal level at their far ends, as shown in Figure 1.9(b). They were then riveted to girder C at piers 3 and 4 and the jacks were removed. Girders B and D acted as hugh cantilevers, and their weight placed them and girder C into reversed bending. Span A was similarly jacked up at pier 1 before it was riveted to girder B at pier 2. The whole procedure had the effect of creating a negative bending moment at piers 2, 3, and 4 roughly equal to the positive mid-span moments. Figure 1.9(c) which were reduced to about half of what they would have been had their girders been connected without jacking.

This was prestressing on a really grand scale, and its aim was to set up a system of internal forces in the structure that would act in opposition to the effect of applied loads, particularly its own weight in this case, and so save material and cost.

In 1849, Clapeyron developed three-moment equations for the solution of con-

(b)

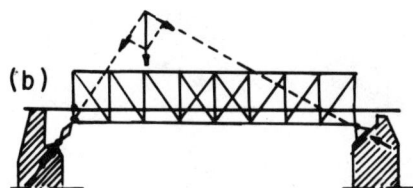

Fig. 1.8 Prestressed truss by Böhrecke.

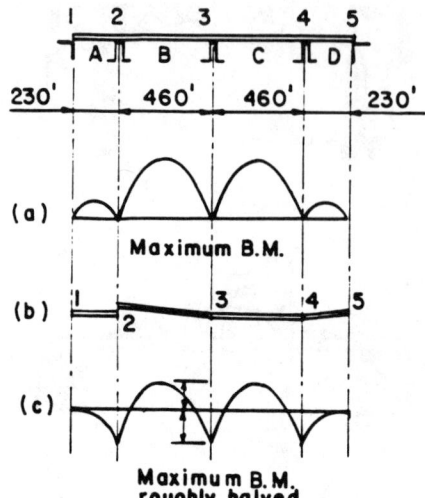

Fig. 1.9 Prestressing of girders of the Britannia Bridge. (a) Dead-load bending-moment diagram before spans were connected. (b) Tubes jacked up at 2 and 5, riveted together at 3 and 4, and then the jacks released. (c) Dead-load bending-moment diagram after prestressing.

tinuous beams by raising the intermediate supports. This method was used by C. Koepcke in 1856 and O. Mohr in 1860.[14,15]

In the second half of the nineteenth century, a number of bridges were prestressed by the use of ballast loads. To this group belong all those structures whose stress distribution was improved by support reactions, which were changed by the influence of permanent load introduction.

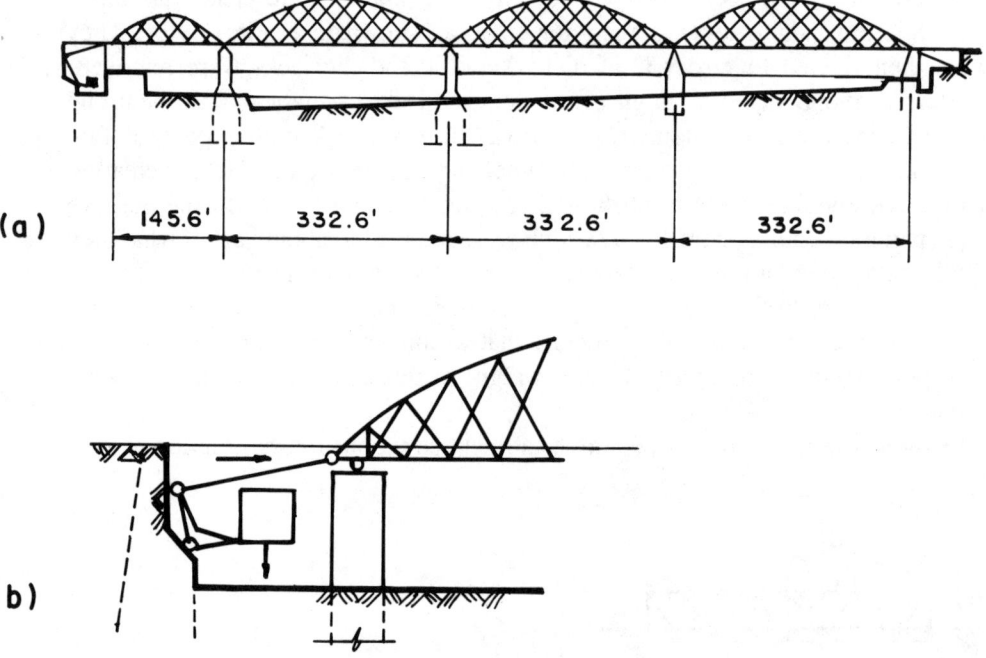

Fig. 1.10 Elbe Bridge.

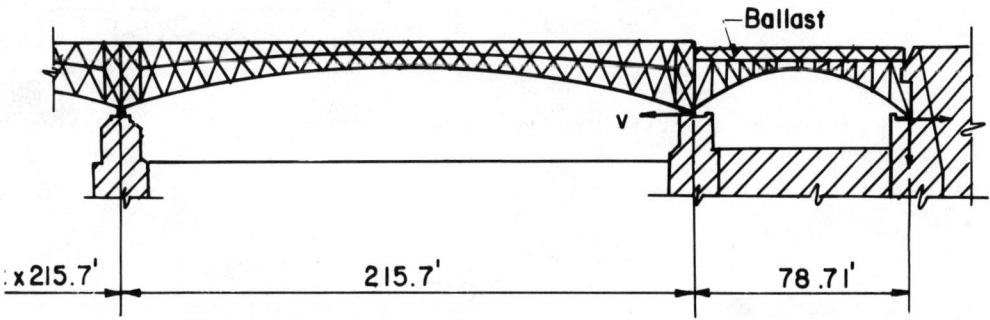

Fig. 1.11 Prestressed railway bridge over the Elbe River at Dresden.

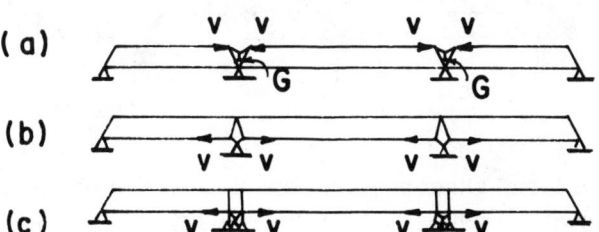

Fig. 1.12 Adjustment of prestressing forces on a continuous truss.

In 1878, after a design by Koepcke, the Elbe Bridge was built in Germany.[16] Its bottom chord was prestressed by an artificial load over which one lever adjustment increased this load by a factor of two in order to produce prestress pressure (Fig. 1.10). The force canceled the tensile force at the bottom chord due to the weight of the bridge. Also, after the project by Koepke, a four-track railway bridge over the Elbe River at Dresden, under an artificial loading over its side spans of 79 ft each, was transferred into a three-hinged each and produced a horizontal force at its abutment, as illustrated in Fig. 1.11.[17]

Figure 1.12 shows a proposal made by Lindenthal in 1883 for a bridge prestressing in Pittsburgh, Pennsylvania.[18] The last alternative was built and the details of its support are shown in Fig. 1.13. Koepke also developed a project for prestressing an arch bridge by force regulation, by using a ballast G as shown in Fig. 1.14.[19]

Some of the principles discussed above may also be applied to suspension bridges, Fig. 1.15(a).[20] The tension in the cable is achieved by the ballast, G,

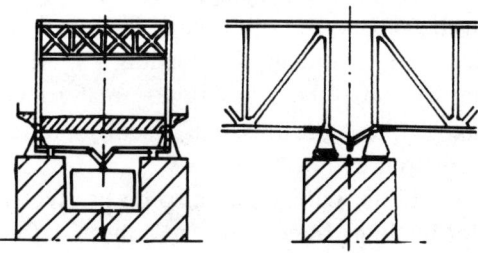

Fig. 1.13 Details of a support for a bridge at Pittsburgh.

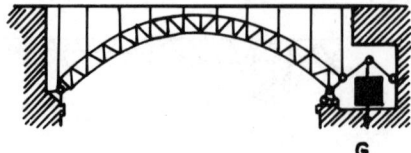

Fig. 1.14 Prestressing of an arch bridge by forced regulation, using a ballast.

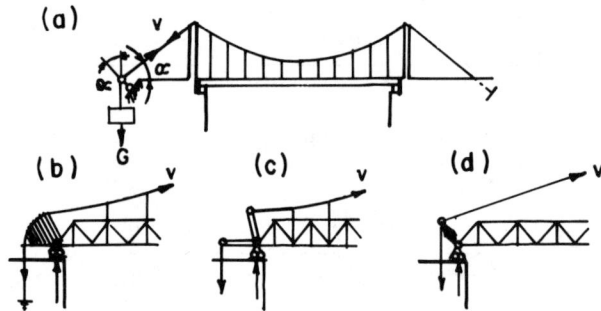

Fig. 1.15 Regulation of stresses of suspension bridges. (a) By ballast. (b) By segments. (c) By a powerful lever. (d) By support.

hanging on the free end of the cable. Other possible applications of prestressing of suspension bridges are shown in Figures 1.15(b), (c), and (d).

1.3 BASIC CONCEPT OF STEEL PRESTRESSING

In prestressed steel structures or their structural elements, the stresses are created artificially and are generally of opposite sign to those stresses due to loadings. Separate examples of the application of prestressed steel structures have been found in the past century, but only during the last decade has this method of increasing the effectivity of structures had wide application.

When in a structure we create prestressing f_0, having an inverse stress sign due to the action of the loading, the stretching of the elastic work of the material is increased (Fig. 1.16). Initially, a preliminary stress, f_0, is created; then a load, P, is applied to bring the stress in the bar to an allowable stress value, F. The tensile force taken by the prestressed bar is greater by a value of f_0A than the force taken by the same bar without prestressing.

It is possible to increase the carrying capacity further by applying multistep prestressing, in which the prestressing and required loading are achieved in a few cycles (Fig. 1.17).

Under applied changeable loading P_1, an initial stress f_0 is produced in the bar until the limiting value F is attained. In the second cycle, the initial stress f_{02} is again given, which decreases the achieved stress and applied loading P_2. After a few such cycles, the summary loading ΣP_2, may be a few times greater than loading P_1, which the structure may take without prestressing. Usually the required loading can be achieved in three or four cycles.

Prestressing is always related to deformations, which may also be of inverse sign compared to the deformations under loading. In certain cases, these deformations may limit prestressing, because of the possibility of stability loss.

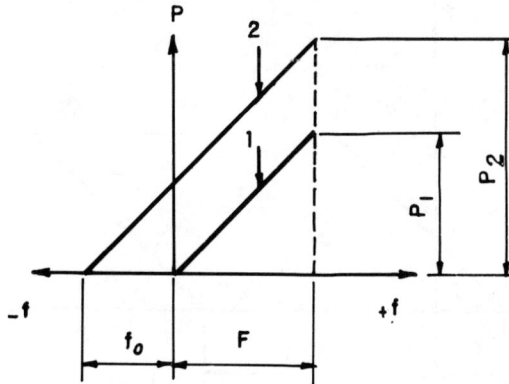

Fig. 1.16 A diagram of the work of a bar (1) without prestressing, and (2) with prestressing.

For the creation of prestressing and in the regulation of the structural forces, a number of different methods exist, each of which may be the most effective under particular conditions. The most highly developed and widely used method at present is the prestressing of separate tensioned and compressed members, and in whole structures (beams, trusses, arches), through squeezing. The rods are made of high-strength steel (steel cables, steel wires, etc.). This type of prestressing is well known theoretically and experimentally, and has found wide practical application. Prestressing by tie members has been applied in bridges.

The tightening of tie members makes it possible to reduce bending moments or a combination of the compression moments within the structure. By applying this type of prestressing, one can effectively and economically use high-strength materials in structures, since the strength of the steel cables or steel wires is four to six times higher than that of medium steel, although the cost is only two to three times higher.

Prestressing, then, makes it possible to use the strengths of wires or cables in

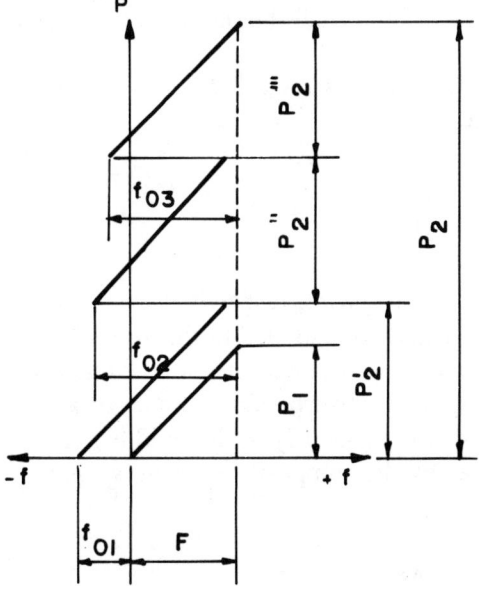

Fig. 1.17 Scheme of multistep prestressing.

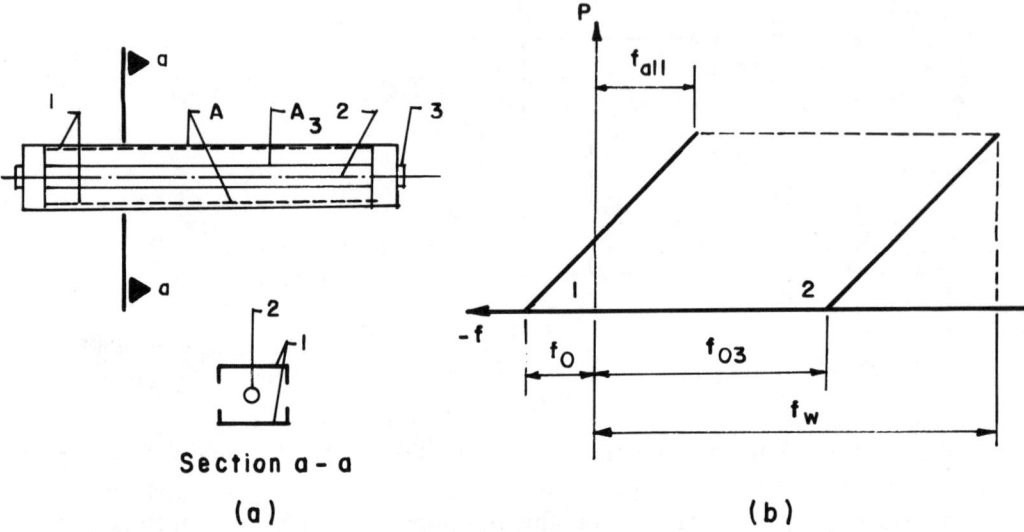

Fig. 1.18 The work of prestressed f_{03} element. (a) Scheme: 1, bar; 2, tie; 3, anchor. (b) Diagram of work; 1, bar; 2, tie.

hybrid-type structures made of medium and high-strength steels. If, to a rigid bar of medium steel having cross section A, we apply prestressing by using high-strength wire having cross section A_w (Fig. 1.18(a)), and obtain in the rigid bar a maximum compressive stress of f_0 (which will be somewhat smaller than the allowable stress, F), the tie bar will then be under prestressed tensile stresses f_{0w}. It will also be smaller than the limited resistance of the wire, F_w, than will the tension of this bar due to external loading. Thus, it is possible to achieve total utilization of the carrying capacity of the basic metal and the wire (Fig. 1.18(b)).

The deformation of this composite bar under tensile loading will be somewhat greater than the deformation of a bar made of mild steel (but no more than twice as great).

In the wire that obtained prestressing f_{0w}, tensile stresses ($F_w - f_{0w}$) will be added which correspond to the bar stress, or are equal to

$$(F + f_0) \frac{E_w}{E}$$

where E and E_w are elastic moduli of the bar and wire, respectively.

The bar receives tensile deformation, which corresponds to the stress ($F + f_0$). Under external loading, the rigid bar will be stressed for ($f_0 + F$) and tie-by the difference of stresses ($F_w - f_{0w}$). By a corresponding choice of areas, the limiting stresses may thus be obtained simultaneously in the rigid bar and in the wire. In this case, the total effort taken by the composite element is equal to the following:

$$P = AF + A_w F_w = AF(1 + \alpha\beta) \tag{1.1}$$

where

$$\alpha = \frac{A_w}{A}$$
$$\beta = \frac{F_w}{F} \qquad (1.2)$$

The multiplier in parentheses indicates an increase in the carrying capacity of the rigid bar at prestressing by the high-strength material. The total deformation of the bar under loading is

$$\Delta l = \frac{(f_0 + F)l}{E} = \frac{(F_w - f_{0w})l_w}{E_w} \qquad (1.3)$$

With the use of the high-strength material, deformations are not 4 to 6 times greater but only 1.5 to 2 times greater than those in the bar of mild steel corresponding to the value f_0.

1.4 METHODS OF PRESTRESSING

At the present time, the methods of applying prestressing forces are as follows: (1) prestressing by tendons made of high-strength steel, and (2) the introduction of prestressing without tendons. The method using tendons is widely applied in bridge engineering and is generally considered the classical method. The method without tendons is somewhat limited and considered secondary to the classical method.

1.4.1 Prestressing with Tendons

Steel bridges that are prestressed with tendons consist of the following three elements: (1) the structure that originally was not prestressed, (2) tendons of high-strength steel, and (3) the anchorages and saddles supporting the tendons. For the introduction of stresses in the tendons, auxiliary installations such as jacks are used.

After the cables, bars, or wires are connected to the anchorages, which are fixed to the structure, the stressing is applied to either the whole structure or part of it. One of the basic conditions to prestressing is to securely attach the tendons to the anchorages and ensure that the connection to the structure is unmovable. The prestressing forces keep the structure in equilibrium, and consequently no vertical reactions arise.

In many cases it it necessary to attach certain elements to the structure that provide direction for the tendons. These elements come in different shapes.

The action of tendons on the structure may be considered in two stages. In the first stage it acts normally, under its own weight, and the prestressing force is one part of the dead load or erection load. The prestressing force should be chosen by taking into consideration the compressive stresses of a single member or a group of members (for a truss), or the flange of a plate girder so that the buckling stresses will be smaller than the permissible stresses. In all cases, the anchorage locations for prestressing forces should be chosen so that the stresses resulting from prestressing are opposite to the stresses due to the dead and live loads. In the first stage the tendon will be tensioned with the prestressing force X. In the second stage, all the loads, namely, the prestressing forces and the dead and live loads, are acting. The total stresses under these loadings have, at all cross sections, the opposite sign of the stresses due to prestressing. The tendons remain tensioned and the tension force in the tendon increases to the value ΔX. The force will appear due to the presence of the tendon as part of the structure, assuming that in the second stage there is loading. The final tension force in the tendon at this stage will be $X_0 = X + \Delta X$. The cross-sectional area of the tendon should be chosen so that even under the influence of this tensioning force, the stress in the tendon will not be greater than the permissible stress. The appearance of force ΔX and its origination in the high-strength steel tendon leads to further economy. The truss members and girder flanges, due to the action of the tendon, must be considered common elements, which, irrespective of their different strengths and elastic values, perform as one unit.

The two working stages of the structure and stresses, as described above, represent one basic scheme as prestressing at each stage is developed. Other prestressing procedures, such as the method of prestressing by stages, are available. The introduction of stress forces and external forces—for example, constant loads—may occur many times between or after the stages. In this case, the explained scheme does not change, except that its numerical application will be complicated during the erection of the structure.

1.5 ECONOMY BY PRESTRESSING

The application of prestressing to steel structures makes it possible to reduce the amount of steel and consequently the cost. By using tendons of high-strength steel we may achieve the following benefits: (1) Utilizing conventional steel for the structural members increases the limit of the elastic deformation under greater loads, so the amount of steel is reduced, (2) The tendons participate in the loading.

To simplify the technical-economical analysis of prestressed steel structures, the following expression may be established:

$$Q = \gamma l \frac{P_x}{\sigma_{\text{all}}} \tag{1.4}$$

under the assumption that the amount of steel is in reverse proportion to the allowable stresses. In the above equation,

Q = weight of the steel in Kp
γ = specific weight of the steel in Kp/cm^3
l = length of prestressed member in cm
P_x = prestressing force in Kp
σ_{all} = allowable stress in Kp/cm^2

This expression is valid only when the member in tension is being prestressed by the tension. Therefore, the equation, in this form, cannot be applied to the whole structure.

The following factors influence the final cost of a prestressed steel structure:

1. The weight of the structure, saddles, anchorages and tendons
2. Construction of the structure at the plant
3. Transportation and loading cost
4. Erection cost, including prestressing
5. Cost of corrosion protection

In all the countries where prestressed steel structures have been built, these structures have proven economical with regard to steel and cost. According to data provided by Brodka and Klobukowski in 1969, the economic advantages of prestressed structures compared to the corresponding nonprestressed structures are as given in Table 1.1.[21]

The values in Table 1.1 were obtained for prestressed steel structures without considering cases of deficiencies due to lack of experience in projects and erection. It is also assumed that the data given in Table 1.1 may be applicable to the design of structures not investigated previously. The economy of steel increases as the difference between the allowable stresses of the steel used for the structure and the high-strength steel for the tendons increases.

When the tendons are prestressed in order to obtain members in tension in a single state, a load-carrying capacity two times greater may be obtained for the member. However, considering the possibility of buckling under prestressing, the bearing capacity may actually be only 70–80%. If the structure is stressed in several stages, under the corresponding loadings, the structure should increase the carrying capacity of the tensioned members from three to five times, and the forces from the main cross section will be transferred to the tendons. It should

Table 1.1

Structures	Economy in Steel, %	Economy in Cost, %
Plate girders	10–12	8–12
Trusses of spans 100 ft	5–10	2–5
Roof trusses of spans 100–130 ft	10–20	5–10
Roof trusses of spans 130–200 ft	10–45	7–20
Frames and arches of spans 100–200 ft	20–50	10–30

be noted that since the time of construction is slower in this case, there will be an increase in the cost. The rationalization of this method is still in the development stage.

1.6 ANALYSIS AND DESIGN OF PRESTRESSED STEEL STRUCTURES

1.6.1 Overloading and Underloading Factors

For the design of prestressed steel structures there are, at present, the following known standards in Czechoslovakia and the USSR.[22,23]

Russian standards for the design of prestressed steel structures use the method of limit state. During the design of prestressed steel structures a number of coefficients are introduced taking into consideration the special character of each prestressed structure's work. With this method the actual value of the prestressing force, considered as constant load, is found by introducing the overload and underload factors, n. The overload factor, $n_1 = 1.1$, and the underload factor, $n_2 = 0.9$, allow for the possibility of the actual prestressing force to respectively exceed and fall below the specific value due to poor accuracy of the prestress measurement methods. These factors are introduced into the calculation if the prestressing force value in a tendon is determined by indirect means, for example, in anchoring the tendon by a bolt, by the force required to tension the bolt; in anchoring the tendon by wedges, by the value of the force necessary to drive or press in wedges, and in other techniques by suitable means.

In the case of reliable direct measurement of the prestressing force by means of pressure gauges on jacks, for the measurement of deflections or stresses in a structure by instruments, and so on, the values of factor n in the calculation of a structure are assumed equal to unity.

The overload factor, $n_1 = 1.1$, is taken into account in two cases: (1) When the structure is being checked during prestressing, the factor is introduced in all the calculated values of cross sections and bars; and (2) When the structure is undergoing external loading, the factor is introduced in the values of members and cross sections in which the external load stresses are of a like sign with that of the stresses due to prestressing or in which the prestresses are greater in value and opposite in sign to the stresses due to external load.

The underload factor, $n_2 = 0.9$, is introduced when checking the structure in the course of loading by an external load, in the values of all the cross sections and bars in which stresses from the external load are greater in value and opposite in sign to the prestressing force values.

The overload and underload factors should be introduced when beam and frame structures, trusses, and similar constructions are being calculated. During the calculation of combined members in tension or compression for service loads, the factors may be omitted, as small deviations in the tendon prestress value have no substantial effect on the load-carrying capacity of such members but merely cause a negligible redistribution of force between the bar and the tendon.

This has been borne out by tests on members in tension. When a bar is checked for buckling in the course of prestressing, the prestress should be assumed with an overload factor of $n = 1.1$.

1.6.2 Controlled Prestressing Force

A prestressing force is generally created in a tendon by means of tensioning devices: The tendon is secured by anchoring to the structure, the tensioning device is removed and the force transferred to the structure, the anchoring is stretched out, the tendon shortens, and the force in it decreases. In addition, tendons from wire ropes or strands of wire are prone, when loaded, to relaxation, which involves a progressive decrease of stresses over the course of time.

Both these factors require allowances for a decrease in the tensioning force and provision for a controlled prestressing force of a value somewhat greater than that indicated by calculations. The value of the controlled force is to be determined in the course of prestressing with the aid of instruments or by any other suitable means.

Similarly to prestressed reinforced concrete structures, the value of the controlled force is found with the aid of the formula

$$X_k = \frac{X}{0.95} + \Delta_a \frac{AE}{l} \tag{1.5}$$

where

X = calculated force in the tendon due to prestressing

0.95 = coefficient of relaxation, which is introduced for tendons of steel wire ropes and bundles of high-strength wires only

$A, E,$ and l = cross-sectional area, modulus of elasticity, and length of the tendon, respectively

Δ_a = yielding of anchors assumed equal to $1/32$ in. (1 mm) for anchors composed of tightly screwed nuts or wedge-shaped plugs and to $3/32$ in. (2 mm) for liner-type anchors

Practically, the yielding of anchors, Δ_a, should be allowed for in shorter tendons only, ones less than 65–98 ft (20–30 m) in length, where for the longer tendons a shortening of $1/32$ in. (1 mm) or $3/32$ in. (2 mm) is of no consequence, because the loss in stress amounts to a mere 4 or 5%.

1.6.3 Protection and Connections of Tendons to Structures

The tendons that are under direct atmospheric influences should be protected against corrosion by painting, by a covering of protective varnish, by plastic sheathing, by zinc, or by other methods. When tendons are placed in the covered parts of the beams or trusses, these covered parts should be filled with cement mortar or bitumen, simultaneously providing independent longitudinal deformations of the tendon and rigid member.

The tendons are fixed to the rigid members of the structure with the help of the gripping wedges, cases, or other anchors that are used during the construction of prestressed structures.

REFERENCES

1. Strub/Roessler, "Technique of Shipbuilding in Ancient Egypt," *Techische Rundshau*, Bern, August 1952 (in German).
2. Landstrom, B. *Sailing Ships*, Doubleday, Garden City, N.Y., 1969.
3. Samuely, F. J., "Structural Prestressing," *The Structural Engineer*, February 1955, pp. 41-52.
4. Gadolin, A. V., "Theory of Barrels Reinforced by Rings," *Artillery Magazine*, no. 12, 1861, pp. 1033-1071 (in Russian).
5. Tyrrell, H. G. *History of Bridge Engineering*, published by the author, Chicago, 1911 pp. 141-142.
6. Jourawski, D. I. *On Bridges of Howe System*, St. Petersburgh, 1885 (in Russian).
7. Kolar, J., "Bridge Construction," vols. I, II, III, Prague, CMT, 1923, 1925, 1926 (in Czechoslovakian).
8. Velflik, A. V., "Main Bridge Trusses," Stavitelsvi Mostri, II, Prague, CMT, 1905 (in Czechoslovakian).
9. Ferjencik, P., and Tochacek, M., "Prestressed Metal Structures," Bratislava, SVTL, 1966 (in Czechoslovakian).
10. Kolar, op cit. p. 19.
11. Ibid.
12. Ibid.
13. Francis, A. J., *Introducing Structures*, Pergamon Press, New York, 1980, pp. 139-140.
14. Tolmachev, K. C., "Regulation of Stresses in the Spans of Metal Bridges," Moscow, Edition Avtotransport, 1960, (in Russian).
15. Mohr, O., "Theory of Timber and Iron Constructions," *Zeitschrift des Architekten und Ingenieur*, nos. 2, 3, 4, 1960 (in German).
16. Kolar, op. cit. p. 19.
17. Ibid.
18. Tolmachev, op. cit.
19. Ibid.
20. Ibid.
21. Brodka, J., and Klobukowski, J., "Prestressed Steel Constructions," Edition by Wilhelm Ernst und Sohn, Berlin-München, 1969, p. 10 (in German).
22. ON 73 1405, "Guidelines for Designing Prestressed Steel Structures," Prague, UNM, 1969, (in Czechoslovakian).
23. *Instructions on the Designing of Prestressed Steel Structures*, Gosstroiizdat, Moscow, 1963, (in Russian).

Chapter 2
Methods of Prestressing

2.1 INTRODUCTION

The basic idea of prestressing steel structures consists in the following: The separate tensile members or the whole structures are subjected to compression by different methods. The tendons, for example, are anchored by their ends onto the structure and remain tensioned, working together with the structure under loading.

Prestressing forces are introduced in structural members in the reverse direction to those which originate in the member or structure under external loading.

The effect of structural prestressing of the metal is, first, the partial substitution of metal required for the structure by high-strength steel in the form of cables or wires. Second, by prestressing, it is possible to create a convenient redistribution of forces in a structure. This reduces the amount of metal needed in the structure but complicates its fabrication and erection.

Prestressing may be applied to either the whole structure or its separate members. Prestressing of steel structures may be performed by different methods, such as:

1. The application of tendons
2. The bending of rolled sections, connecting them by welding and reinforcing them with cover plates
3. The application of a predeflection technique known as "Preflex"
4. The redistribution of the bending moments at continuous beams by regulation of their support levels

2.2 THE PRESTRESSING OF STEEL MEMBERS BY TENDONS

2.2.1 Members in Axial Tension

In order to understand the basic structural behavior of a prestressed steel structure, we shall first investigate the simplest case, that of a steel bar of rectangular

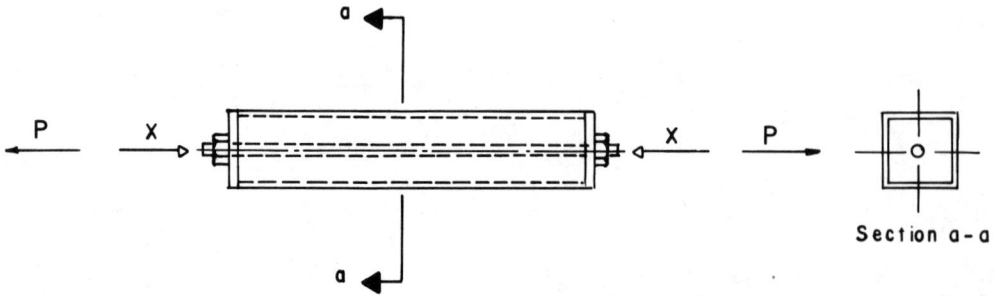

Fig. 2.1 A member in tension under prestress.

cross section under working load conditions (Fig. 2.1).[1] The member is subjected to a tensile force, P, and a compression force, X, due to the initial prestressing force. The stresses in the members and cable are:

$$f_m = \frac{P - X}{A} \qquad (2.1)$$

and

$$f_t = \frac{X}{A_t} \qquad (2.2)$$

where

f_m = member stress
f_t = tendon stress
P = tensile force
X = compression force
A and A_t = cross-sectional areas of the member and tendon, respectively

When we exert an additional tensile force, P_{LL}, onto the member, this force will cause an elongation of Δl in the member as well as in the tendon. A Δl elongation of the tendon will add an increment of prestressing force ΔX (Fig. 2.2).

$$\frac{P_{LL} - X}{AE} = \frac{\Delta X}{A_t E_t} \qquad (2.3)$$

The final tendon stress is:

$$f_t = \frac{X + \Delta X}{A_t} < F_t \qquad (2.4)$$

The final member stress is:

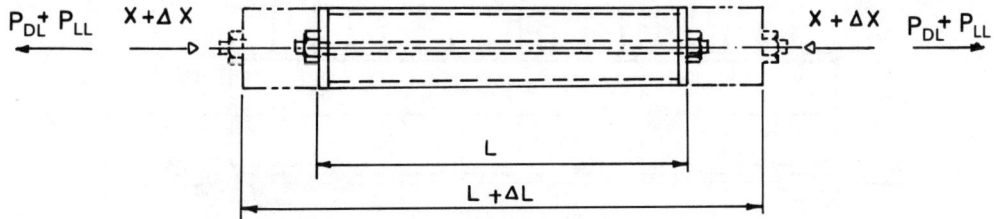

Fig. 2.2   The effect of member elongation.

$$f_m = \frac{(P + P_{LL}) - (X + \Delta X)}{A} < F \qquad (2.5)$$

From the above simultaneous equations, the unknowns X, ΔX, A_t, and f_m can be calculated. If $X > F_{DL}$, the equation is:

$$f_m < \frac{F}{\Psi}$$

where Ψ is a safety factor. In other words, the elastic stability of buckling must also be investigated.

2.2.2 Prestressing of a Symmetrical I Beam

Let us now consider a symmetrical I beam prestressed by tendons placed under the bottom flange (Fig. 2.3) and loaded by external loads. Due to prestress X, a uniform stress of

$$f = -\frac{X}{A} \qquad (2.6)$$

will be produced across the A-sectional area.

Due to an eccentric prestress, the beam is subject to a moment as well as a direct load. The moment produced by the prestress is Xec, and the stresses due to this moment are:

$$f = \pm\frac{Xec}{I} \qquad (2.7)$$

If M is the external moment due to the load on a section and the weight of the beam, then the stress at any point across the section due to M is:

$$f = \pm\frac{Mc}{I} \qquad (2.8)$$

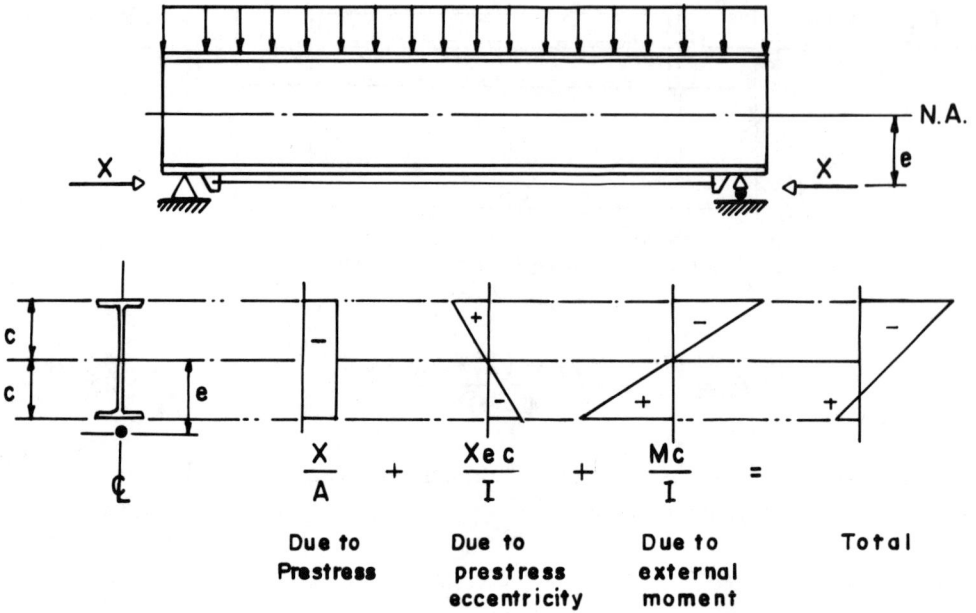

Fig. 2.3   The stress distribution due to prestress and load on a symmetrical steel I beam.

The resulting stress distribution is given by

$$f = \frac{X}{A} \pm \frac{Xec}{I} \pm \frac{Mc}{I}$$

(2.9)

Considering this symmetrical I beam, as shown in Fig. 2.3, the resulting top and bottom fiber stresses are not equal; the top fibers are more stressed than the bottom ones.

2.2.3 Prestressing of an Asymmetrical I Beam

Consider now a beam with its top flange larger in cross section than its bottom flange. In Fig. 2.4, the same three component-stress diagrams are shown added together to give a combined stress distribution.

2.2.4 Increment of Prestressing Force on an Eccentrically Stressed Beam

The basic stress equations derived above are to be modified in such cases as beams subjected to bending. Figure 2.5 shows a simply supported prestressed beam under the influence of its working load.

Under a live load, the tendon length S is increased by ΔS and X by ΔX, in a manner similar to that for the tendon behavior. But for the determination of ΔX, the equations of the statics

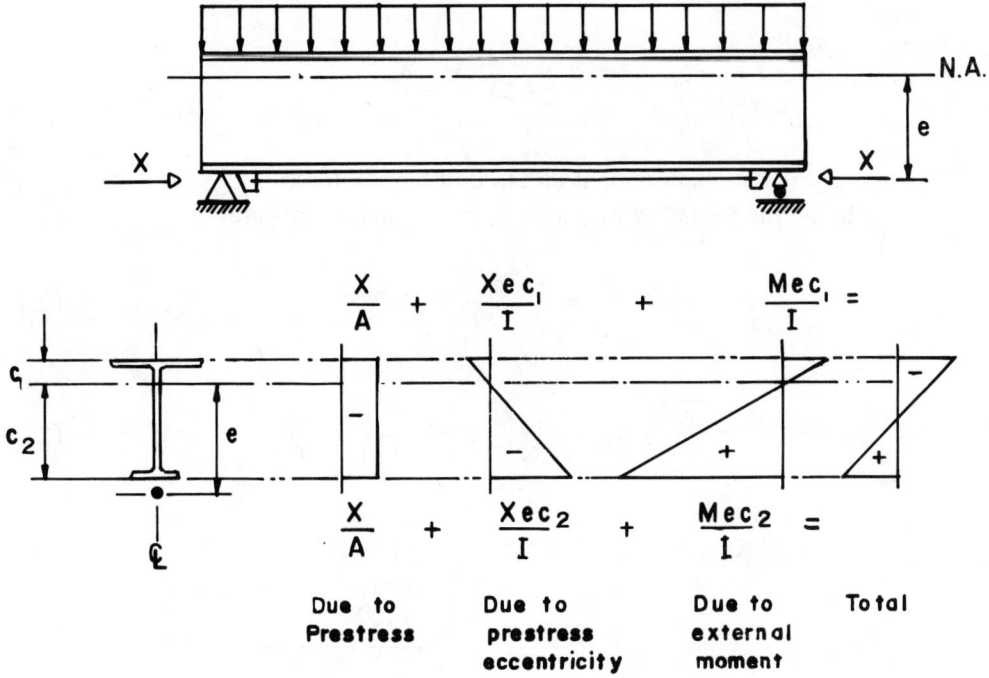

Fig. 2.4   The stress distribution due to prestress and load on an asymmetrical steel I beam.

$$\Sigma M = 0 \qquad \Sigma V = 0 \qquad \Sigma H = 0$$

are not sufficient, as the structure is statically indeterminate, similar to that of a simply supported tied arch. Thus, one additional equation of elasticity must be used.

Cutting the cable vertically, we may express in equation form that the horizontal displacement due to live load (δ_{ip}) and the force increment ΔX is 0. Therefore,

$$\delta_{ip} + \Delta X \, \delta_{11} = 0 \qquad\qquad (2.10)$$

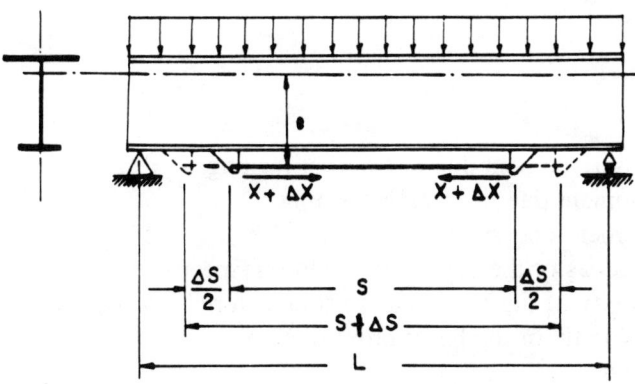

Fig. 2.5   A simply supported prestressed beam under the influence of its working load.

or

$$\Delta X = -\frac{\delta_{ip}}{\delta_{11}} \tag{2.11}$$

where δ_{11} is the displacement due to unit tendon force.

Utilizing the virtual work method, the displacements are:

$$\delta_{ip} = \int_0^l \frac{Mxm}{EI} \, dx \tag{2.12}$$

$$\delta_{11} = \int_0^l \frac{m^2}{EI} \, dx + \frac{l}{E_t A_t} + \frac{l}{EA} \tag{2.13}$$

or

$$\Delta X = -\frac{\displaystyle\int_0^l \frac{Mxm}{EI} \, dx}{\displaystyle\int_0^l \frac{m^2}{EI} \, dx + \frac{l}{E_t A_t} + \frac{l}{EA}} \tag{2.14}$$

Uniformly distributed loading (Fig. 2.5(a)), Fig. 2.5(a) prestressed (beam under uniformly) distributed loading.

$$\Delta X = -\frac{\displaystyle\int_a^{l-a} mM_p dx}{\displaystyle\int_a^{l-a} m^2 dx + \frac{Il_a}{A} + \frac{EIl_a}{E_t A_t}} = -\frac{\displaystyle\int_a^{l-a} eM_p dx}{\displaystyle\int_a^{l-a} e^z dx + \frac{Il_a}{A} + \frac{EIl_a}{E_t A_t}}$$

$$= -\frac{2M_p e(l^3 - 6l_a^2 + 4a^3)}{3l^2(l - 2a)\left(e^2 + \dfrac{I}{A} + \dfrac{EI}{E_t A_t}\right)} = -\frac{2M_p e}{3\left(e^2 + \dfrac{I}{A} + \dfrac{EI}{E_t A_t}\right)}\left(2 - \frac{l_a}{l}\right) \tag{2.14a}$$

where

M_p = maximum bending moment due to external loading
M = that bending moment due to external loading
m = the bending moment due to unit load
A, A_t = cross-sectional areas of the beam and tendon, respectively
E, E_t = moduli of elasticity of the beam and tension materials, respectively
I = the moment of inertia of the beam cross section
l = the tendon length

2.2.5 Draped Tendons

The stress distribution in a member due to prestressing is generally composed of an axial stress and a bending stress due to eccentricity, and is given by the equation

$$f = -\frac{X}{A} \pm \frac{Xe}{S} \qquad (2.15)$$

where

X = the horizontal component of the prestressing force
e = the eccentricity of the force
S = the sectional modulus of the member

The stress due to the applied design loads is

$$\pm \frac{M}{S}$$

where M is the bending moment.

Since the purpose of the prestressing steel is to counteract the stress M/S due to design loads, it is seen that the ideal system is one that would cause stress distribution similar but opposite in character to that of the design loads, or equal to $\pm Xe/S$, and this can be obtained only by eliminating the axial stress X/A. In this case, the tensioning steel is draped in a predetermined polygon depending on the applied loads, tensioned to a desired amount, and anchored into a compressed part of the girder (Fig. 2.6).

In a simply supported girder, the vertical components of the tensioning steel produce upward forces at their point of anchorage. The upward forces are designed to completely counteract the downward dead load or to reduce it by any desired amount.

The draped tendon produces moments and bending stresses that are similar but opposite in direction to those produced by their design loads.

The reduction or elimination of the dead-load moments permits their increased dead and live loads to act upon the structure, increasing its sectional modulus. For example, the beam in Fig. 2.6 supports three concentrated dead loads, P. By draping the tendon to show their ordinates, the forces in this structural system are as designated, with X being the horizontal tension component in the tendon. The vertical component at the center of a force X is found assuming that $X = M/e$, and substituting the value of the moment

$$M = \left(\frac{3P}{2} \cdot \frac{L}{2}\right) - \left(\frac{PL}{4}\right) = \frac{PL}{2} \qquad (2.16)$$

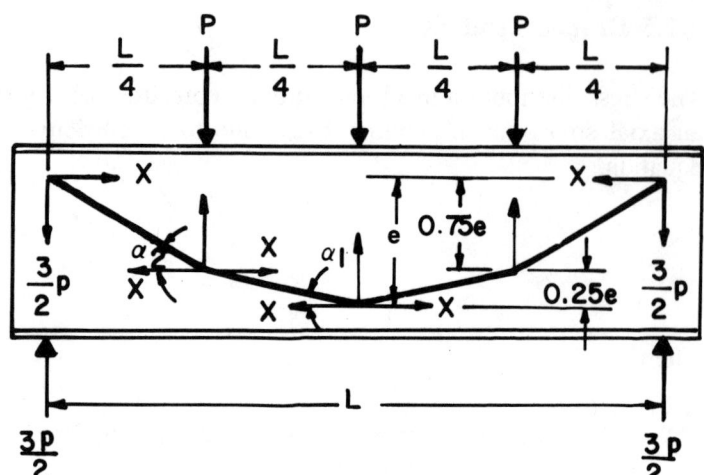

Fig. 2.6   A draped tendon.

then

$$X = \frac{PL}{2e} \tag{2.17}$$

The vertical component of the cable tension is

$$V = 2X \tan \alpha = 2X \frac{0.25e}{0.25L} \tag{2.18}$$

and by substituting the value for X, we obtain

$$V = \frac{PL}{2e} \cdot \frac{2e}{L} = P \tag{2.19}$$

Similarly, the vertical component of the tensile force at each quarter point, considering that $X = PL/2e$ and the vertical component of the cable tension is

$$V = X\frac{0.75e}{0.25L} - X\frac{0.25e}{0.25e} = \frac{2Xe}{L} = P \tag{2.20}$$

Therefore, when the tendons are tensioned, the vertical components are equal to P, and the beam under the three concentrated loads has a zero shear moment and deflection at any section and is completely without stress under its dead loads. The total beam load may, therefore, be increased to $3P$, as shown in Fig. 2.6, without increasing its sectional modulus. The bending moment at any section of the beam due to tensioning may be obtained from a loading diagram or from the equation

$$M = Xy$$

where y is the tendon sag at the section considered.

2.2.6 Composite Prestressed Structures

The stress analysis of a prestressed composite steel structure depends on its construction method, when temporary supports are used and removed after the composite section has been prestressed (Fig. 2.7). The structure is statically indeterminate in its prestressed form.

The fundamental assumptions used in analysis are: The composite structure, due to its effective connection between the steel and concrete provided by means of its shear connectors, acts as a homogeneous section following Hooke's law and Navier's hypothesis.[2]

Considering a typical prestressed composite section (Fig. 2.8), the sectional properties can be determined based on the transformed steel section. The following notations are used:

A_s = the net cross-sectional area of steel
A_c = the cross-sectional area of concrete
A_t = the cross-sectional area of the tendon
I_s = the moment of inertia of steel with respect to the neutral axis of steel
y_1, y_2 = the extreme fiber distances of steel
E_s = the modulus of elasticity of steel
E_c = the modulus of elasticity of concrete
b = the effective width of a concrete slab
t_c = the thickness of concrete
$n = E_s/E_c$, the modular ratio
N_{st} = the neutral axis of steel
N_c = the neutral axis of concrete
A_{tr} = the transformed cross-sectional areas of the composite section

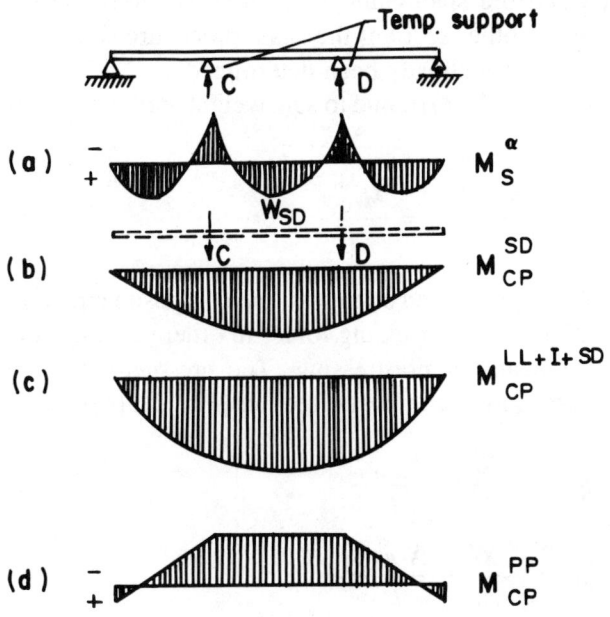

Fig. 2.7 Moment diagrams.

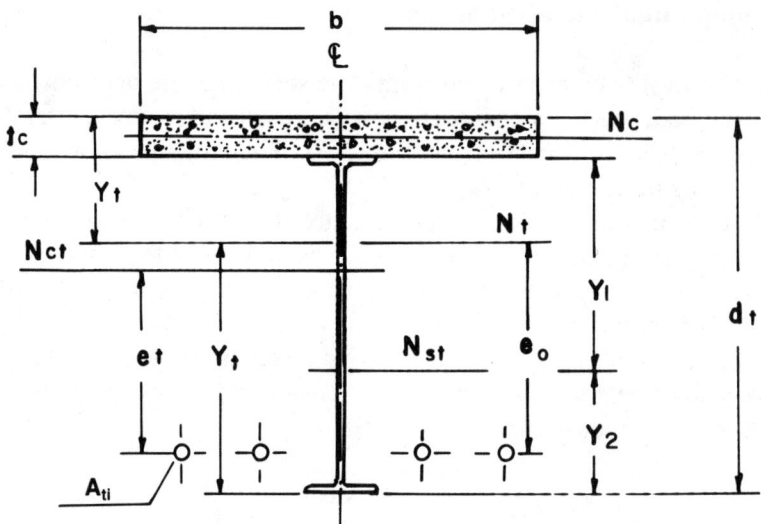

Fig. 2.8 A typical prestressed composite section.

I_{tr} = the transformed moment of inertia with respect to the neutral axis of the transformed section

N_t = the neutral axis of the transformed section

y_t, y_b = the extreme fiber distances of the transformed section from the top and bottom, respectively

e_0 = the distance from the cables to the neutral axis of a transformed section

d_t = the depth of the composite girder

Knowing all the sectional properties, it is thus necessary to investigate the stresses of the critical section and the span camber in order to determine the required prestressing force. For the other section, the same procedure should be followed, after a prestressing force has already been determined.

The stresses in the steel girder (Fig. 2.9(a)), due to self-weight and wet weight of the concrete, are

$$f_b = \frac{M_{DL}y_2}{I_s} \qquad f_t = \frac{M_{DL}y_1}{I_s} \tag{2.21}$$

After the concrete has reached its prescribed strengths, the composite structure will then be prestressed by an X_0 initial prestressing force, in order to reduce the final bottom or top stresses by means of prestressing. The prestressing force would cause the following stresses in the composite section (Fig. 2.9(b)):

$$f_{tb} = -\frac{X_0}{A_{tr}} - \frac{X_0 e_0 y_b}{I_{tr}} \tag{2.22}$$

$$f_{tt} = -\frac{X_0}{A_{tr}} - \frac{X_0 e_0 y_t}{I_{tr}} \tag{2.23}$$

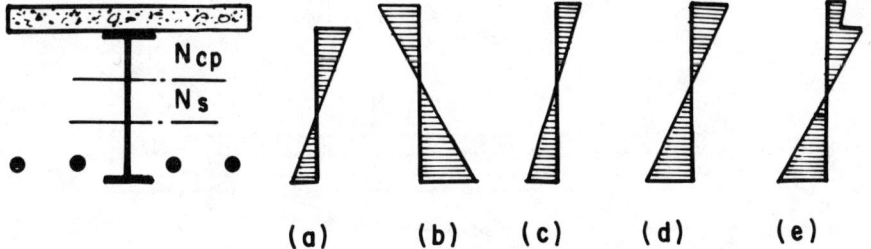

Fig. 2.9 Stress diagrams of a prestressed composite girder.

where a minus sign indicates a compressive stress and a plus sign indicates a tension stress.

The prestressing force lifts the girder from its temporary supports. Consequently, the composite section will carry dead and superimposed dead load moments M_{DL+SD}, producing the following stresses (Fig. 2.9(c)):

$$f_{tb} = +\frac{(M_{DL+SD})\, y_b}{I_{tr}} \tag{2.24}$$

$$f_{tt} = -\frac{(M_{DL+SD})\, y_t}{I_{tr}} \tag{2.25}$$

Similarly, the live load and impact stresses are (Fig. 2.9(d)):

$$f_{tb} = +\frac{(M_{LL+I})\, y_b}{I_{tr}} \tag{2.26}$$

$$f_{tt} = -\frac{(M_{LL+I})\, y_t}{I_{tr}} \tag{2.27}$$

The external live load and superimposed dead load acting on the prestressed structure produce a statically indeterminate prestressing force increment, ΔX_0, which can be determined by the virtual work method. Assuming that the cables are cut at point 1 (Fig. 2.10), the horizontal displacement of the cable at that point is expressed as zero on that basic structure which has been made statically determinate. Therefore,

$$\delta_{01}^{LL+I+SD} + \Delta X_0\, \delta_{11} = 0 \tag{2.28}$$

and

$$\Delta X_0 = -\frac{\delta_{01}^{LL+I+SD}}{\delta_{11}} \tag{2.29}$$

where $\delta_{01}^{LL+I+SD}$ is displacement due to total external load.

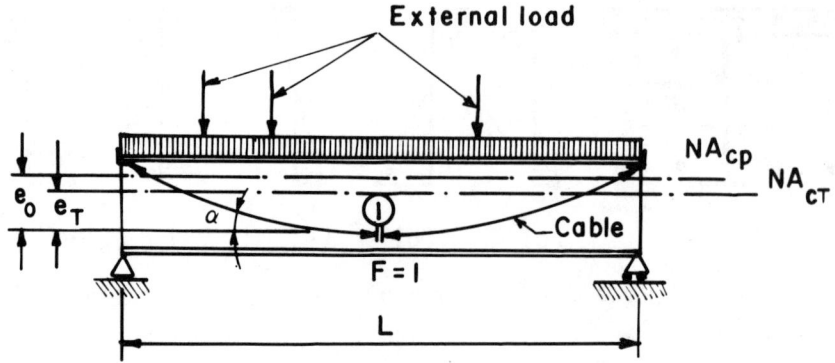

Fig. 2.10   A diagram related to virtual work.

Applying the virtual work method

$$\delta_{01}^{LL+I+SD} = \int_0^l \frac{(M_{LL+I+SD})}{E_s I_t} e_0 dl + \sum \frac{N_{LL+I+SD}}{E_s A_t} \Delta l \qquad (2.30)$$

where

I_t = moment of inertia of transformed composite cross section
$N_{LL+I+SD}$ = axial force in the beam under total external load

and the second term is zero or negligible.
   In a similar manner,

$$\delta_{11} = \int_0^l \frac{e_0^2 \, dl}{E_s I_t} + \sum \frac{\Delta l}{E_s A_t} + \sum_0^1 \frac{\sec^2 \alpha}{E_{cb} A_{cb}} \Delta l \qquad (2.31)$$

The prestressing force increment produces additional stresses (Fig. 2.9(e))

$$f_{tb} = -\frac{\Delta X_0}{A_{tr}} - \frac{\Delta X_0 e_0 y_b}{I_{tr}} \qquad (2.32)$$

$$f_{tt} = -\frac{\Delta X_0}{A_{tr}} + \frac{\Delta X_0 e_0 y_t}{I_{tr}} \qquad (2.33)$$

where

E_{cb} = modulus of elasticity of concrete beam
A_{cb} = cross section of the concrete beam

2.2.7 Loss of Prestress

The value of the required prestress force X_0 is reduced by the following factors.

1. ΔX_{CC}—Loss Due to Creep of Concrete

The effect of creep is to relieve stresses in the concrete and to increase the steel stresses as well as any loss due to plastic flow. The AASHTO specifications[3] compensate for creep and shrinkage by requiring threefold values of the concrete modulus, or that

$$n = \frac{E_s}{3E_c}$$

2. ΔX_{sr}—Loss Due to Shrinkage of Concrete

The shrinkage of concrete varies widely. For ordinary prestressed concrete, an average shrinkage strain value of 0.0003 may be used.

3. ΔX_{fr}—Frictional Loss

The frictional loss is small and almost insignificant in the design of simply supported prestressed composite girders, due to the slight directional tendon changes that occur only at the saddles. Also, the coefficient of friction existing between the lubricated steel saddles and greased cables is of little magnitude.

4. ΔX_t—Loss Due to Unequal Temperature Changes

Since the coefficient of expansion in steel is nearly the same as that for concrete, there is practically no loss of prestress due to temperature drop. The heat in setting the cement is almost entirely dissipated within the first week after placing.

For pretensioning work, if the steel is tensioned at one temperature and the concrete sets at a higher temperature, there will be a loss of prestress. Consequently, the total prestress loss may be considered as the summary of separate factors, or

$$\Sigma \Delta X_{\text{loss}} = -\Delta X_{cc} - \Delta X_{sr} - \Delta X_{fr} - \Delta X_t \qquad (2.34)$$

and the required initial and final prestressing forces are

$$X_{\text{final}} = X_0 + \Delta X_0 - \Sigma \Delta X_{\text{loss}} \qquad (2.35)$$

The magnitude of such losses can be expressed as those given by Lin in the percentage of prestress.[4]

For the average steel and concrete properties cured under average air conditions, the percentage may be taken as shown in Table 2.1. However, when con-

Table 2.1 Average Prestress Loss Factors

Loss of Prestress	Pretensioning (%)	Posttensioning (%)
Elastic shortening and bending of concrete	3	1
Creep of concrete	6	5
Shrinkage of concrete	7	6
Creep in steel	2	3
Total	18	15

ditions deviate from the average, alternate allowance should be made accordingly.

2.2.8 Shear

1. General Equation

The general equation for a shear force under working load (Fig. 2.11) is

$$V = V_0 - X \sin \alpha \tag{2.36}$$

where V_0 represents those shearing forces on a statically determinate structure at the section under consideration.

The shear stresses in the web are

$$\tau = \frac{VQ}{It_w} \tag{2.37}$$

where

Q = the static moment of steel and concrete that lies above the fiber under consideration, about the neutral axis

t_w = the web thickness

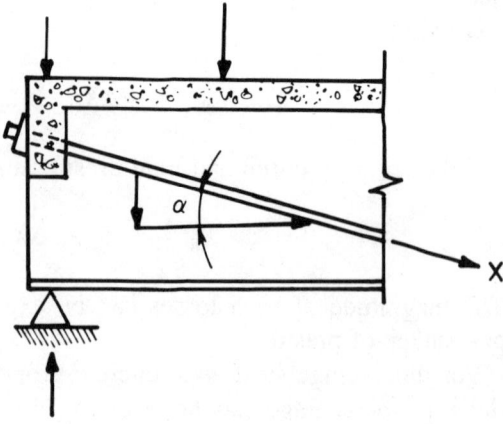

Fig. 2.11 Shear forces.

2. Shear Connectors

The design of shear connectors is basically the same as that used for conventional composite steel structures. It is recommended, however, that they be slightly overdesigned, since their cost represents a small fraction of the total cost but their performance is very important in order to have a composite section that acts monolithically.

2.3 BEAMS PRESTRESSED BY BENDING

2.3.1 Bending of Rolled Sections

A rolled section may be bent within its elastic range by jacking and welding together at this stage, and it will remain prestressed after the load is removed (Fig. 2.12). When such a beam is stressed by a load in which the beam is bent in a reverse direction to that caused by prestressing, the stresses in the extreme fibers of the cross-sectional area will be opposite in sign to that of the prestresses. This will result in a uniform distribution of the stresses throughout the beam cross section and will thus ensure a more effective utilization of the material.

Symmetrical beams

Consider a prestressed beam composed of two members, symmetrically shaped, with respect to their horizontal axis (Fig. 2.13) and bent to cause tensile stresses at their top edges and compressive stresses at their bottom edges. Those stresses in the outside fibers of separate rolled sections occurring before welding are

$$\sigma_0^1 = \frac{M_0}{I_0} \times \frac{h_0}{2} = \frac{M_0}{S_0} \tag{2.38}$$

where

M_0 = the bending moment of one member
I_0 = the moment of inertia of the member
S_0 = the sectional modulus of the member

After the members are welded, the bending load is removed. This is equivalent to an application to the built-up beam of the bending moment $2M_0$, of opposite sign. Then, the stresses in the extreme top and bottom fibers of the beam are

$$\sigma_{01}^1 = \frac{2M_0}{S_w} = \frac{2\sigma^1 S_0}{S_w} \tag{2.39}$$

where S_w is a sectional modulus of the welded beam.

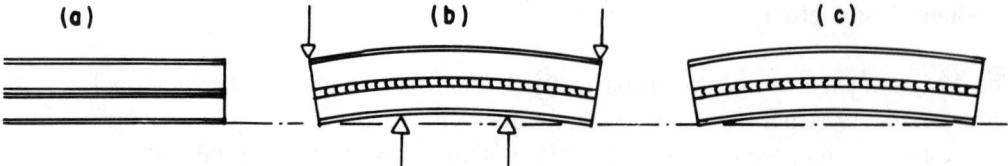

Fig. 2.12 Prestressing by initial bending. (a) Initial configuration. (b) Bending and welding of the two components. (c) A prestressed member.

When the beam is under a working load in an opposite direction to that of the prestress, the stresses at the extreme fibers are

$$\sigma_p^1 = \frac{M_L}{I} \times \frac{h}{2} = \frac{M_L}{S} \tag{2.40}$$

The resulting stresses at the top and bottom edges are, respectively

$$\sigma_t^1 = \frac{M_0}{S_0} - \frac{2\sigma_0^1 S_0}{S_w} - \frac{M_L}{S_w} \tag{2.41}$$

$$\sigma_b^1 = \frac{M_0}{S_0} + \frac{2\sigma_0^1 S_0}{S_w} + \frac{M_L}{S_w} \tag{2.42}$$

2.3.2 Asymmetrical Beams

Let us now consider two asymmetrical beams, bent to cause tensile stresses at their top edges and compression stresses at their bottom edges (Fig. 2.14). As in the case of symmetrical beams, the resulting stresses at the top and bottom edges are, respectively

$$\sigma_t'' = \frac{M_0}{S_t} - \frac{2M_0}{S_w} - \frac{M_L}{S_w} \tag{2.43}$$

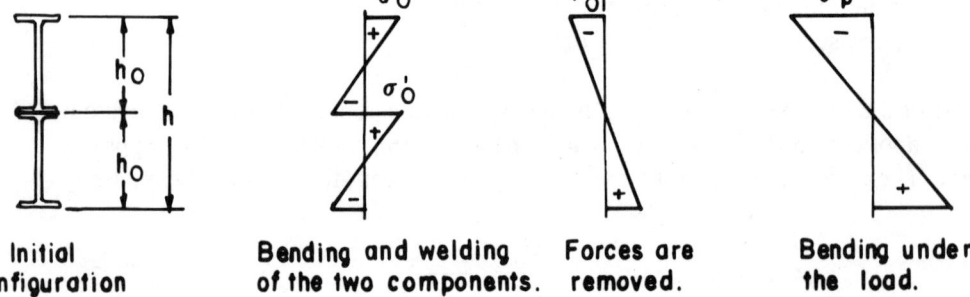

| Initial configuration | Bending and welding of the two components. | Forces are removed. | Bending under the load. |

Fig. 2.13 Diagrams of normal stresses in symmetrical beams formed by the connection of bent components.

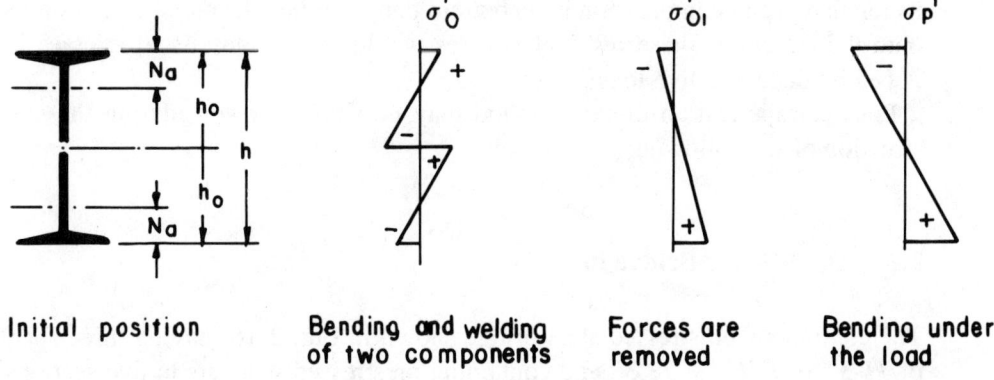

Fig. 2.14 Diagrams of normal stresses in asymmetrical beams by the connection of bent components.

$$\sigma_b'' = \frac{M_0}{S_t} + \frac{2M_0}{S_w} + \frac{M_L}{S_w} \tag{2.44}$$

where

S_t = the sectional modulus of a single member with respect to its flange

S_w = the sectional modulas for the welded members

2.3.3 The Bending and Welding of High-Strength Cover Plates

It is possible, by prebending the rolled beams, to obtain a prestressed hybrid beam having high-strength steel plates welded to the flanges (Fig. 2.15).[5] Prestress is applied by deflection of the beam in an upward direction, at a predetermined amount, and by welding high-strength cover plates onto the top and bottom flanges before releasing its deflection forces.

Upon removal of the loads, we obtain a deflected member having tension in

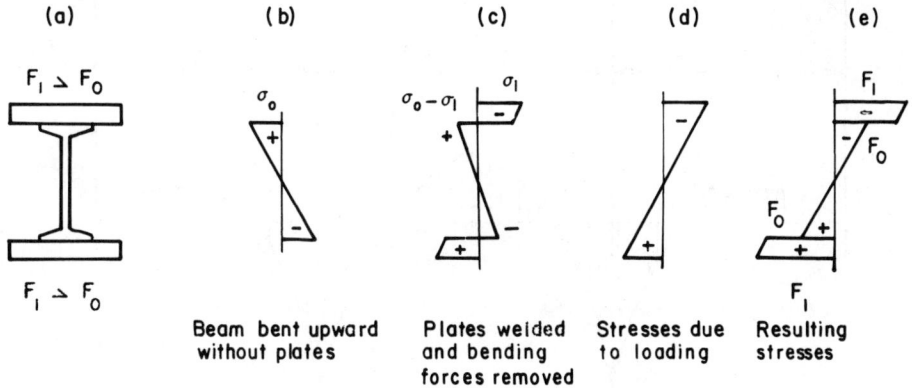

Fig. 2.15 Stresses in a prestressed beam with flanges having welded high-strength plates.

its top flange and compression in its bottom flange. When the member is installed onto the structure, these residual stresses would oppose any usual stresses induced by dead and live loads.

Practical application of this method may be shown by considering the construction of the following two bridges.

2.3.4 Des Moines Bridge in Iowa

The continuous prestressed steel bridge shown in Fig. 2.16, having three spans of 73'3" + 93'6" + 73'3" and containing prestressed stringers in five sections, was built in the state of Iowa.[6] In cross section, there are four lines of stringers at 9'8" spacings carrying a 30-ft roadway and 3-ft curbs (Fig. 2.17). For each line of stringers there are two 51'3" end sections, two 44' pier sections, and one 49'6" center section, a total of 20 prestressed stringer sections.

The prestressed steel stringers were composed of A36 wide flange sections with T 1 steel cover plates for prestressing. The rolled beams of A36 steel have an ultimate strength of 60,000–80,000 psi at a design stress of 20,000 psi. The cover plates of T1 steel have an ultimate tensile strength of 115,000–135,000 psi, at an allowable design tensile stress of 54,000 psi. The prestressing for each beam was performed by placing it on a platform and jacking it into predetermined deflections. Temporary stiffeners were added as a necessity at its jacking and bearing points. The lateral supports at the jacking points prevent lateral buckling.

The jacking load at the center of the spans varies from 60,000 lb for an end section of an exterior stringer (Fig. 2.18 (a)) to 100,000 lb for a section of an

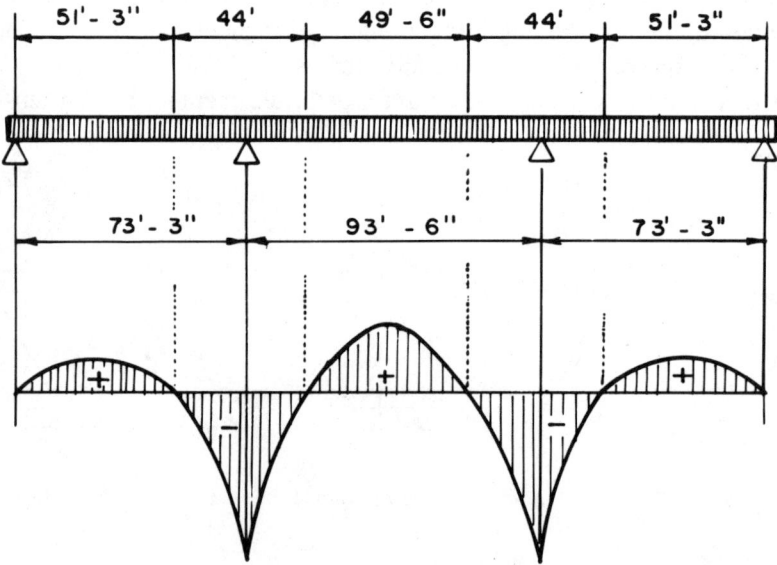

Fig. 2.16 Des Moines Highway Bridge. Distribution of prestressed parts. Elevation.

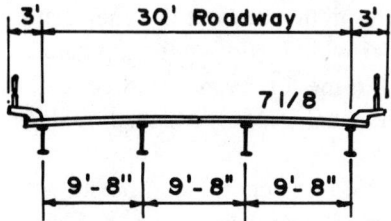

Fig. 2.17   Des Moines Highway Bridge. Cross section.

interior stringer over a pier (Fig. 2.18 (b). The estimated deflections after the release of a jacking load are 7/16 in. and 11/16 in., respectively.

Welders tack the cover plates while the beam is deflected under the jacking load. After the prestressing load is released, the cover plates on the pier sections are then welded onto the top and bottom flanges of the beam with 5/16-in. continuous fillets. Continuous 3/16-in. fillet welds are used at the end and center sections.

The stringers are spliced at their points of contraflexure. Field splices were made with high-strength bolts, using A36 steel for all splice material.

Designers estimated a savings in the weight resulting from prestressing to be as much as 25% as compared to a nonprestressed continuous steel bridge of comparable length.[7]

2.3.5 Bentalon Bridge

In the designing of a deck girder bridge carrying Bentalon Avenue in Baltimore, prestressed welded girders were adopted.[8] The prestress was applied by deflecting the girder upward a predetermined amount and welding a high-strength cover

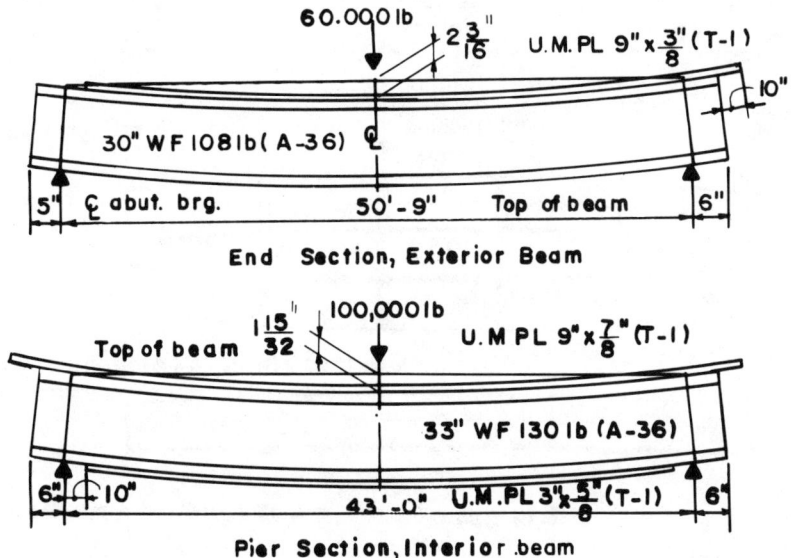

Fig. 2.18   Stringers have one cover plate for end sections (top) and two for pier section.

plate onto the bottom flange before releasing the deflected forces. The result, upon removal of the load, was a deflected member with tension in the top flanges and compression in the bottom flanges. When the member was installed in the bridge, these residual stresses were opposing the usual stresses induced by dead and live loads.

The bridge was designed with eight stringers, 7'4" on centers, composite with a 7" slab having a 2" wearing surface. Each stringer consists of an A36 steel 36WF230 rolled girder, with a 15" × 3/4" cover plate of T1 steel welded to the bottom flange and a 15" × 1/2" cover plate of A441 steel welded to the top flange (Fig. 2.19).

To apply the prestress, the beams were restrained at 31 ft on either side of the center, and then jacked against each other at the third point in between (Fig. 2.19(a). Temporary stiffeners were placed at the restraints and jacking joints, the ends of the beams rotating freely. The jacking force was held at 105 kips and T1 steel plates were tack-welded onto the bottom flanges. As shown in Fig. 2.19(b), after the jacking force was removed, the T1 plates were fully welded, and A441 steel plates were welded onto the top flanges.

Figure 2.20 shows a camber beam layout diagram. The actual deflection compared favorably with that of the computed value. The specifications called for design stresses of 55,000 psi in the T1 steel and 27,000 psi in the A441 steel, respectively. The computed stresses in the prestressed girder under dead and live loading were 53,000 psi in the T1 steel, 20,000 psi in the rolled girder, and 23,000 psi in the A441 top cover plate. These figures reflect the efficient used of different types of steel in this hybrid welded girder.

It should be observed that only the 62-ft central length of each 106-ft girder

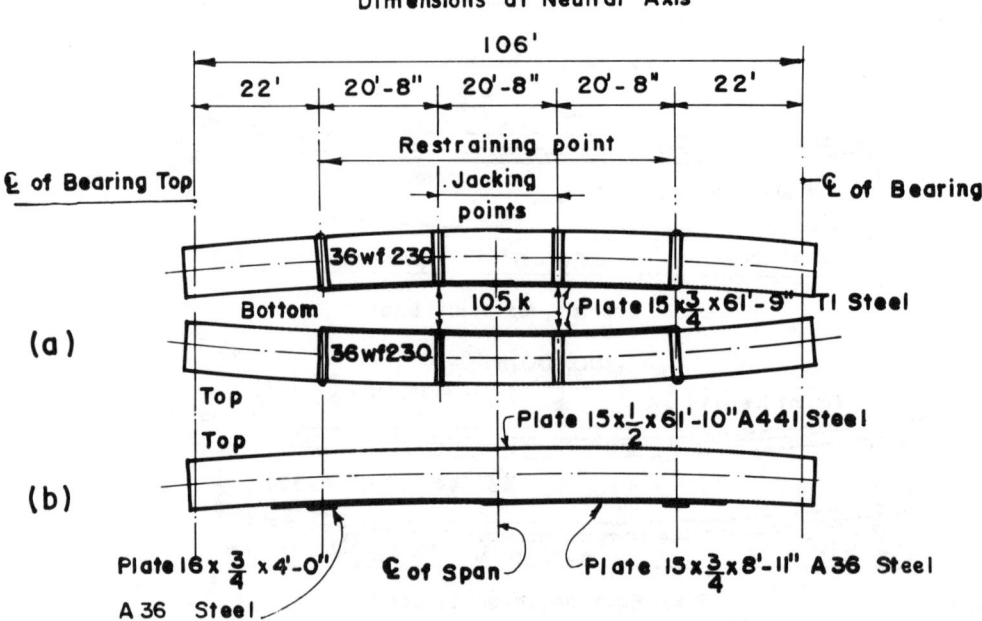

Fig. 2.19 Bentalon Bridge. Prestressing of stringers.

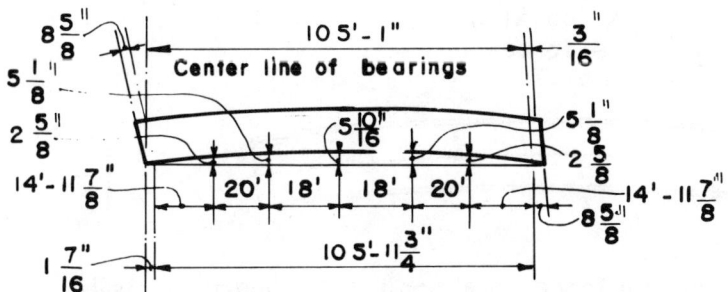

Fig. 2.20 Bentalon Bridge.
A camber diagram.

was subjected to prestressing. By the introduction of prestressed rolled girders, the saving in weight was about 20% in the steel and the reduction in overall cost about 5%.

2.4 PREDEFLECTION TECHNIQUE

The predeflection technique was developed in 1949 by Lipski and is known under by the term *Preflex* beam.[9] This technique enables the use of concrete-encased high-strength steel beams in cases where deflections or cracking of concrete, or both, would otherwise be excessive. The system is generally used when shallow construction depths are required.

During fabrication, a rolled form, having a cover plate if necessary, or a welded section of high-strength structural steel, is deflected in the direction of the design loads. While the section is maintained in a deflected and stressed condition, the tension flange is encased in high-strength reinforced concrete. When the concrete casting achieves the required strength, the predeflection is released, and this precompresses the encasing concrete of the tension flange (Fig. 2.21). Prior to deflection, the fabrication included precambering to the desired shape and the addition of shear connectors for composite action.

The bending stresses and their properties of the cross-section at four separate stages of fabrication or construction are shown in Fig. 2.22.[10]

Stage 1: The steel beam is under the jacking forces.
Stage 2: The concrete is placed while the jacking forces are maintained. Reinforcement in the casting may be included. The bending moment due to self-weight may be deduced from the predeflection moment provided the beam is supported at its ends.
Stage 3: Stresses occur after casting of the concrete slab forming the top flange of the beam, including concrete encasement of the web. Some loss of prestress will occur between the time of release and the casting of the top slab, due to creep and shrinkage of the concrete.
Stage 4: Superimposed dead and live loadings are applied. Separate calculations should be carried out for these two types of loading, using appropriate modular ratios.

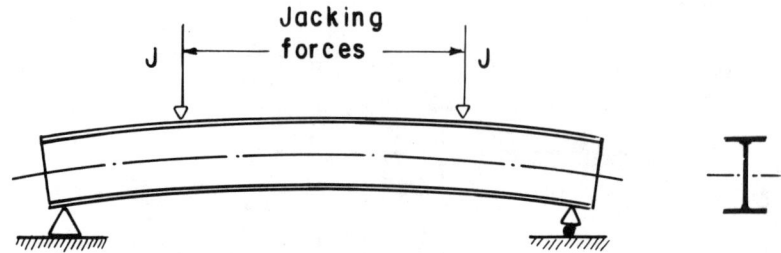

(A) Step -1, jacking forces are applied to beam furnished
by mill with predetermined camber.

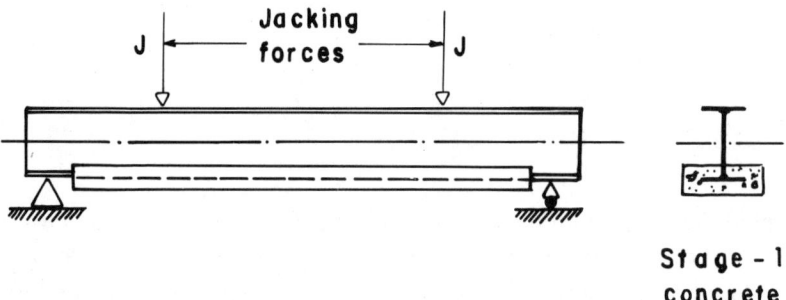

(B) Step - 2, stage-1 concrete is placed while jacking
forces are maintained.

Fig. 2.21   Diagrams illustrating the Preflex technique.

The advantages to be gained by preflexing a steel beam are as follows:

1. An increase in the stiffness and a corresponding decrease in deflections over
   the range of the working load
2. Permanent encasement of the steel beam by a high-strength concrete in
   which there are fewer cracks at the working load than in a cased beam of
   the same size
3. An increase in the span/depth ratio compared to that of an uncased composite beam.

The predeflected beam is transported and erected in a manner similar to that
of a steel beam. The web and top flange are then encased in concrete, usually
monolithically with the floor slab.

2.5 REDISTRIBUTION OF BENDING MOMENTS

By changing the support levels of continuous girders, one can redistribute the
bending moments and consequently redistribute the stresses in cross sections of

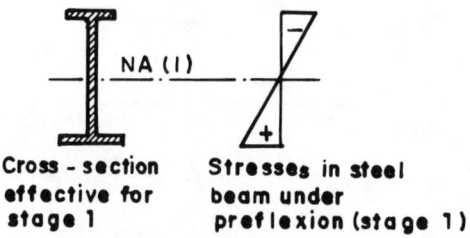

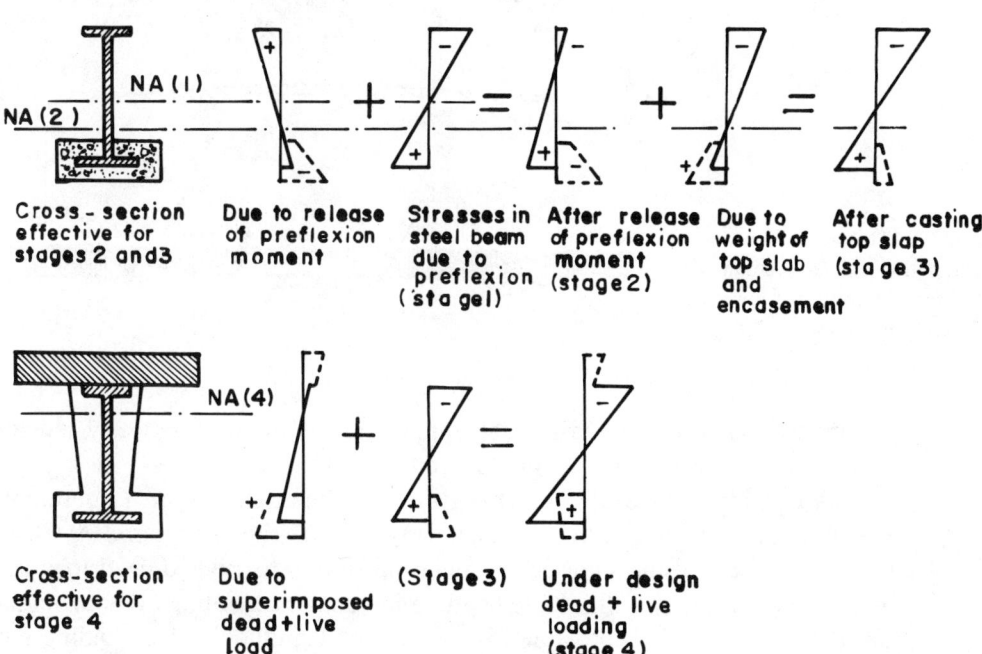

Fig. 2.22   Cross-sectional properties and their stresses at various stages of loading a Preflex beam.

the girder. By the application of additional forces and by changing the support levels and other measures, it is always possible to achieve an expedient distribution of the forces in the main carrying members of a bridge or to redistribute their stresses in cross sections under consideration.

This artificial method of allowing the most useful or expedient combination of stresses in the carrying members of a girder may be called the *regulation of stresses*. The regulation of stresses is, therefore, the rational use of this method in creating the most favorable state of stress in the members of a structure.

The regulation of stresses may be achieved by several methods, all of which basically posses a common concept, namely, an artificial creation in the member under consideration of a force in reverse direction to that force originated in the member under the loading.[11]

The regulation method of changing support levels leads to the artificial creation of additional bending moments, which are added algebraically to those moments existing in a given section. It is possible to change the bending moments

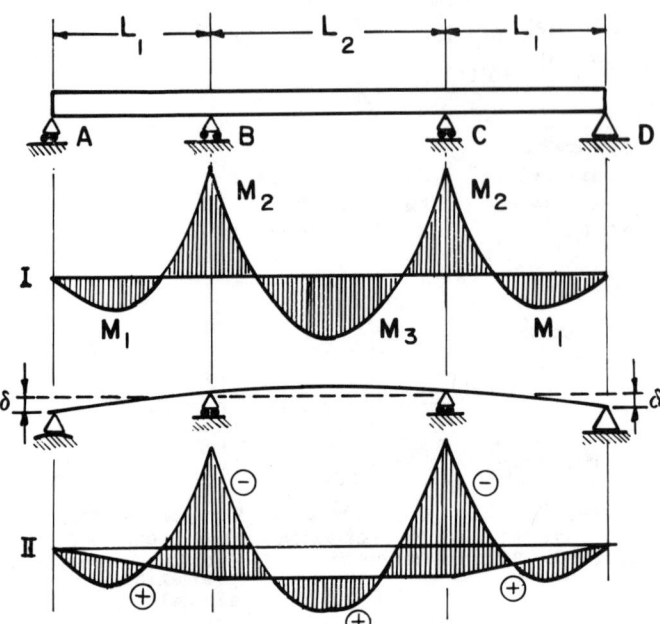

Fig. 2.23 The bending-moment redistribution of a continuous three-span girder by the lowering of its end supports.

by artificially lowering or raising the support levels in such statically indeterminate structures as continuous girders and frames.

Let us consider, for example, changing the support levels of a continuous three-span girder having different spans (Fig. 2.23). When supports A, B, C, and D are situated at the same level under uniformly distributed dead load in a continuous girder, they will initiate bending moments, according to moment diagram I. In this case, it is possible to change the magnitude of the bending moment by lowering support levels A and D. If each of the supports A and D are lowered onto a value δ, then in the girder above supports B and C, negative moments M_B and M_c will be initiated, and after adding them to the bending moments under dead load, the resulting moment diagram will be as shown under moment diagram II.

In this way, by artificial redistribution, it is possible to achieve the equalization of M_1 and M_3, which are approximately equal to

$$L_2 : L_1 = 1.3$$

However, it is also necessary to consider that the equalization of these span bending-moment magnitudes by the lowering of their supports will lead to an increase of the support moments. By this, a reduction in the mid-span girder depth will be achieved by increasing the depth above its supports, and this will require a girder of changeable depth.

An advantage of support moment increase due to span moment reduction consists in the following consideration. The section modulus for an I girder may be taken as

$$S = \frac{Ah}{2} - \frac{A_w h}{3} \qquad (2.45)$$

or

$$S = \frac{Ah}{6}(3 - 2m) \qquad (2.46)$$

where

A = area of girder cross section

h = the girder depth

$m = A_w/A$ = the ratio of the cross-sectional area of a web to the total cross-sectional area of the girder

By substituting $S = M/f$, we obtain

$$\frac{A}{M} = \frac{6}{fh(3 - 2m)} \qquad (2.47)$$

This expression represents a specific cross-sectional area of the girder or that area per unit of bending moment.

Equation (2.47) indicates that a specific area of the girder is in inverse proportion to its depth. Therefore, accepting m as the constant value, it is possible to conclude that it is advantageous to transfer the moment to act over the higher depth of the girder toward its supports.

However, it is necessary to note that the use of girders having changeable depth complicates the technological process of their fabrication and therefore increases their cost. For this reason, the advantage in applying this method of moments regulation may be justified only for bridges having large spans, when the economy of metal obtained due to moment regulation exceeds that of the cost increase due to more complicated fabrication.

For medium bridge spans, continuous girders having parallel chords will be most advantageous. However, such a solution is not practical in cases of large differences of absolute values of span and support moments.

2.5.1 The Bending-Moment Redistribution in Two-Span Continuous Girders

By applying the regulation method for the vertical displacement of supports, it is possible to obtain the equivalence of absolute values of the span and support bending moments (Fig. 2.24). In the moment diagram shown in Fig. 2.24(b),

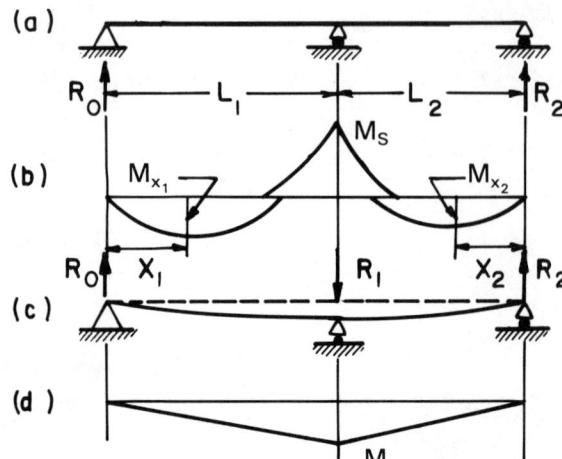

Fig. 2.24 A scheme for the regulation of bending moments of a two-span continuous girder.

under dead- and live-load action, we designate:

M_{x_1} = the maximum positive bending moment in the left span at a section located at a distance x_1 from the left support

M_{x_2} = the maximum positive bending moment in the right span at a section located at a distance x_2 from the right support

M_s = the maximum (by absolute value) support moment

For a two-span continuous girder, we have

$$M_{x_1} < M_s > M_{x_2} \tag{2.48}$$

The problem of moment regulation consists in obtaining the equality by absolute value of a maximum support moment to the maximum span moment. If, by lowering the middle support at a value δ_1 (Fig. 2.24(c)) we introduce an additional positive moment, m_s, above the middle support, (Fig. 2.24(d)), then we may express the problem of regulation as follows:

$$M_{x_1} + \frac{m_s x_1}{L_1} = M_s - m_s \tag{2.49}$$

Solving this equation, we have

$$m_s = \frac{(M_s - M_{x_1})L_1}{L_1 + x_1} \tag{2.50}$$

The value of a calculated support moment, considering the lowering of the middle support is

$$-M_s + m_s = -M_s' \tag{2.51}$$

The calculated span moment will then obtain an increment, or

$$M_{x_1} + \frac{M_s x_1}{L_1} = M'_{x_1} \qquad (2.52)$$

and there will be equality between the support and the span moments:

$$-M_s = M'_{x_1} \qquad (2.53)$$

The maximum moment in a second span, considering its initial bending, is

$$M'_{x_2} = M_{x_2} + m_s \frac{x_2}{L_2} \qquad (2.54)$$

To initiate moment m_s at the middle support, it is necessary to lower the middle support to δ_1. After lowering the middle support to δ_1, reaction R_0, R_1, and R_2 will originate at its supports, the direction of which is shown in Fig. 2.24(c).

Evidently, the additional moment at the middle support, in this case, will be

$$m_s = R_0 L_1 = R_{01} \, \delta_1 \, L_1 \qquad (2.55)$$

where R_{01} is the unit reaction at the end support 0 due to vertical displacement of the middle support for $\delta_1 = 1$. Equalizing expressions (2.50) and (2.55), we have

$$\frac{(M_s - M_{x_1}) \, L_1}{L_1 + x_1} = R_{01} \, \delta_1 \, L_1 \qquad (2.56)$$

Therefore, the required displacement of the middle support, applying the moment M_s, will be

$$\delta_1 = \frac{M_s - M_{x_1}}{R_{01} \, (L_1 + x_1)} = \frac{M_s}{R_{01} \, L_1} \qquad (2.57)$$

We determine the unit reactions F_{01} and R_{ii} by applying the virtual work method using the general formulas

$$R_{ik} = \Sigma \int \frac{M_i M_k}{EI} \, ds \qquad (2.58)$$

$$R_{ii} = \Sigma \int \frac{M_i^2}{EI} \, ds \qquad (2.59)$$

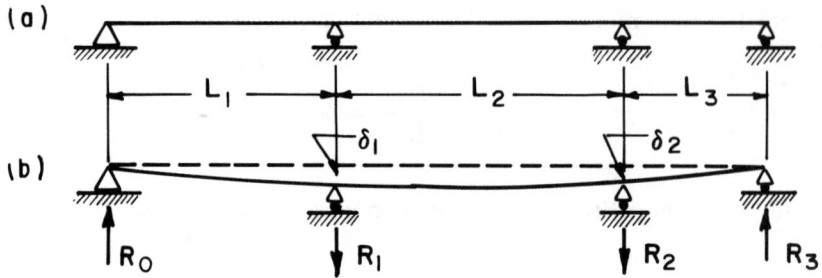

Fig. 2.25   A scheme for the regulation of bending moments of a three-span continuous girder.

2.5.2 The Bending-Moment Redistribution in Three-Span Continuous Girders

Let us assume that in applying the previous method, the values of the additional support bending moments m_{s1} and m_{s2} are determined (Fig. 2.25). To obtain bending moments m_{s1} and m_{s2}, it is necessary to lower supports 1 and 2 correspondingly for δ_1 and δ_2 (Fig. 2.25 (b)).

These additional bending moments at the middle supports may be expressed as follows:

$$m_{s_1} = R_0 L_1 \quad \text{and} \quad m_{s_2} = R_3 L_3 \tag{2.60}$$

For the end supports, the reactions are

$$R_0 = R_{01} \delta_1 + R_{02} \delta_2$$
$$R = R_{31} \delta_1 + R_{32} \delta_2 \tag{2.61}$$

where

R_0, R_3 = Reactions at supports 0 and 3
$\quad R_{01}$ = unit reaction at support 0 due to the vertical displacement of support 1, due to $\delta_1 = 1$
$\quad R_{02}$ = unit reaction at support 0 due to the vertical displacement of support due to $\delta_2 = 1$
$\quad R_{31}$ = unit reaction at support 3 due to the vertical displacement of support 3, due to $\delta_1 = 1$
$\quad R_{32}$ = unit reaction at support 3 due to the vertical displacement of support 2, due to $\delta_1 = 1$

By substituting expression (2.61) into expression (2.60), we obtain

$$m_{s_1} = (R_{01} \delta_1 + R_{02} \delta_2)L_1$$
$$m_{s_2} = (R_{31} \delta_1 + R_{32} \delta_2)L_3 \tag{2.62}$$

A solution of these equations gives the required values for the vertical displacements of the supports

$$\delta_1 = \frac{R_{02}\, m_{s_2}\, L_1 - R_{32}\, m_{s_1}\, L_3}{L_1\, L_3\, (R_{31} R_{02} - R_{32} R_{01})}$$

$$\delta_2 = \frac{R_{31}\, m_{s_1} L_3 - R_{01}\, m_{s_2}\, L_1}{L_1 L_3\, (R_{31} R_{02} - R_{32} R_{01})}$$

(2.63)

REFERENCES

1. Szilard, R., "Strengthening Steel Structures by Means of Prestressing," *The Engineering Journal*, October 1955, pp. 1379–1381.
2. Szilard, R. "Design of Prestressed Composite Steel Structures", *Journal of the Structural Division, Proceedings of the ASCE*, November 1959, pp. 97–123.
3. *The AASHTO Standard Specifications for Highway Bridges*, 12th ed., Washington, D.C., 1977, p. 194.
4. Lin, T. Y., *Design of Prestressed Concrete Structures*, 2nd ed., Wiley, New York, 1963, p. 113.
5. Subcommittee 3 on Prestressed Steel of Joint ASCE-AASHO Members, Committee on Steel, "Development and Use of Prestressed Steel Flexural Members," *Journal of the Structural Division, Proceedings of the ASCE*, September 1968, pp. 2033–2060.
6. "Prestressed Steel Beams Replace Concrete," *Steel Construction Digest*, vol. 18, no. 2, Second Quarter, 1961, p. 10.
7. "Prestressing Steel Stringers Reduce Bridge Weight by 25%," Engineering News-Record, October 19, 1961, pp. 32–33.
8. Levine, L., and McLoughlin, J., "Prestressed Steel Stringers Solve Bridge Clearance Problem," *ASCE—Civil Engineering*, April 1967, pp. 42–43.
9. Baes, L., and Lipski, A., *Preflex Beam—Principles, Notes on Calculation, and Description Notes*, I, II, and III, Preflex S.A., Brussels, June 1953, May 1954, April 1958.
10. Johnson, P. P., and Buckby, R. J., *Composite Structures of Steel and Concrete*, vol. 2, Granada, London, 1979, pp. 284–289.
11. Tolmachev, K. H., *Regulation of Stresses at Metal Bridge Structures*, Scientific and Technical Edition of Automobile Transportation, Moscow, 1960 (in Russian).

Chapter 3
Tendons and Anchorages

3.1 INTRODUCTION

3.1.1 Basic Definitions

Prestressing is the imposition of a state of stress upon a steel structure before it is placed into service. Prestressing can enable structures to withstand any forces and loads imposed on them in service. When prestressing is applied to a structure that is already in service to increase its carrying capacity, it is known as post-stressing or posttensioning.

The prestressing of steel structures is generally performed by tendons connected at both ends to anchorages that are fixed to those members being pre-stressed. Tendons are tensioned or stretched elements that are used to create compression in the structure. Generally, high-strength steel tendons are stretched between two anchorages and jacked to about three-fourths of their ultimate strength.

Tendons of high-tensile steel used for prestressing usually take one of the following forms: wires, strands, wire ropes, cables, and bars. Configuration of the tendons, strung along the structural member being prestressed, may be one of the following: rectilinear, curvilinear, and polygonal.

Tendons are usually connected to a main structural member at diaphragms along its length. Diaphragms assure stability of the members during prestressing, and the tendon should have close contact with the diaphragms. However, diaphragm design should ensure free longitudinal displacement of the tendon during its tensioning.

Internal tendons are the tendons contained within the cross-sectional area of the steel member. They may later be bonded to it by concreting or grouting. External tendons are those tendons which lie outside the cross-sectional area of the steel member. The external tendons may be connected by stirrups to the steel member at certain spacings.

The prestressing may be applied in a sequence of two or more steps and is then called stage-stressing.

46

3.1.2 Mechanical Properties of Tendons

The following mechanical properties of tendons are important.

1. *Ductility:* Tendons should possess ductility in order to prevent brittle failure during installation and service. Ductility is measured by bend and elongation tests. The elongation should be 2% or more.

2. *Stress relaxation:* This is an irreversible plastic steel flow that occurs under applied high stress and loads to cause a reduction in the tendon stress, which may range from 6% to 13%.

3. *Corrosion:* Tendons should be protected from corrosion, which may affect the ductility and fatigue strength and reduce the cross-section. Protection may be provided by galvanizing and epoxy coating, which should be continuous.

4. *Fatigue:* With a variation of live load, prestressed tendons undergo only a very small range of stress change. Therefore, fatigue is generally not a problem.

5. *Temperature variation:* Tendons are not substantially affected by very low temperature, except to increase the modulus of elasticity and decrease the ductility. At elevated temperature, the rate of stress relaxation, as well as the ductility, increases substantially.

3.1.3 Steel Wires

Steel wires to be used in tendons for prestressing are drawn from a cold rod in a simple continuous length and produced in diameters of up to 0.276 in. (7 mm).[1] The tensile strength and minimum yield strength, measured by a 1.0% total-elongation method, are given in Table 3.1 for the common wire sizes.

3.1.4 Structural Strands

Structural strands for prestressing are manufactured by a wire arrangement to produce a symmetrical section. A parallel wire strand is obtained when individual wires are arranged in a parallel configuration without a helical twist.

The static mechanical properties of a structural strand are stated in the American Society for Testing and Materials (ASTM) Specifications.[2] These specifications contain various information on the physical requirements, tests for zinc coating, weight, data on those wires used to make the strand, strength tables, sampling, testing, inspection, and packaging. The mechanical properties of some strands consisting of Class A zinc-coated wires are given in Table 3.2.

3.1.5 Prestretching

Prestretching is applied to a structural strand or rope in order to remove any looseness (constructional stretch) inherent to the manufacturing process. The prestretched wire or rope becomes an elastic material within the limits of the

Table 3.1 Prestressing Wire

Nominal Diameter (in.)	Weight (lb/ft)	Area (in.²)	Ultimate Strength (psi)	Minimum Yield (psi)
0.192	0.098	0.0289	250,000	200,000
0.196	0.10	0.0302	250,000	200,000
0.250	0.17	0.0491	240,000	192,000
0.276	0.20	0.0598	235,000	188,000

prestretching operation, which enables the prediction of that elongation under load, to a high degree of accuracy.

The removal of the constructional stretch is affected by the repeated application of tension load to the strand or rope, which forces the component wires to arrange themselves into closer contact. The prestretching load applied to a cable does not usually exceed 55% of the rated minimum breaking strength of a strand, or 50% for a rope, which essentially eliminates the constructional stretch of a cable or rope.

Strands shall be prestretched to the necessary loading value and held long enough to attain the minimum modulus of elasticity. Such loading shall be approximately 50% of the breaking strength.

When construction stretch is eliminated, any given working tension or load of predetermined relation thereto can be applied, and overall lengths and fitting positions can be measured and located within close tolerances.

3.1.6 High-Tensile Steel Bars

High-strength steel bars up to 2 in. in diameter are used primarily for prestressing and are produced from medium-carbon steel with the addition of several alloy elements. This steel is presently obtained with an ultimate strength of about 150,000 psi (ASTM A722 bars) and a yield point of approximately 70% of its ultimate strength. Its high quality is usually attained by heat treatment. Because

Table 3.2 Steel Strands

Diameter (in.)	Weight (lb/ft)	Area (in.²)	Minimum Breaking Strength (tons)	Minimum Modulus of Elasticity (psi)
1/2	0.52	0.150	15	
3/4	1.18	0.338	34	
1	2.10	0.600	61	
1 1/2	4.73	1.35	138	24×10^6
2	8.40	2.40	245	
2 1/2	13.1	3.75	376	
3	18.9	5.40	538	
3 1/2	25.7	7.35	724	23×10^6
4	33.6	9.60	925	

Table 3.3 Steel Bars

Bar Size (in.)	Weight (lb/lin ft)	Area (in.²)	Initial Tensioning Load at 80% of Ultimate Strength	Final Design Load at 60% of Ultimate Strength	Minimum Guaranteed Strength (lb)
3/4	1.5	0.442	51,300	38,500	64,100
7/8	2.04	0.601	69,700	52,300	87,100
1	2.67	0.785	91,000	68,300	113,800
1 1/8	3.38	0.994	115,300	86,400	144,100
1 1/4	4.17	1.227	142,300	106,700	177,900
1 3/8	5.05	1.485	172,200	192,200	215,300

of the low ductility of this steel, slight bending is necessary for the fabrication of prestressed reinforcing, which must be performed very carefully to avoid rupturing the steel.

The high strength needed for prestressed tendons is achieved by using a specially selected hot-rolled alloy steel. In their hot-rolled state, the alloy bars have neither uniform elastic properties nor the required ductility. These properties are obtained by subjecting the bars to a stress-relieving treatment at 600°F for 8 hr in a furnace. They are left closed in the furnace until cooled to approximately room temperature.

After stress relieving, the bar has all the desirable properties except for the stress-strain curve and high yield point typical of that for prestressed tendons. These are developed by cold stretching each bar to at least 90% of its ultimate strength. The cold-stretching process also serves as a proof stressing which eliminates any bars having surface imperfections or metallurgical defects.

Table 3.3 gives the physical properties for stress-steel bars, present values for ASTM A722 bars.

3.2 ANCHORAGES

The prestressing system comprises essentially a method of steel stressing combined with a method of anchoring the tendon(s) onto the structure. Anchorages are mechanical devices used to transmit the tendon force to the steel structure. They include the means of gripping and securing the tendon installation to the steel member.

There are a number of different systems, some of which are patented, for tensioning and anchoring the tendons to the structure.[3] Detailed knowledge of the end anchorages and tensioning jacks is necessary in order to design or specify a system, including design of the ends of the posttensioned members so as to install these anchorages and jacks.

The methods of mechanical posttensioning can be classified into three groups, as follows:

1. Posttensioned anchorages for bars
2. Posttensioned anchorages for strands
3. Posttensioned anchorages for wires

3.3 ANCHORAGES FOR BARS

3.3.1 Anchorages Having Threaded Joints

In the threaded-anchorage type of system, the anchoring of bar tendons is based on their threaded joints (Fig. 3.1). The bearing capacity of a bar tendon is calculated, considering its threaded joint, and the bar cross section is decreased 30–35%.

In nut-type anchoring, bar tendons are generally tensioned by hydraulic jacks, whose pulling device is connected to the threaded end of this bar.

3.3.2 Lee-McCall System

One common approach to end anchorages for steel bars is known as the Lee-McCall system in England, or stress-steel system in the United States. The bar ends are threaded and anchored with nuts onto washers and bearing plates. With the use of tapered threads, about 98% of the bar strength is developed (Fig. 3.2). A short length of the bar is threaded at the untensioned end, sufficient to receive the nut resting on a washer. However, for the jacking end, a long threaded end is required.

The hexagonal nuts for the bars have a diameter equal to about twice the bar diameter and a thickness about 1.6 times that of the bar diameter. The standard washers are made of 3/16-in. and 14-gage metal. The anchorage plates differ in size and can accommodate one to three bars per plate. The plates have a thickness of 5/8 to 1 1/2 in.

3.3.3 Wedge-Type Anchorage

Bar tendons may also be anchored by means of wedges (Fig. 3.3 (a)) and chuck grips (Fig. 3.3 (b)). The thread for connecting the jack pulling device is located at that end of the bar beyond its grip, and the bars are therefore not weakened by threading between the two end points of tendon anchorage. The advantage of the wedge anchorage is its convenience in gripping the bar at any point along its length.

3.4 ANCHORAGES FOR STRANDS

3.4.1 Sleeve-Type Anchorage

Tendons formed from wire ropes have sleeve-type anchoring at their ends (Fig. 3.4). The sleeves are manufactured in the form of a hollow cylinder with an

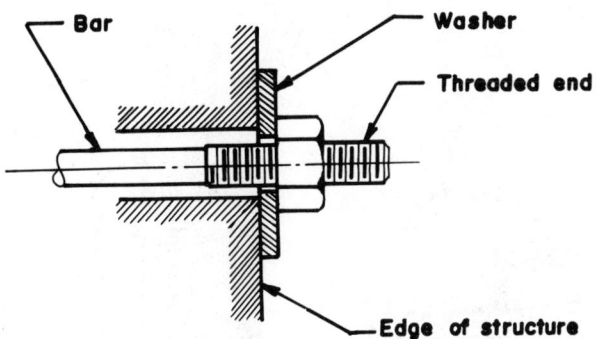

Fig. 3.1   Threaded anchorage.

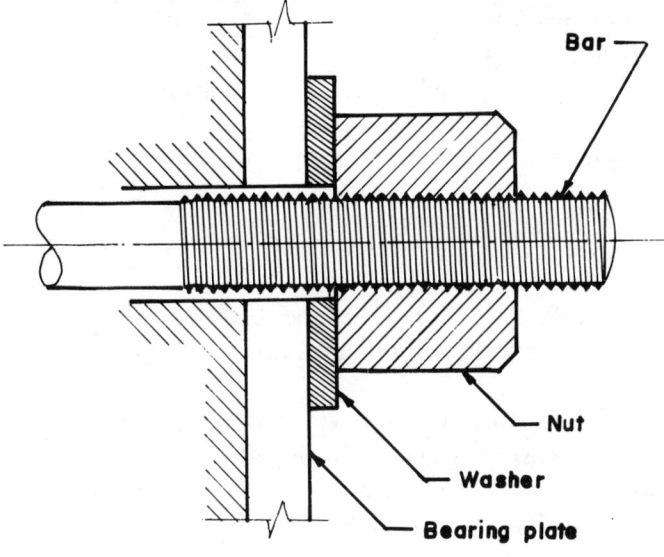

Fig. 3.2   An end anchorage for the Lee-McCall or stress-steel system.

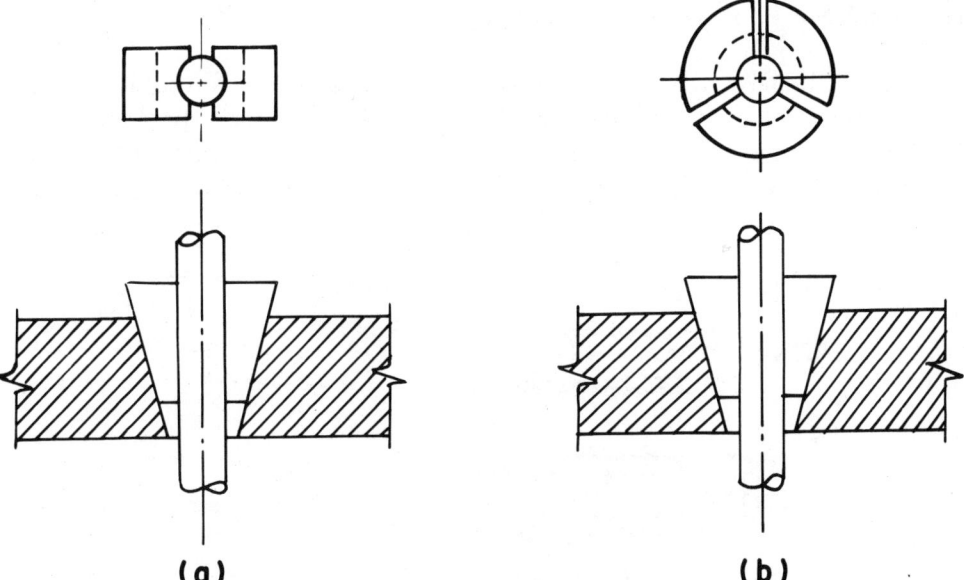

Fig. 3.3   Wedge- and chuck-type bar tendon anchors.

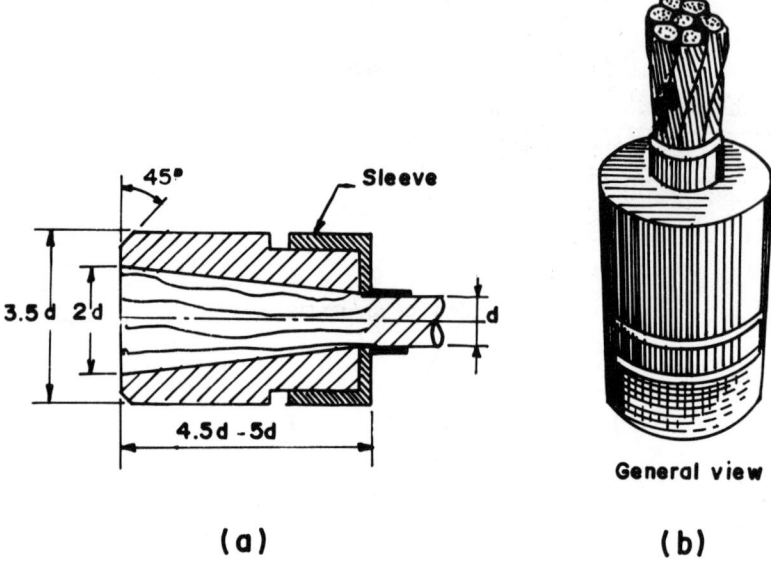

Fig. 3.4   Sleeve-type anchor for steel wire rope.

inside tapered or cylindrical surface which receives the uncoated end of a pre-pared wire rope. Generally, the sleeve dimensions are a function of the wire rope diameter.

The outside of the sleeve carries a thread for screwing the jack grips. It is good practice to fit the sleeves of the tensioning anchors with a stop nut that transmits the tightening force to the structure. The placing of sleeve-type anchors into service requires pouring them with alloys and is thus a complicated procedure.

3.4.2 Shell-and-Wedge Anchorage

The shell-and-wedge anchorage, shown in Fig. 3.5, is relatively long, 8 to 10 diameters of the wire rope, and is only suitable for relatively slight breaking stresses in the wire ropes. The sleeves are manufactured in the form of hollow cylinders, with an inside tapered or cylindrical surface, which receives the un-

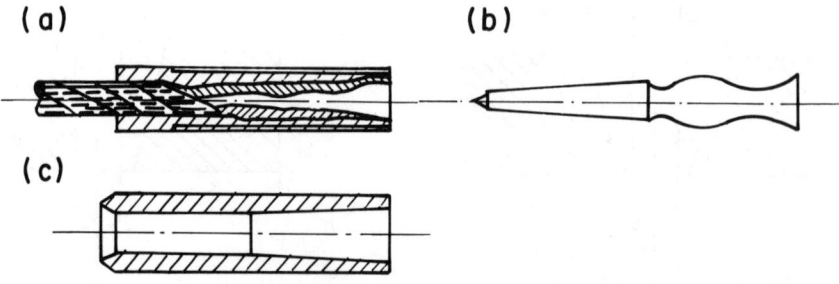

Fig. 3.5   Shell-and-wedge anchorage.

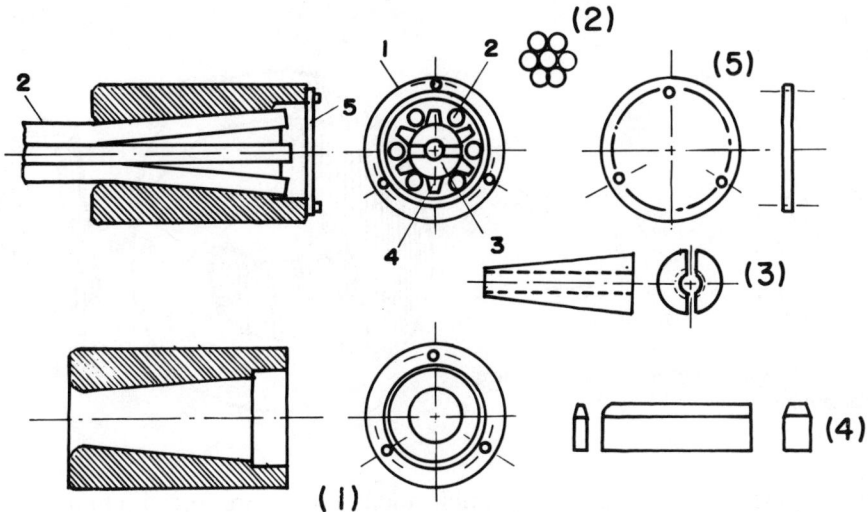

Fig. 3.6   An anchor with wedge grips for a seven-strand wire rope.

coated end of a prepared wire rope. The sleeve dimensions are a function of the wire rope diameter.

3.4.3 Anchorage with Wedge-Type Grips

An anchor with wedge-type grips, for a seven-strand wire rope, 45 mm in diameter, was developed in bridge-building practice (Fig. 3.6). The uncoated wire rope strands are inserted into the sleeve and secured by driving the wedges in manually. To provide protection against corrosion, grease is packed inside the sleeve and a cover is screwed on.

Anchors under static and repeated loads have proven a sleeve-type anchor with driven-in wedges adequately serviceable. The fitting of the wire rope inside the sleeve permits a somewhat greater deformation of the wire rope as compared to an anchor that is poured with an alloy.

3.4.4 Barrel-and-Plug Anchorage

Strands built as rectilinear tendons of tubular cross section from wires 4–8 mm in diameter are generally secured by means of an anchorage consisting of barrels with plugs (Fig. 3.7). The wire ends are inserted into the conical orifice of the barrel and, after the tendon is tensioned, are secured by driving in the plug. Tensioning the tendon and driving in the plug are performed by double-action hydraulic jacks. The barrels are manufactured from quality structural carbon steel, and the plugs from the same grade steel. Along its side, the plug carries a triangular or trapezoidal thread (Fig. 3.7(b)). As the jack exerts a longitudinal force upon the plug by driving it in, it slides in closely between the wire ends.

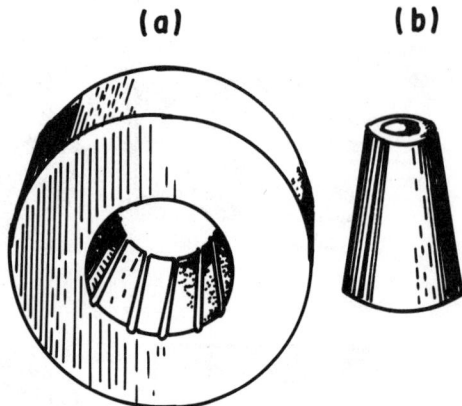

Fig. 3.7 Anchoring a strand in a barrel containing a plug.

The plug thread projections then crumble and the wires are forced into the softer plug steel to be wedged dead between the barrel and the plug. To avoid the wires slipping into the barrel under a load, the force exerted on the plug as it is driven in should be as close to the tendon prestressing value as possible.

A softer steel barrel anchors reliably all the wires of the strand as various deviations in anchor dimensions and in the shape of the wire are compensated for (as the plug is driven in) by those differences in the values of areas over which the wire is pressed into the barrel's conical orifice surface (Fig. 3.7(a)).

Barrel-and-plug anchors are also used for solid-section strands. The simplicity of manufacture and minimal cross-sectional area of this type of tendon make their use practicable.

3.4.5 Shell-and-Bar Anchorage

The strand end is inserted into a shell and clamped between the shell and a bar inserted along the shell length between the wire ends (Fig. 3.8). As tests indicate, this type of anchorage ensures a bearing capacity of only 75–80%. Its advantages are its relatively low cost and ready installation.

3.5 ANCHORAGES FOR WIRES

The Magnel or Magnel-Blaton system uses rectangular sandwich plates of steel which have tapered notches to receive the wedges (Fig. 3.9). Wires of 0.196 or 0.276 in. are gripped between the grooves of the wedges and the sandwich plate. The number of wires for the Magnel cable varies from 2 to 64 per cable for both sizes of wires. In addition to Magnel sandwich plates, anchorage may be considered the following current type anchorages for wires, such as VSL, Freyssinet, BBRV, etc., as used in prestressed concrete.

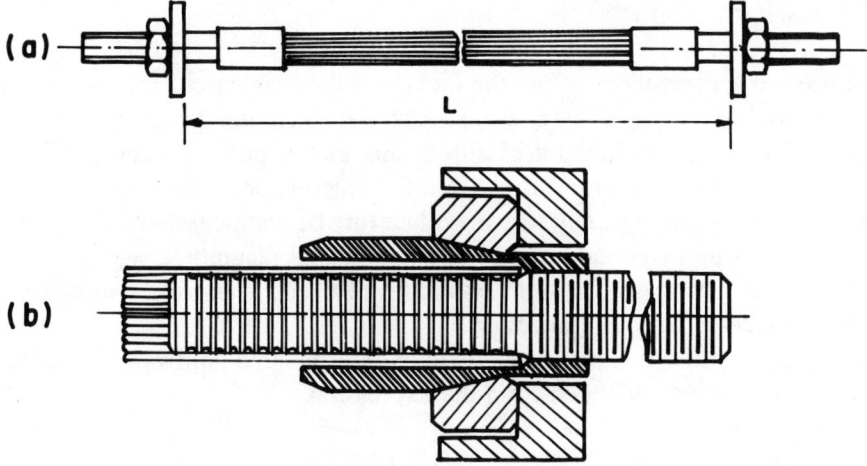

Fig. 3.8   A shell-and-bar anchor.

3.6 LOSS OF PRESTRESS

3.6.1 Introduction

The losses of a prestressing force can be grouped into two categories:

1. Those that occur immediately during the prestressing of a member. The prestress jacking force may be immediately reduced by losses due to friction and anchorage slip. This is called the initial prestress force.

2. Time-depending losses, especially due to relaxation. Then the prestressing jacking force is known as the effective prestress force.

In general, losses do affect such service load behaviors as the deflection or camber, as well as any deformations occurring during construction.

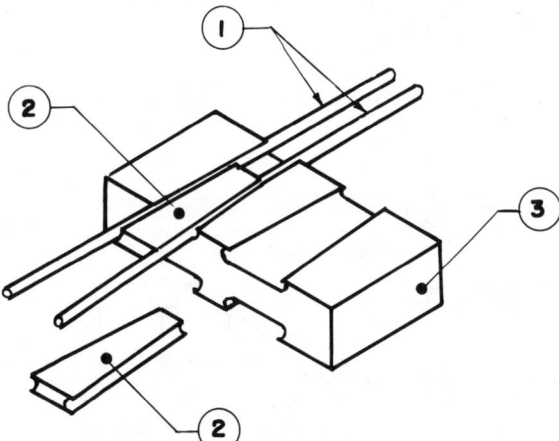

Fig. 3.9   A Magnel sandwich plate with wedges. 1, Wires; 2, wedges; 3, anchor plate for eight wires.

3.6.2 Anchorage Slip

In posttensioned members, when the jacking force is released, the steel tension is transferred to the girder by a special anchorage of one type or another. Inevitably, there is a small amount of slip at the anchorages upon transfer, as the wedges seat themselves more firmly into the tendons or as the anchorage hardware deforms. Anchorage slip loss may therefore be compensated by overstressing, providing its magnitude can be anticipated. Its magnitude will depend on the particular prestressing system or hardware used. The wide variety of anchorages preclude any generalization.

Given the amount of slip characteristic of any specified hardware, the anchorage slip loss can be calculated from the expression

$$\Delta_s = \Delta_a \frac{E_t}{l_t} \tag{3.1}$$

where

Δ_a = the amount of slip
l_t = the tendon length
E_t = the elastic modulus of the prestressing steel

The importance of anchorage slips generally depends upon the length of a member. For very short tendons, the anchorage set will produce high slip losses. For long posttensioned members, slip losses become insignificant.

3.6.3 Losses Due to Friction

The connections between the deflected tendons and structures, apart from their anchorages, include such guides as rollers and shoes. These are introduced to minimize the frictional loss at deflected tendon points. Typical details for plate girder shoes are shown in Fig. 3.10. For continuous girders, tendons may be placed into boxes, Fig. 3.11.

For prestressed steel members, the tendons are usually anchored at one end

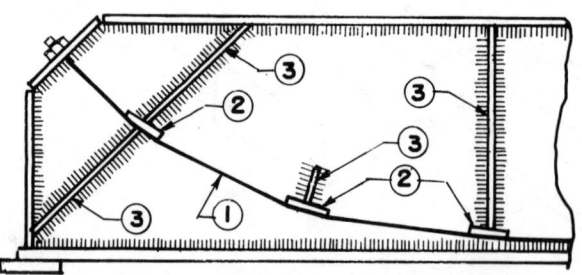

Fig. 3.10 Shoes at turning points
for the tendon of a plate girder.   **I - Tendon   2 - Support plate   3 - Stiffening plate**

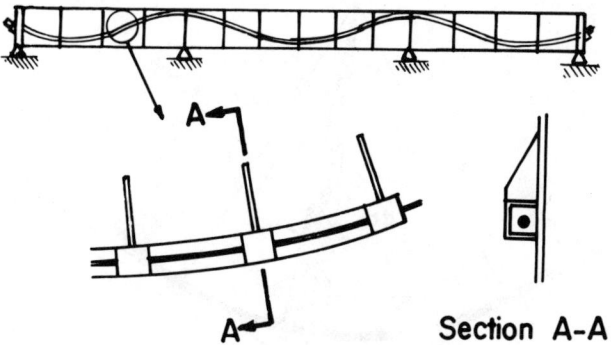

Section A-A

Fig. 3.11 Curved tendons passing through boxes.

and stretched with jacks at the other end. At those support locations where a tendon will change direction, friction resistance is developed during prestressing, with the result that the tension at the anchored end is less than that of the jack tension. Figure 3.12 shows the detail of a typical cylindrical-shaped support for a curvilinear-strung tendon. The prestress loss is due to the curvature friction of the tendon.

Considering an infinitesimal length dx of a prestressed tendon supported by a cylindrical surface of radius R, the angle change of the tendon (Fig. 3.13), is

$$d\theta = \frac{dx}{R} \tag{3.2}$$

The normal pressure component produced by a force, F, acting in the tendon bent around an angle, $d\theta$, is given as

$$N = fd\theta = \frac{Fdx}{R} \tag{3.3}$$

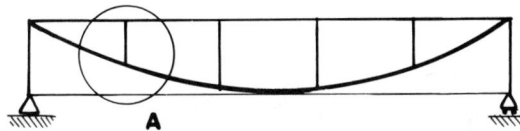

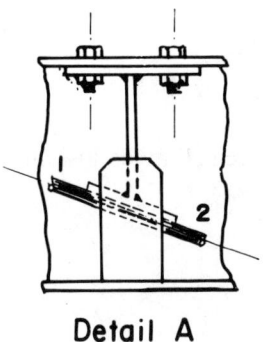

Detail A

Fig. 3.12 A prestress loss due to friction at the location of change in a tendon's direction.

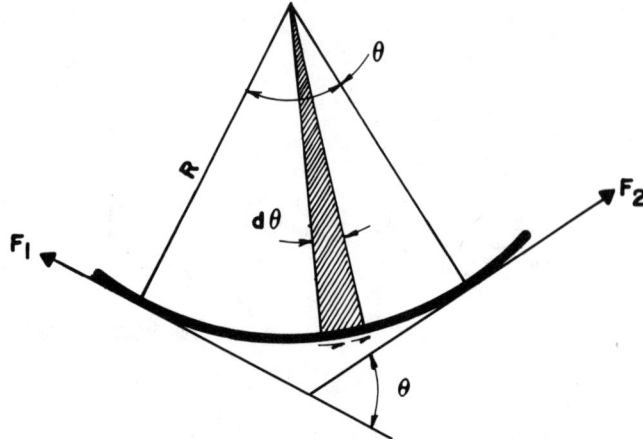

Fig. 3.13 Friction loss along
the tendon.

The magnitude of friction loss, dF, around the length, dx, is

$$dF = -\mu N = -\mu \frac{F dx}{R} = -\mu F d\theta \tag{3.4}$$

or

$$\frac{dF}{F} = -\mu d\theta \tag{3.5}$$

where μ is a coefficient of friction.

By integrating both sides of equation (3.5) between the limits F_1 and F_2, we have

$$\int_{F_1}^{F_2} \Leftarrow \frac{dF}{F} = -\mu \int_0^\theta d\theta$$

or

$$\log_e\left(\frac{F_2}{F_1}\right) = -\mu\theta \qquad F_2 = F_1 e^{-\mu\theta} \tag{3.6}$$

Therefore, loss in the tendon force is

$$\Delta F = F_1 - F_2 = F_1(1 - e^{-\mu\theta}) \tag{3.7}$$

Dividing the friction loss dF by the tendon area A_t gives loss in stress due to curvature friction

$$\Delta F = f_s(1 - e^{-\mu\theta}) \tag{3.8}$$

where f_s is the tendon stress at the jack.

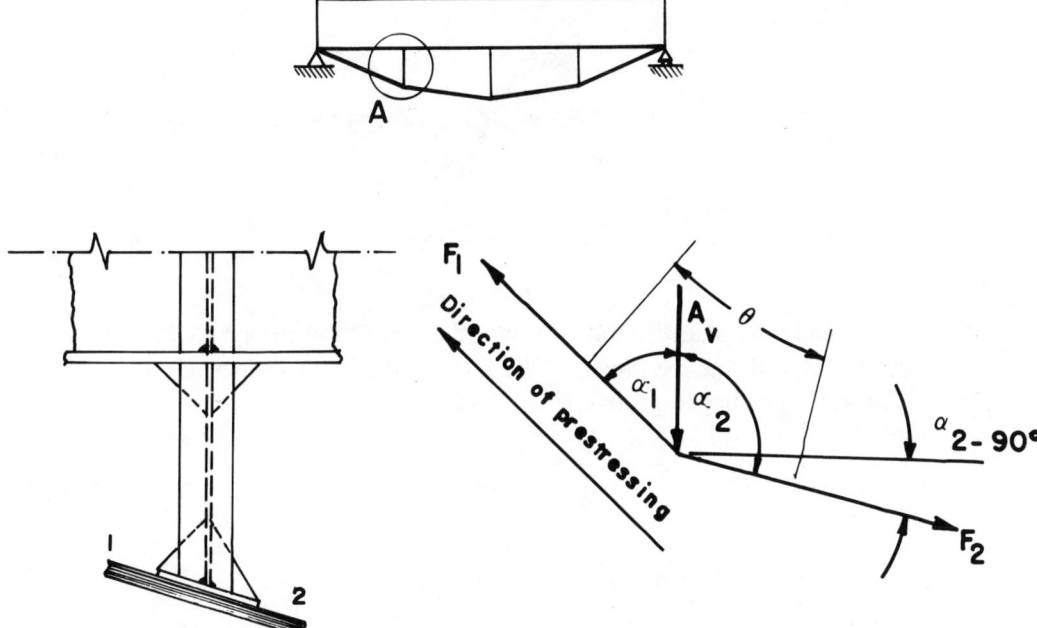

Fig. 3.14 Equilibrium of forces at their turning point.

3.6.4 Friction at Supports of a Polygonal Tendon

We now consider the general case of a tendon draped in a polygonal configuration (Fig. 3.14), where at sections 1 and 2, due to prestressing, forces F_1 and F_2 are acting in equilibrium with reaction A_v.[4]

We assume that the tendon is supported at the turning point by a fixed cylindrical support and that the tendon is not placed in the groove. Reaction A_v is considered uniformly distributed over the length of the tangential surface along the tendon. For an equilibrium condition, by projecting forces F_1 and F_2 on the horizontal axis, we have

$$F_1 \sin \alpha_2 = F_2 \cos (\alpha_1 - 90)e^{-\mu\theta} = F_2 \sin \alpha_1 e^{-\mu\theta}$$

or

$$F_1 = F_2 \frac{\sin \alpha_1}{\sin \alpha_2} e^{-\mu\theta} \qquad (3.9)$$

When there is no friction, $\mu = 0$, and formula (3.9) gives

$$F_1^0 = F_2^0 \frac{\sin \alpha_1}{\sin \alpha_2} \qquad (3.10)$$

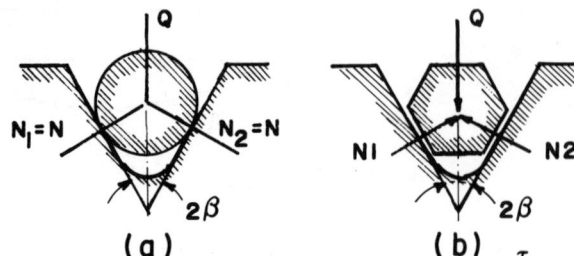

Fig. 3.15 Groove-type supports
for the tendons.

If the tendon is placed in one groove, there are more friction surfaces and the friction loss will thus be greater. The increased friction values are as follows:
(1) for a wedge-shaped groove (Fig. 3.15(a) and (b))

$$\mu' = \frac{\mu}{\sin \beta} > \mu \tag{3.11}$$

(2) for a semicircular groove (Fig. 3.16)

$$\mu'' = \frac{\mu}{\sin \beta} \beta > \mu' > \mu \tag{3.12}$$

3.6.5 Friction at Rotating Cylinder Support

When a rotating cylinder is used as a support for a tendon, there is friction around the pin. The influence of the pin friction is expressed through the moment as follows:

$$M_p = A_p r \tag{3.13}$$

and the product

$$r = \rho_p \mu_p$$

is designated as the radius of the friction circle, where

A_p = reaction on the pin

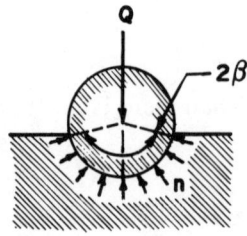

Fig. 3.16 Semicircular groove.

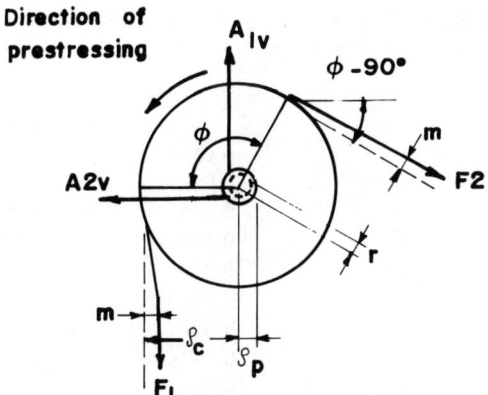

Fig. 3.17 Rotating cylindrical support for a tendon.

ρ_p = radius of the pin
μ_p = friction coefficient of the pin

The stiffness of the tendon results in its deviation outside of the support for a value m, as shown in Fig. 3.17.

From the equilibrium conditions, it follows that

$$F_1 = F_2 \frac{\rho_c + m + r(\sin \psi - \cos \psi)}{\rho_c - m - r} \tag{3.14}$$

where

ρ_c = radius of cylinder
m = deflection of the tendon from straight direction due to tendon rigidity

and from Equation (3.14), we obtain a loss in prestress

$$\Delta F_2 = F_2^0 \left[1 - \frac{\rho_c - m - r}{\rho_c + m + r(\sin \Psi - \cos \Psi)} \right] \tag{3.15}$$

and a supplementary value

$$\Delta F_1 = F_1^0 \left[\frac{\rho_c - m + r(\sin \Psi - \cos \Psi)}{\rho_c - m - r} \right] \tag{3.16}$$

For a tendon profile composed of a combination of straight and curved segments, the losses may be calculated progressively, beginning at the jacking end. For each segment, the end force nearest the jack is equivalent to F_1 and is equal to a reduced force F_2 calculated at the end of the preceding segment.

3.6.6 Values for the Calculation of Prestress Loss

Tables 3.4 and 3.5 give values for use in the calculation of prestress loss.

Table 3.4 Values for Formula $\varphi = 1 - e^{-\mu\theta}$

μ / θ	0.15	0.20	0.25	0.30	0.35
10°	0.026	0.034	0.043	0.051	0.059
20°	0.051	0.067	0.084	0.099	0.115
30°	0.076	0.099	0.123	0.145	0.167
40°	0.099	0.130	0.160	0.189	0.221
50°	0.123	0.160	0.196	0.230	0.263
60°	0.145	0.189	0.230	0.270	0.307
70°	0.167	0.221	0.263	0.307	0.348
80°	0.189	0.244	0.295	0.342	0.387
90°	0.210	0.270	0.325	0.376	0.423

3.6.7 Relaxation of Steel

Prestressed tendons are stressed at an essentially constant length during the lifetime of a member. Relaxation is defined as the loss of stress in a stressed material held at constant length. From the available evidence, it appears that relaxation continues almost indefinitely, although at a diminishing rate. It must be accounted for in the design as it produces a significant loss in the prestress force.

The amount of relaxation depends on the intensity of steel stress, as well as time. For conventional stress-relieved steel, the ratio of the relation reduced stress f_p to an initial stress f_{pi} can be estimated using the following expression:

$$\frac{f_p}{f_{pi}} = 1 - \frac{\log t}{10}\left(\frac{f_{pi}}{f_{py}} - 0.55\right) \tag{3.17}$$

Table 3.5 Values for the Friction Coefficient μ

Type of Support	Frictional Support Surfaces	Type of Prestressed Tendon		
		Smooth Wire Strand Locked Coil Cable	Spiral Wire Rope: Six-Strand and Multiple-Strand Wire Ropes	Wire Ropes Woven Together
(curved support)	Free from rust	0.25	0.30	0.35
(grooved support)	Covered by Vaseline, graphite, paraffin	0.20	0.25	0.30
(V-support)	Free from rust			
(flat support)	Covered by Vaseline, graphite, paraffin	0.15	0.20	0.25

where

f_{py} = the effective yield stress
t = the time in hours occurring after stressing
$log\ t$ = the base 10
f_{pi}/f_{py} = not less than 0.55

For present purposes, this relation may be restated in terms of that loss of steel stress resulting from relaxation:

$$\Delta f_{\text{rel}} = f_{pi}\frac{\log t}{10}\left(\frac{f_{pi}}{f_{py}} - 0.55\right) \qquad (3.18)$$

Special low-relation wires and strands are available. According to ASTM Specifications A416 and A421, such steel shall exhibit relaxation after 1000 hr of not more than 2.5% when initially loaded to 70% of a specified tensile strength, and not more than 3.5% when loaded to 80% of a specified tensile strength.

3.6.8 Controlled Prestressing Force

When a prestressing force is created in a tendon and is transferred to the structure, the anchorage is stretched out, the tendon shortens, and the force is decreased.

Also, the tendons from the wire rope or strands are prone, when loaded, to relaxation which involves a progressive decrease of stresses during that time.

These factors require allowances for a decrease in the tensioning force and the provision for a controlled prestressing force of a value somewhat greater than that indicated by its calculations. The values of the controlled force may be found by the formula

$$X_c = \frac{X}{0.95} + \Delta_a\frac{A_t E_t}{l_t} \qquad (3.19)$$

where

X = a calculated force in a tendon due to prestressing
0.95 = a coefficient of relaxation which is introduced for those tendons of steel wire rope and bundles of high-strength wires only
A_t, E_t, and l_t = the cross-sectional area, modulus of elasticity, and length of a tendon, respectively
Δ_a = the yielding of anchors assumed approximately 1/32 in. for those anchorages composed of tightly screwed nuts or wedge-shaped plugs and to 1/16 in. for gasket-type anchors

Practically, the yielding of anchors Δa should be allowed for in shorter tendons only (less than 65–100 ft long), while for the longer tendons, a shortening of 3/8 in. or 3/4 in. is of no consequence.

3.7 ARRANGEMENT OF TENDONS, DIAPHRAGMS, JOINTS, AND ANCHORAGES

3.7.1 Introduction

Prestressed structures are generally composed of a rigid member from medium-strength steel and tendons from high-strength steel and anchorages. It is advisable to make the cross section of the rigid member symmetrical with respect to the main axes of inertia (Fig. 3.18). These members may be fabricated of two channels, a pipe, an I beam, two or four angles, sheets, and so on, in continuous and solid types.

In heavy members it is good practice to design a tendon from several branches each composed of a wire rope, a single strand of wire, or a high-strength steel bar. Such an arrangement helps to place the tendons conveniently throughout the cross section. Also this facilitates anchoring and reduces prestressing per branch when the tensioning force intensity is limited by the available equipment. Tendon branches should be symmetrically installed in order to obtain coincidence of the center of gravity of tendon branches and the member.

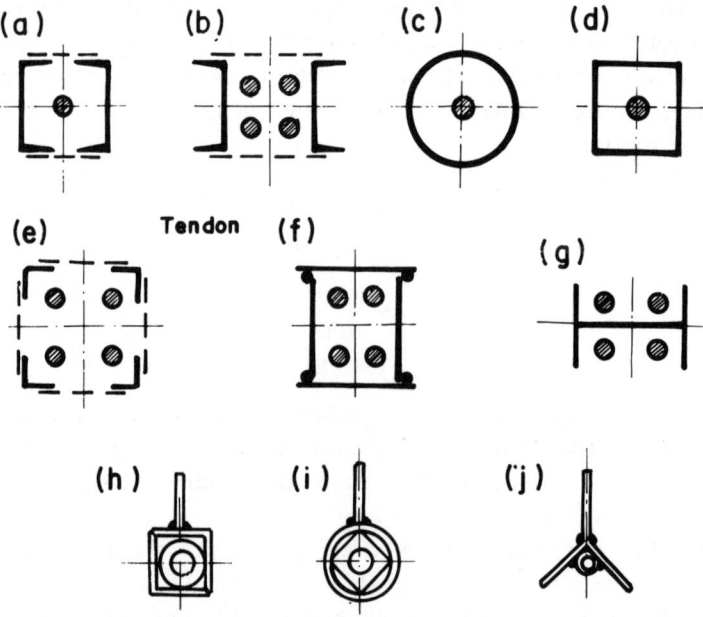

Fig. 3.18   Types of cross sections of members with tendons.

3.7.2 Diaphragms

Throughout its length the rigid member is connected to the tendon by diaphragms spaced at intervals of (40-50) r_{min} of minimum radius of gyration of the member chord cross section during prestressing. The tendon should be in solid contact with the diaphragms. The member acts through the diaphragms on the tendon and remains straight under compressive loads. At its end edges the member has anchors that connect the tendon to the member.

Cross-sections of members may be either open, as in Fig. 3.18 (a), (b), (e), and (g), or closed, as in Fig. 3.18(c), (d), (f), (h), and (i). Members with open cross-section are more convenient for installation of diaphragms, checks on the tendon prestressing value and supervision of the structure during service.

Diaphragms are welded into a cross section as transverse sheets with holes for passing the tendons. The gaps between the edges of holes and tendons should allow free longitudinal movement of the tendon under tensioning to avoid friction at the points of contact. Generally, the gap value is 1/32 in.

In members having closed cross sections, the installation of diaphragms is always a problem; therefore, this type of section is suitable for shorter members which require no reinforcing diaphragms. In this case, the inside of such members is filled with cement grout or concrete to connect the member to the tendon and ensure the stability of the member. Filling the inside of the member also protects the tendon against corrosion.

3.7.3 Joints for Long Members[6]

Sometimes it is necessary to install tendon branches of wire rope or bars for long members consisting of several lengths (Fig. 3.19). Joints are also required when tendons are short or when the member is subdivided into separate units. Should the tendon be too short, it can be jointed together by diaphragms, as in Fig. 3.19(b) and (c). The diaphragms should be strong enough to resist deformation

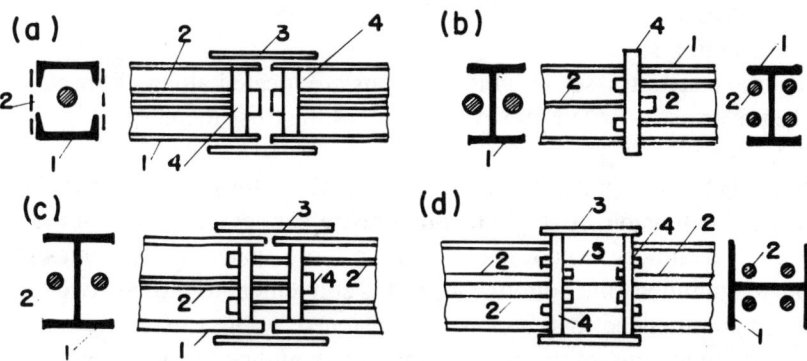

Fig. 3.19 Types of joints of long members. 1, Member; 2, tendon; 3, butt strap; 4, end bearing plate with anchors fastening of tendon; 5, high-strength bolt.

under the action of concentrated forces from the branches of the tendon. The joints of prefabricated components prestressed at the manufacturing plant are provided with butt straps that overlap the member only, as in Fig. 3.19(c). The butt straps then absorb all the stresses from the service load acting in the joint. In heavy members, all the force in a joint or part of it can be transmitted through high-strength bolts (Fig. 3.19(d)), which are installed in place during erection.

3.7.4 Plate Girder Anchorages

For plate girders, the possible arrangements of the anchorages for tendons may be of different types.[7,8,9] The arrangement of tendons at anchorages may be either for single, double or multiple tendons. Large concentrated forces are transmitted to the girder at the location where the tendon is anchored, causing substantial local stresses in the web and chord of the girder. Therefore, it is important to reinforce the anchorage connection to the web of the girder. Usually auxiliary ribs are provided in the anchorage zone to contribute to a uniform transmission of forces and reinforce girder components. The girder web may be reinforced at this location by welding stiffeners on both sides or inserting a thicker plate. (See Figs. 3.20, 3.21, 3.22, and 3.23.)

3.7.5 Truss Anchorages

By the location of tendons and their effect on the behavior of the structure, prestressed trusses may be divided into three main types as follows.[10]

1. Simple span trusses having tendons used for prestressing of most stressed members in tension only, as shown in Fig. 3.24(a).

In such trusses each member is stressed in compression by its individual tendon. Prestressed trusses of this type are effective only for large spans and loads and when each of the prestressed members is an individual prefabricated unit. These members are prestressed either during the fabrication or during preassembly at the erection site.

2. In simple span trusses in which tendons are located along all or part of the span, the prestress is carried for several or all of the truss members (Fig. 3.24(b), (c), (d)).

With this pattern of prestressing one or several tendons are installed along the bottom chord in tension. A single tendon prestresses several panels of the chord, but the other members remain unstressed. In large spans having forces of considerable magnitude at the bottom chord, it is convenient to provide two tendons. Also it is recommended that tendons be tensioned before trusses are erected.

3. In continuous trusses the straight tendons should be located along the chord panels in tension as in Fig. 3.25. Cross sections of typical prestressed members used for trusses are shown in Fig. 3.26.[11,12]

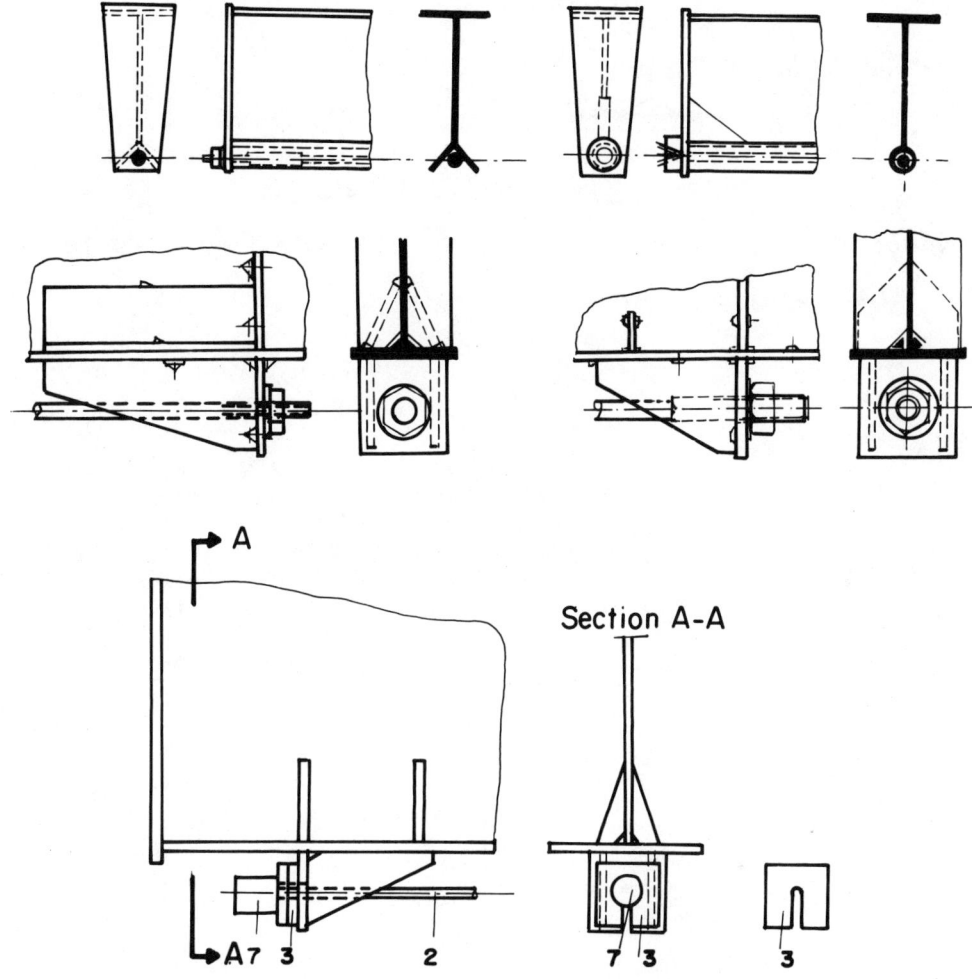

Fig. 3.20 Different arrangements of single tendon at anchorage.

Generally, such members should have cross sections symmetrical with respect to the main axes of inertia. Typical arrangements of the anchorages at trusses are shown in Fig. 3.27.

3.7.6 Protection of the Tendons and Anchorages Against Corrosion

Prestressed steel structures, especially those stressed under tension, should be protected against corrosion, mechanical damage, and very high temperatures.

Due to possible corrosion prestressed steel structures should not be used in areas having high agressive moisture. Rigid elements are protected against corrosion in a manner similar to that used for nonprestressed structures. A covering against corrosion may be used either before or after prestressing. Steel is initially covered by prime, then by one or two layers of the final cover.

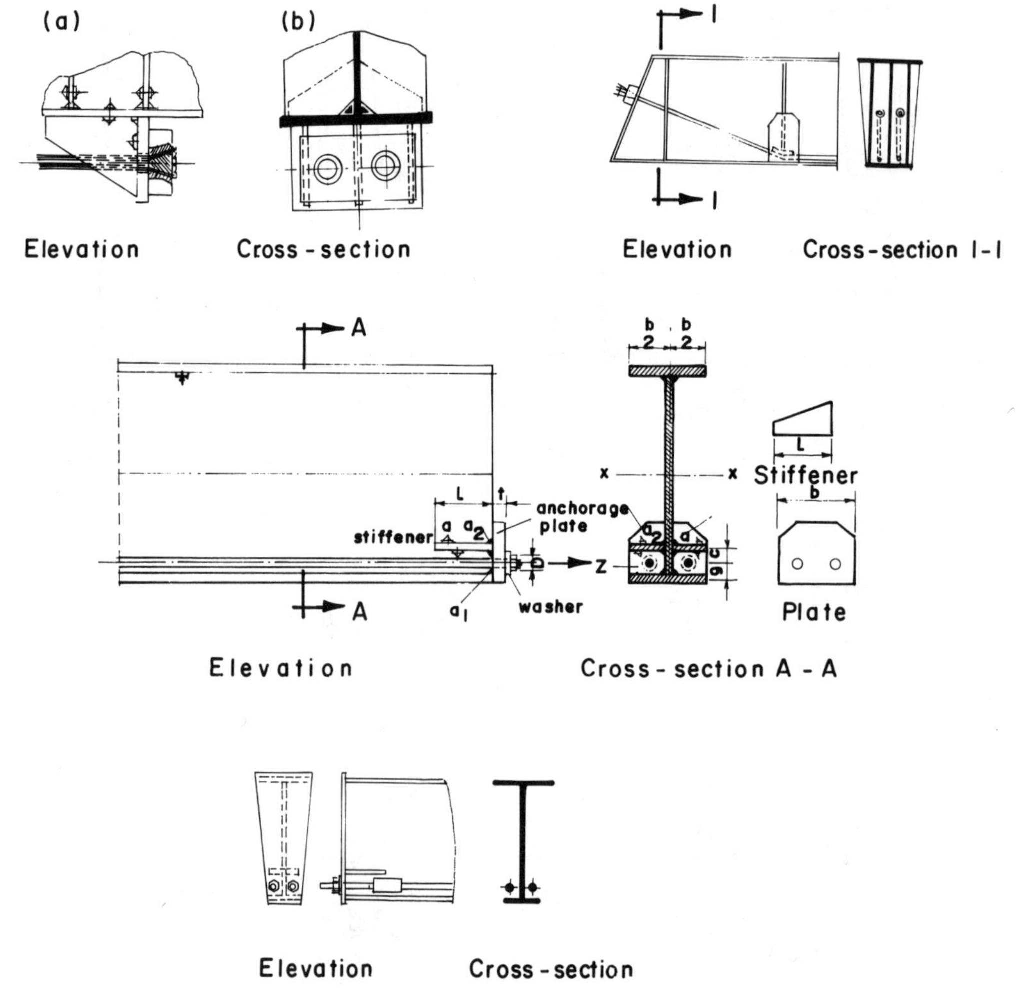

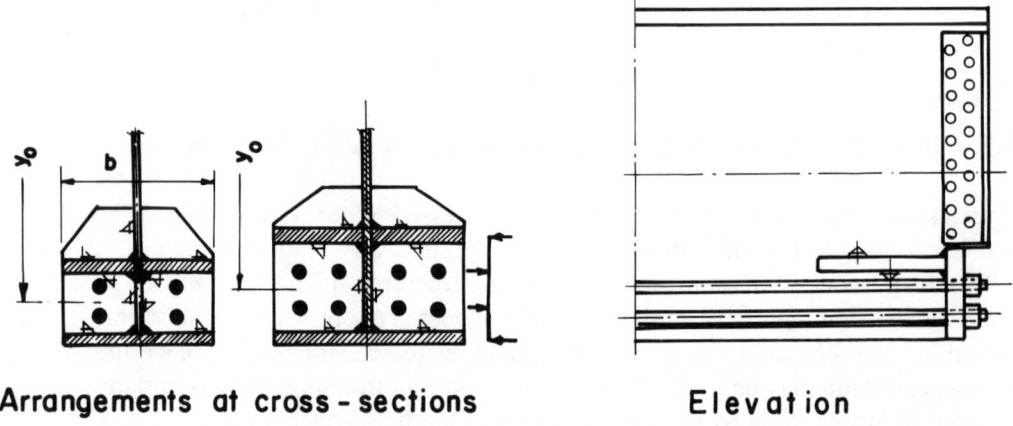

(a)

Elevation   Cross-section

(b)

Elevation   Cross-section I-I

Elevation   Cross-section A-A

Stiffener

Plate

Elevation   Cross-section

Fig. 3.21   Different arrangements of double tendons at anchorage.

Arrangements at cross-sections

Elevation

Fig. 3.22   Arrangement of multiple tendons at anchorage.

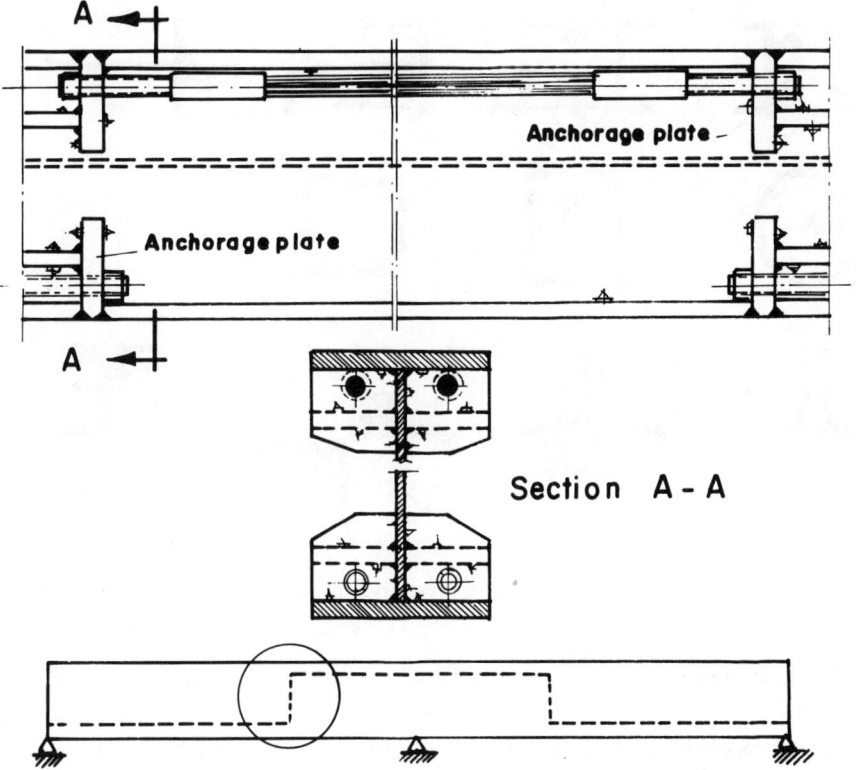

Fig. 3.23 Arrangement of straight tendons at continuous girder.

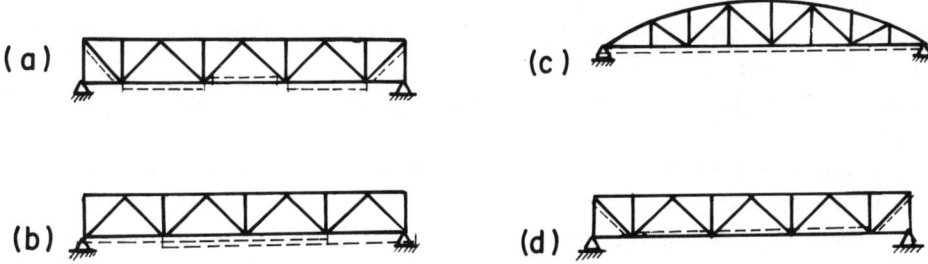

Fig. 3.24 Trusses having different locations of tendons.

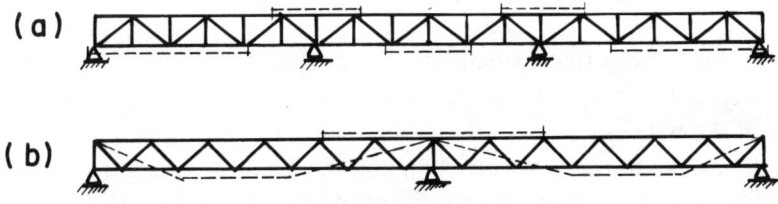

Fig. 3.25 Location of tendons in continuous trusses.

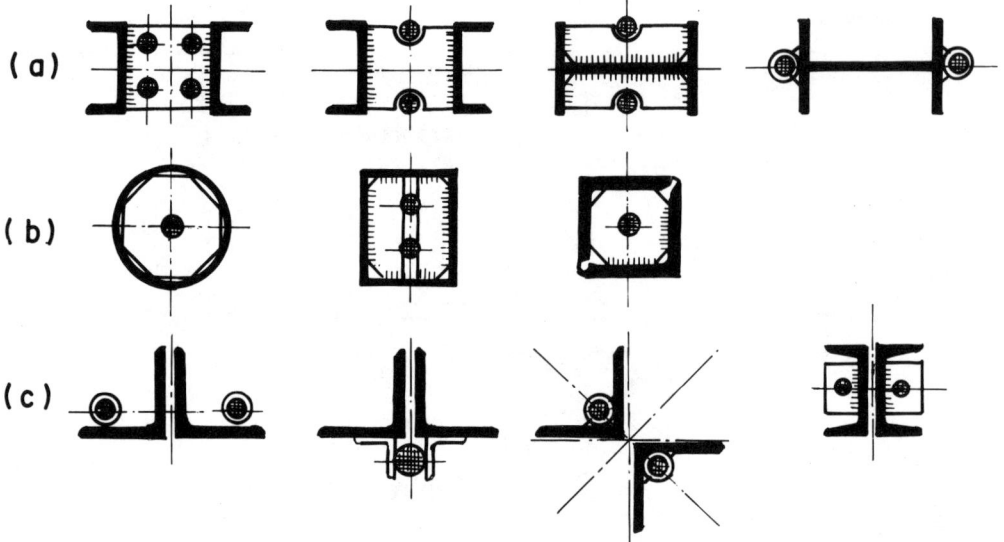

Fig. 3.26   Types of cross sections of prestressed members for trusses.

It should be noted that corrosion must not be allowed to weaken tendon cross sections.

Inside parts of the wire rope are covered by red lead, and external protection is accomplished by applying a corresponding covering. The covering is applied on the wire rope after prestressing of the structure, to prevent the formation of cracks after large deformations. To improve protection against corrosion it is possible to zincify the wire and simultaneously use a coating of zinc chrome over the prime. Wire rope may also be protected against corrosion by the spray application of a protective layer of plastic or by the winding of water-resistant plastic film around the wire rope. In this case it is convenient to insert inside of shell preventors that will absorb moisture that may penetrate into the protected space.

The fabrication of wire ropes having Z-shaped wire covered by plastic film has been proposed.

Cables made from unprotected wires are best for placing inside of closed shapes such as pipes and box-shaped cross sections, which may be filled with asphalt, concrete, grout, and so on. Wire of large diameter, for example, 9/32 in. (7 mm), may be protected with the help of the same covers used for rigid elements; however, the covered part of the wire should not be inserted into the anchorage, because this will reduce friction and increase the slip of the wire.

Special types of wire covered by zinc do not require additional protection from corrosion. Apart from this special wire for cables, other kinds of wire covered by zinc may also be used. It must be noted, however, the use of zinc lowers the strength of the wire.

Considering the hazards of corrosion and mechanical damage, it is necessary to satisfy the requirements of the specifications and standards regarding minimal permissible diameter of the wire. For the wire ropes, which are under atmo-

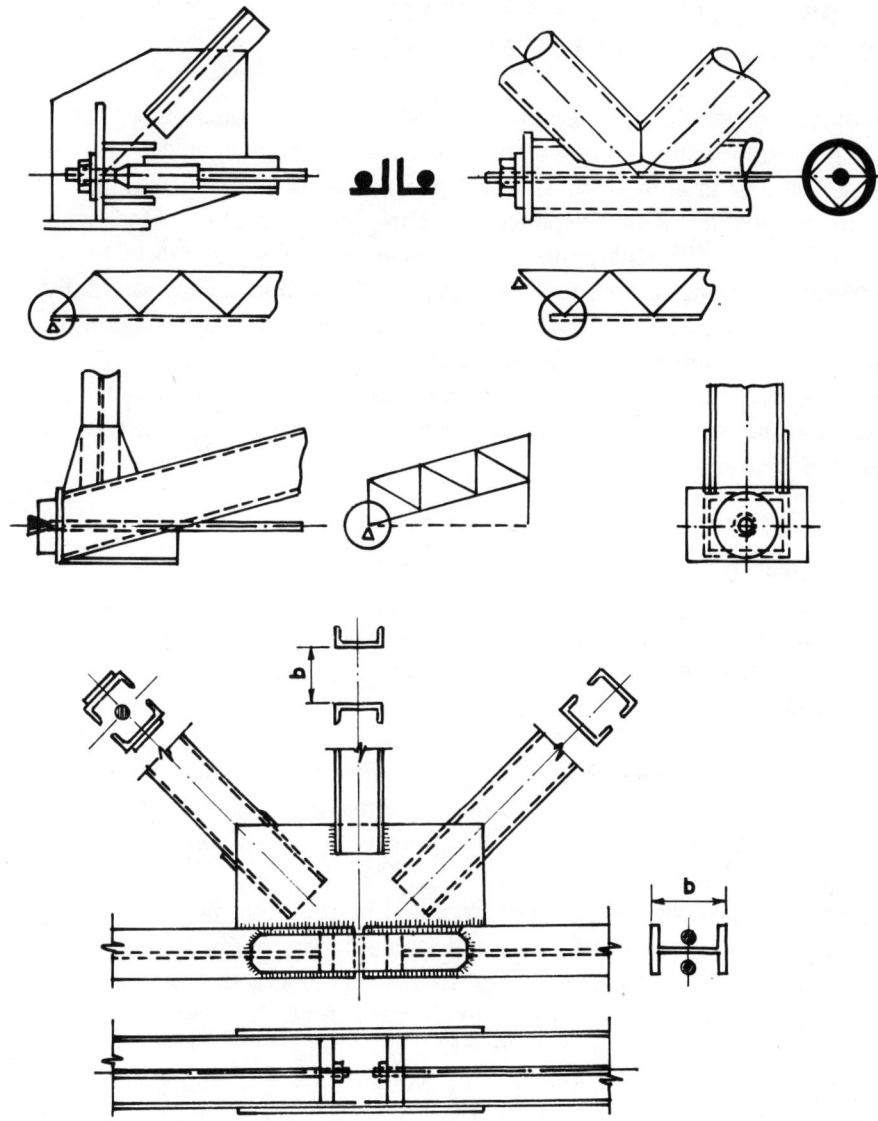

Fig. 3.27 Typical arrangements of anchorages at prestressed trusses.

spheric influences, the diameter should be 3/32 in. (2 mm), or a minimum of 1/32 in. (1 mm). Single high-strength steel wire should have a diameter not less than 1/8 in. (3 mm).

Wire ropes, cables, special wire, and wire for the fabrication of rope should not be exposed to temperatures greater than 180°C, and tendons fabricated from other kinds of steel, 350°C. For brief exposures to higher temperatures elements may be protected by the appropriate heat insulation. Welding operations should not be carried out near unprotected tendons.

Prestressed steel structures that will constantly be exposed to temperatures greater than 100°C should be designed with the unfavorable influence of these temperatures on the properties of the material in mind.

3.8 ABBREVIATED RECOMMENDATION FOR CORROSION PROTECTION FOR TENDONS*

The durability of unbonded tendons in prestressed steel structures has proven to be of great importance. Unbonded tendons are usually comprised of individual strands, bar, or groups of wires that are permanently prevented from bonding and are allowed to move independently. High-tensile steel used for prestressed tendons is particularly susceptible to corrosion in the presence of water or more aggressive substances. In some applications the unbonded tendon, fully protected from corrosion, is inserted into a preformed duct and protective material is subsequently pumped into the duct to fill it.

The abbreviated information presented here is based on the following FIP (The Fédération Internationale de la Précontrainte) recommendations: *Corrosion Protection of Unbonded Tendons.*[13]

3.8.1 Research

Three series of exposure tests relating to, or including, the corrosion protection of unbonded tendons have been carried out in Japan, the United States, and the Netherlands.

1. In Japan the Shinko Wire Co. Ltd. reported in 1978 on unbonded tendon specimens with three years of exposure to normal environments and one year of exposure to maring environment. The unbonded tendons appeared to be "perfectly protected from corrosion."[14]

2. In the United States Schupack reported in 1980 that in 11 years, the prestressing steel (which included one beam with a paper-wrapped wire tendon) was not structurally damaged by exposure to a severe freeze-thaw environment, although there was corrosion to various degrees in all the tendons (bonded and unbonded) of all the beams in the test.[15]

3. In 1981 a report prepared by the Netherlands Committee for Corrosion Protection of Unbonded Tendons was published by the Netherlands Committee for Research, Codes and Specifications for concrete, following a six-year study of the problem, during which long-term exposure tests and accelerated laboratory tests were carried out. The report placed great emphasis on the need to exclude moisture from all parts of the tendon at all times and from the anchorage in service conditions.[16]

3.8.2 Basic Requirements for Corrosion Protection

Considering the above research, the basic requirements for protection of unbonded tendons are as follows:

---

*Courtesy of the Institution of Civil Engineers. Thomas Telford Publications.

1. The main cause of corrosion is the penetration of moisture to the tendon, either through the sheathing and coating, or at the anchorage. Therefore, the sheathing should be watertight, free from pinholes, and highly resistant to the penetration of water vapor. It should be continuous throughout the length of the tendon and strong during installation.

2. The tendon should be protected inside during the sheathing by grease that contains corrosion inhibitors.

3. Probably the most vulnerable part of the tendon, considering corrosion, is at the end anchorages. Special measures need to be taken to ensure that moisture cannot penetrate to the tendon.

3.8.3 Protective Material

Unbonded tendons should have the prestressing steel permanently protected against corrosion by a properly applied coating. It is recommended that the protective compound take the form of a grease, which also assists the free movement of the tendon during stressing.

Various types of grease are not impervious to moisture, so harmful elements can migrate through the grease to reach the tendon unless the outer covering is completely waterproof. Ordinary lubrication greases, however, give sufficient protection in most circumstances, if they satisfy the following requirements:

1. The sheathing should be completely waterproof and continuous for the full length of the tendon.

2. The coating material should be continuous over the entire tendon length and should completely fill the sheathing without air pockets.

3. The coating should remain ductile and free from cracks and should not become fluid.

4. The coating material should not contain harmful impurities, such as chlorides, sulfides, or nitrates.

Special proprietary anticorrosive greases are available with the following additional properties.

1. They provide a barrier to moisture and air.

2. They provide a self-healing film and displace water.

3. They have reserve alkalinity for long-term acid neutralization.

3.8.4 Testing

The protective material is considered to give the required degree of protection if it meets the performance specification given in Table 3.6. It is not considered necessary to carry out these tests frequently. It should generally suffice if (1) the supplier of the protective material guarantees that the requirements of the spec-

Table 3.6 Performance Specification for Protective Material

Test	Criterion	Test Method
Dropping point	Minimum 373 K	ASTM D-566
	(100 C)	ISO-2176
Water-soluble ions	Maximum 20 ppm	ASTM D-512
(a) chlorides	Maximum 20 ppm	ASTM D-992
(b) nitrates	Maximum 20 ppm	ASTM D-1255
(c) sulfides		
Oil separation test period	Maximum 5% by weight, but	DIN 51-817
7 days at 40°C	lower value (say 3%) preferred	
Corrosion	Grade 7 after 1000 h	ASTM B-117
	Grade 0	DIN 51-802
Oxidation	Maximum 0.06 MPa after 100 h	ASTM D942-70
Stability	Maximum 0.2 MPa after 1000 h	DIN 51-808

ifications are being met, and (2) the tendon fabricator produces evidence that the protective material has been ordered in accordance with this guarantee.

3.8.5. Application

If the coating is not applied at the steel wire mill, temporary corrosion protection is required until the permanent protective material can be applied. The temporary protection can take the form of special strand pack wrapping, vapor phase inhibitors (VPI), or a combination of both.

If the coating is applied at the mill, it is advisable for the strand to pass through a bath of de-watering oil after the cooling operation to remove excess cooling water and to apply a temporary protective film.

The permanent protective material must be applied uniformly to all surfaces and should penetrate the interstices as far as possible. The tendon is usually passed through a container which is kept full of the compound at a slight pressure. The diameter of the outlet orifice controls the residual thickness of the protective film, which should be the minimum necessary to allow the tendon to move freely within the sheathing.

Excessive clearance between tendon and sheathing allows the sheathing to be distorted during coiling and placing, while inadequate clearance results in too much friction during stressing.

3.8.6 Sheathing

For Class A exposure, the sheathing must be completely watertight throughout its length, up to and including the anchorages. Since the corrosion protection of unbonded tendons depends very largely on the exclusion of water from the steel, it follows that the sheathing must be completely watertight throughout its length, up to and including the anchorages. Although various types of sheathing have been used, it has been found so far that plastic is the only material that is entirely suitable.

3.8.7 Plastic Material

The plastic material from which the sheathing is made should be of a type that will not react with grease or steel. It should be durable, stable, and flexible during handling and storing on site and in service for the range of temperature that is likely to be experienced.

While abrasion and splitting of the plastic sheathing can be repaired on site, it may not be possible to discover all the damage, so the plastic sheathing should be of a type that is most resistant to damage of any kind.

Deviation in profile causes the tendon to exert constant pressure on the sheathing which may cause the plastic to flow and to become thinner along the line of contact. The plastic should therefore have relatively low creep properties over the anticipated range of temperature.

It is recommended that the plastic material be either high-density polyethylene or polypropylene. Both materials are tough, durable, and nonreactive. High-density polyethylene is more flexible and less liable to become brittle at extremely low temperatures, while polypropylene is more stable at high temperatures. Both materials have high resistance to abrasion and creep, although polypropylene is slightly superior in these respects.

3.8.8 Methods of Application

Class A Exposure

The most satisfactory way of ensuring a continuous seam-free, waterproof tube of uniform thickness is to use the method of extruding the plastic over the coated tendon. This provides a minimum or total absence of air pockets and is most suitable for coiling, storing, uncoiling, and placing the tendon with the least likelihood of damage.

The extrusion head should be adjusted to give the required radial thickness of plastic, which may be between 1/32 in. (0.7 mm) and 3/32 in. (2.00 mm). For general applications a thickness of between 1/32 in. (0.7 mm) and 3/64 in. (1.00 mm) is recommended so that the tube does not deform to the pattern of the strand. The diameter of the tube should be such that, after coiling, it is entirely filled with the protective coating, with sufficient clearance to permit the free longitudinal movement of the tendon.

To ensure a continuous thickness of plastic tube, a continuous and uniform rate of travel through the extrusion head must be ensured. After extrusion, the protected tendon should be rolled before being coiled. In bulk production, coils 3.28 ft (1 m) in diameter are frequently used.

Another method of application which appeared to give satisfactory results and which is in common use is the longitudinal heat sealing of the plastic tube. If this method is employed it is important to ensure that the seams are properly made since it is difficult to detect burst seams after the tendon is uncoiled. The coils should be of sufficiently large diameter to ensure that seams do not burst open when the tendon is being coiled.

Class B Exposure

For Class B exposure, the complete watertightness of the tendon-anchorage assembly is less important. Either of the above methods is recommended for Class B exposure. Alternatively, the sheathing can consist of preformed plastic tubing into which the greased strand is inserted.

3.8.9 General Note

Tests have shown that the anchored end of a tendon is the most vulnerable location for corrosion attack, especially as the interstices between wires, or the wires forming a strand tendon, form capillaries that allow moisture to gain access to the most highly stressed parts of the tendon.

Cutting to Length

After it has been stressed, the tendon should be in accordance with the anchorage manufacturer's instructions, and the amount of cover to the end of the tendon should be the minimum specified for corrosion protection, depending on the degree of exposure of the structure.

The tendon can be cut to length with an abrasive disk cutter, with spark erosion or hydraulic shears, or by oxy acetylene cutting.

As soon as possible after the ends of the tendons have been cut to length, the stressing pockets should be filled with a low-shrink mortar after the sides of the pockets have first been coated with a resin bonding agent.

Under no circumstances should the grout or mortar used for pocket filling contain chlorides or other chemicals known to be deleterious to the prestressing steel.

Before the stressing pocket is filled with mortar, the end of tendon and the gripping part of the anchorage should be completely sealed against moisture. Either of the following methods can be used:

1. The exposed portion of the tendon and the gripping part of the anchorage should be coated with a material that will give permanent protection against the entry of moisture. Suitable materials include epoxy-resin compounds.

2. The vulnerable parts should be coated with the same corrosion protective material that is being used elsewhere on the tendon. In this case, however, it is

necessary to prevent the material from being displaced or damaged during subsequent operations, and a custom-made covering of stout metal or plastic should be fixed to the end of the anchorage to encapsulate the tendon and grips completely.

REFERENCES

1. ASTM A421-74, Standard Specifications—For Uncoated Stress-Relieved Wire for Prestressed Concrete.
2. ASTM A586-68, Standard Specifications—For Zinc-Coated Steel Structural Strand.
3. Belenya, E., *Prestressed Load-Bearing Metal Structure*, Mir Publishers, Moscow, 1977, pp. 56–76.
4. Ferjencik, P., and Tochacek, M., "Prestressing of Steel Structures: Theory and Construction Practice," Edition by Wilhelm Ernst und Sohn, Berlin, 1975, pp. 183–371, (in German).
5. Belenya, op. cit., pp. 79–80.
6. Ibid., p. 80.
7. Hample, E., "Prestressed Constructions," vol. II, VEB Verlag für Bauwesen, Berlin, 1965, pp. 417–418, (in German).
8. Brodka, J. and Klobukowski, J., "Prestressed Steel Constructions," Edition by Wilhelm Ernst und Sohn, Berlin, 1969, pp. 102–103, (in German).
9. Hoyer, W. et al., "Handbook for Steel Structures," vol. III, Verlag für Bauwesen, Berlin, 1974, pp. 334; 338–339, (in German).
10. Belenya, op. cit., pp. 229–230.
11. Streletzkii, N. S., *Metal Structures, Edition of Literature on Construction*, Moscow, 1965, p. 57 (in Russian).
12. Hoyer, W. et al, op. cit., 9, pp. 335.
13. *FIP Recommendation, Corrosion Protection of Unbonded Tendons*, Thomas Telford Ltd., London, 1986, 6.
14. Shinko Wire Co. Ltd., Japan, "Evaluation of Corrosion Protection of Unbonded Tendons," Post-Tensioning Institute, Phoenix, Arizona, 1978.
15. Schupack, M., "The Behavior of Twenty Post-Tensioned Test Beams Subject to up to 2200 Freeze-Thaw Cycles in the Tidal Zone at Trent Island, Maine. Performance of Concrete in Marine Environment," American Concrete Institute, Detroit, 1980, ACI SP-65, pp. 133–152.
16. The Netherlands Committee for Corrosion Protection of Unbonded Tendons (Committee C26A), "Corrosion Protection of Unbonded Tendons," *Heron*, vol. 26, no. 3, 1981, 74 pages.

Chapter 4

Prestressed Plate Girders

4.1 INTRODUCTION

The most highly developed method for the prestressing of plate girders is by the use of different types of tendons made from high-strength steel. The tendons are placed at the bottom flange which is under tension. During prestressing in the girder, axial stresses originate having opposite values to those stresses due to loading. The economic effect of prestressing results from the participation of the high-strength-steel tendons and the consequent reduction of girder cross section. Prestressing increases the elastic stage in the work of steel in a beam, and this permits the high strength of the tendon to be utilized.

The prestressing of steel beams is one of the methods used to effectively increase the use of structural material. By prestressing, it is possible to reduce the amount of metal by 10–20% and the cost of the structure by 5–12%, to lower the construction depth of the beam, and to achieve the rational distribution of material along beam. The advantage of prestressing consists in the fact that during the erection of the structure, the initial prestressing is of opposite sign to the stresses under loading.

4.2 TYPE AND LOCATION OF TENDONS

The installation and shape of the girder tendons are the basic construction problems. There are different types of installations, considering the placement of tendons in simple and continuous spans.[1,2,3] For simple spans, there are the following types of tendons.

1. *Straight tendons:* Such tendons installed at the bottom chord level of a girder are widely used for that part of the girder which is under a maximum bending moment (Fig. 4.1(a)). To install a tendon along the whole length of a single-span girder (Fig. 4.1(b)) is not rational, because at its supports, where the moments under external loading are small, the tendons are not only unnecessary but create an unwanted state of stress. Such a placement may be justified

78

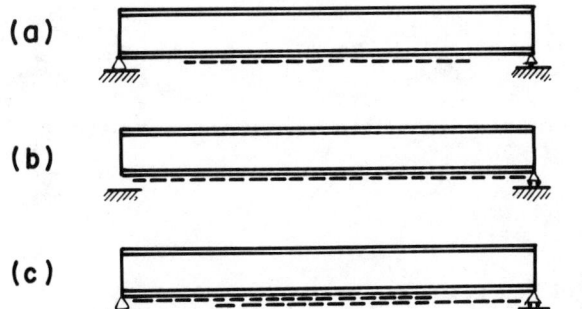

Fig. 4.1 Installation schemes for a straight tendon in a girder.

only by the simplicity of its anchorages and in the jack location used for prestressing.

Along the length of the span, shortened tendons must be placed in such a way that at their connection locations, their cross section will be totally utilized to take their corresponding bending moment.

For better conformity of the beam prestressing diagram to the configuration of the bending moment diagram (Fig. 4.1(c)), the tendons may be placed overlapping and by this increasing their cross-sectional area following increase of the bending moments.

Generally, straight tendons are applied to take only the bending moments, and for relatively small spans. This type may also be applied to simplify the erection procedure.

For girders of variable cross section, an economy in steel may be achieved. In this case, the location of the tendons is shown in Fig. 4.2. The length of the straight parts in the tendon is equal to $0.6\,l - 0.8\,l$ and the girder height above the bearings is $0.45\,h - 0.6\,h$, where l is the span length and h is the height of the girder at mid-span.

In general, for simple-span girders the location of a tendon along the overall length of the girder is impracticable, because the tendon is unnecessary near the bearings, where the bending moments are small. This arrangement may be used where it greatly simplifies the anchorage design and tensioning devices.

2. *Curvilinear and trapezoidal tendons:* Curvilinear or trapezoidal tendons (Fig. 4.3(a) and (b)) have the advantages that they create a prestress whose values vary along the length of a girder. The prestress is larger in the cross section of a greater bending moment. In a single-span girder, the tendon comes

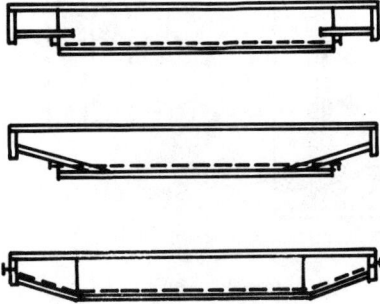

Fig. 4.2 Location of tendons for a girder of changeable height.

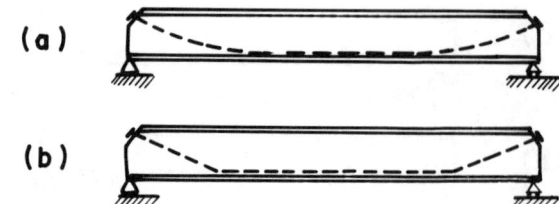

Fig. 4.3 Curvilinear and trapezoidal tendons.

closer to the center of gravity of that cross section above its bearings and the bending moment due to prestressing is reduced.

The installation of curvilinear tendons is somewhat complicated and requires the use of special guides or saddles to profile the tendons as shown in Fig. 4.4. These guides are to be welded onto the girder web and reinforced by short ribs or aligned with the vertical girder stiffeners. Also, friction is created between the tendon and the guide during prestressing, therefore increasing the tendon prestress force.

The reduction of stresses in a girder due to prestressing is greatest as its maximum moment location, because the distance between the tendon and center of gravity of the section is increased. However, at its supports, the tendon approaches the center of gravity of its cross section and the prestressing moment is reduced. But here, the inclined tendon begins to take the shear force, and thus reduces the shear web stresses.

3. *Different tendon locations:* For a pair of girders, it is rational to attach the tendons at the transverse horizontal diaphragms between the girders (Fig. 4.5). In this case, it is possible to achieve tendon tensioning by connecting them together in the span, as shown in Fig. 4.5(b). By applying relatively small forces to the transverse ties, one can thus create substantial prestressing in a longitudinal direction.

For structures of relatively great constructional height, it is possible to construct girders with their tendons placed outside the girder (Fig. 4.6). In terms of use of steel, such girders are more economical than girders with tendons installed within the limits of their height, due to greater effectivity in the work of a tendon.

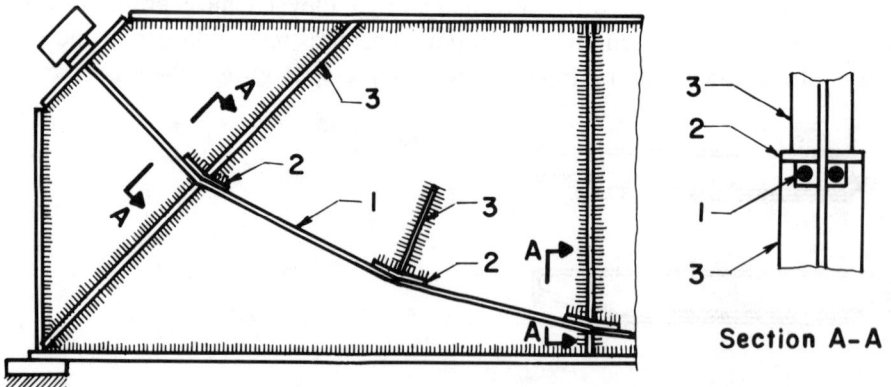

Fig. 4.4 Guides for tendons of curvilinear outline. 1, Tendon; 2, guide; 3, rib.

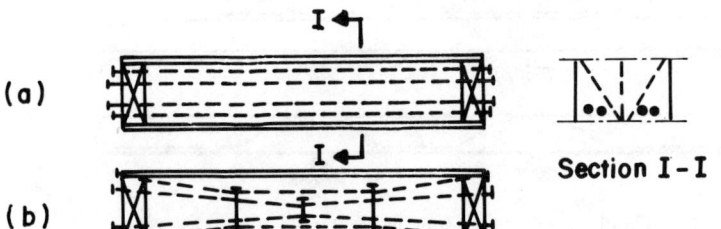

(a)

(b)

Section I-I

Fig. 4.5 Placement of the tendons for a pair of girders.

However, they do have a number of structural deficiencies: The compressed bottom chord is not connected to the tendon and may therefore lose its stability. Also, the transportation of such girders is somewhat complicated.

For a single concentrated mid-span load, it is advisable to use a system with only one strut (Fig. 4.6(a)) and, under a uniformly distributed loading, two struts (Fig. 4.6(b)).

4. *Continuous girders:* In continuous girders, straight tendons are usually located where girders are under the maximum bending moments on their tension sides (Fig. 4.7(a)) or which have a single curvilinear tendon that replaces several straight tendons (Fig. 4.7(b)).[4] This permits a reduction in the number of anchors and the number of jacking operations.

4.3 TYPES OF GIRDER CROSS SECTIONS

When, in a single-span girder, the tendons are located at its bottom chord level, the tendons are stressed in tension and relieve the bottom chord by taking up a greater part of their tensile force due to the moment. In a case where a prestressed girder is designed as a symmetrical cross section, the bottom chord will remain understressed when the compressive stresses in the top chord reach a design value of F.

Therefore, to utilize the girder cross section material to its fullest capacity, the cross section should be designed as an asymmetric I girder, and in medium-size girders the bottom chord may be designed as a rolled section, using a pipe, angles, channels, and other shapes (Fig. 4.8).

The optimum asymmetry of a cross section depends on the following factors: the type of load, the nature of girder stress, and the physical characteristics of the material.

Actually, analysis indicates that economical girder design calls for greater

(a)

(b)

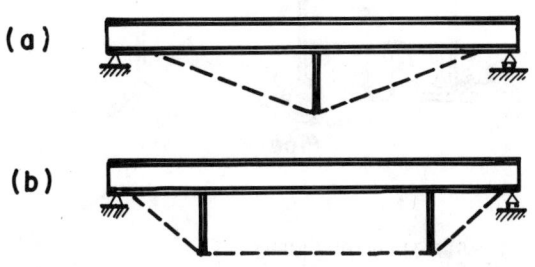

Fig. 4.6 A girder with tendons installed outside.

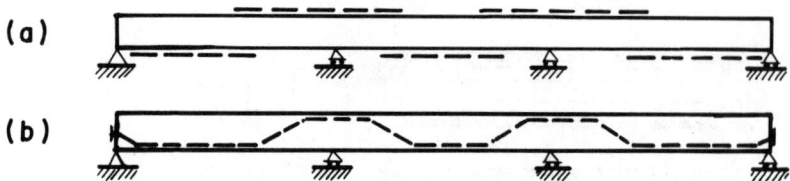

Fig. 4.7   Configuration of tendons in continuous girders.

height and smaller web thickness. However, the girder height may be limited by its design considerations and web stability.

The basic parameters of a cross-section are its asymmetry

$$a = \frac{h_2}{h_1} \quad \text{and} \quad m = \frac{A_w}{A}$$

the ratio of the cross-sectional areas of the web to the total section, respectively (Fig. 4.8). The optimum values of the above parameters for asymmetric steel girders are:

$$a = 1.7 - 2 \qquad m = 0.5 - 0.6$$

This analysis indicates that the economy of the girder increases with an increase of its height and a reduction of its web thickness, to a greater degree than that for conventional girders without prestressing. However, the height increase is limited by construction reasons and the need to ensure the stability of the web.

Asymmetric cross sections may also be designed as a box from four plates (Fig. 4.9). For prestressed plate girders, the safety margin may be calculated by checking the bottom flange for buckling by means of the formula

$$f_b \leq \Psi F_b \tag{4.1}$$

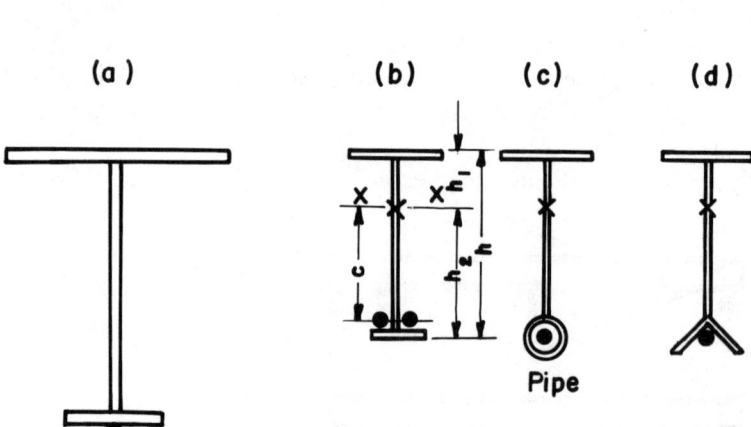

Fig. 4.8   Various types of asymmetric I girders.

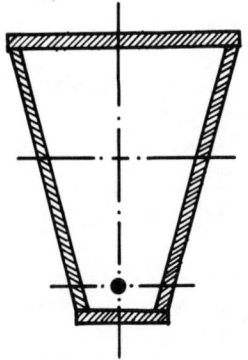 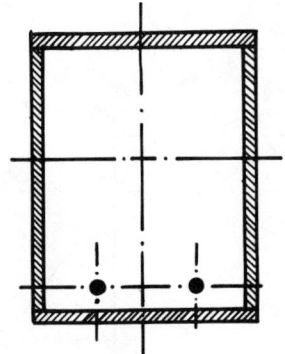

Fig. 4.9 Box-type cross sections.

where

Ψ = the coefficient of buckling determined considering the flexibility of the bottom girder flange with respect to the vertical axis, with the free length of the flange equal to that of the spacing between the points of connection of the tendon to its bottom flange

F_b = the allowable buckling stress

and

$$f_b = \frac{X}{A_b} + \frac{Xe}{S_b} \qquad (4.2)$$

where

A_b = the cross section of the bottom flange

S_b = the section modulus of the bottom flange with respect to the vertical axis

The above formula indicates the compressive stresses in the chord due to the prestressing force in the tendon.

Therefore, a maximum possible tendon prestressing force is

$$X = \frac{\Psi F_b S_b A_b}{S_b + eA_b} \qquad (4.3)$$

4.4 ANALYSIS AND DESIGN OF A PRESTRESSED GIRDER

The introduction of a tendon transfers a girder to a statically indeterminate system. When the tendon is located on the side of those girder fibers in tension balanced by compressive girder stresses, they provide an additional moment of internal forces.

The behavior of a girder in an elastic range, considering its cross section, may

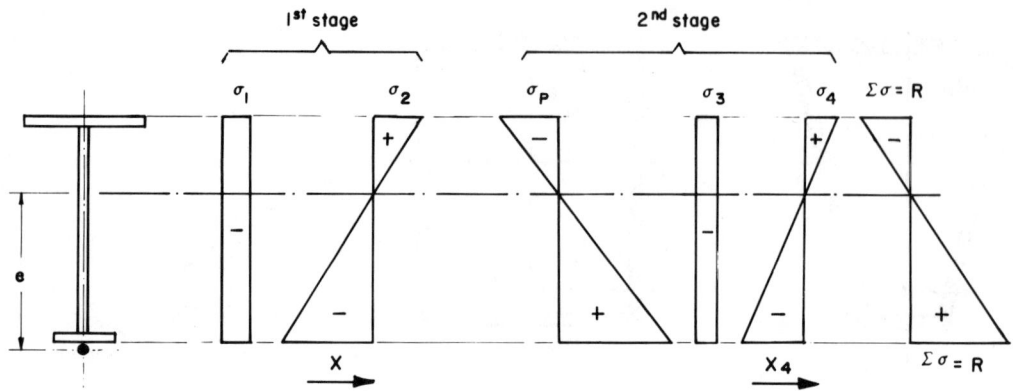

Fig. 4.10   A prestressed beam.

be divided into two stages (Fig. 4.10). At its first stage, a prestressing force X creates stresses

$$f_1 = -\frac{X}{A} \quad \text{and} \quad f_2 = \pm\frac{Xey}{I} = \pm\frac{M_p}{S} \tag{4.4}$$

across the girder, where M_p is the moment due to prestress Xe. At the second stage, the external load is applied until the stresses in the upper and bottom edges attain a yield point. At this stage, the tendon is under an increment of prestressing force ΔX, which induces across the girder the stresses

$$f_3 = -\frac{\Delta X}{A} \quad \text{and} \quad f_4 = \pm\frac{\Delta Xey}{I} \tag{4.5}$$

of an opposite sign to those stresses under external loading.

$$f_p = \frac{My}{I} \tag{4.6}$$

The resulting stresses for the compression edge are

$$f_c = -\frac{M}{S_1} - \frac{X + \Delta X}{A} + \frac{(X + \Delta X)e}{S_1} < F \tag{4.7}$$

and for an edge in tension

$$f_t = \frac{M}{S_2} - \frac{X + \Delta X}{A} - \frac{(X + \Delta X)e}{S_2} < F \tag{4.8}$$

and for a tendon

$$f_{td} = \frac{X + \Delta X}{A_t} < F_t \tag{4.9}$$

where

X = a prestressing force

ΔX = an increment of tendon force

M = a bending moment due to external load ($DL + LL + SDL$)

S_1 = a sectional modulus for a compressed edge

S_2 = a sectional modulus for a tensioned edge

A = a cross-sectional area of a girder

A_t = a cross-sectional area of a tendon

e = eccentricity of a tendon with respect to the centroid of a girder cross section

f_c = a compressive stress

f_t = a tensile stress

F_t = a permissible stress of tendon material

F = an allowable stress of girder material

It should be noted here that a girder may lose its bearing capacity under prestressing if the strength of the compressed flange is inadequate or

$$f_c = -\frac{X}{A} - \frac{Xe}{S_2} > F \tag{4.10}$$

and if the local stability condition is not met, which will be investigated in the section treating buckling.

4.5 BEAMS PRESTRESSED BY TENDONS

4.5.1 Optimum Parameters of a Symmetrical I Beam

Considering geometrical cross-sectional characteristics of a symmetrical I beam, Fig. 4.11, the following dimensionless parameters of the cross section may be developed.[6]

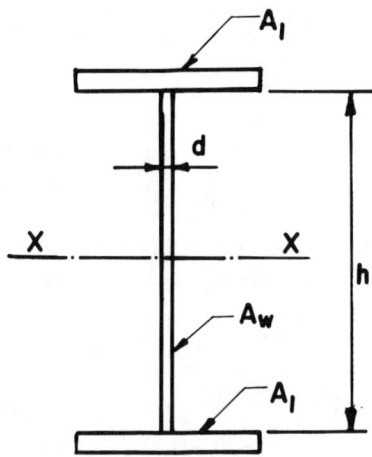

Fig. 4.11 Symmetrical I beam.

1. A parameter characterizing height and thickness of the web

$$n = \frac{h}{d} \qquad (4.11)$$

2. A parameter characterizing the distribution of material in a cross section

$$\kappa = \frac{A_w}{A} \qquad (4.12)$$

where

$$A = 2A_1 + A_w \qquad (4.13)$$

Equation (4.13) = the total cross-sectional area, and A_w = the area of the web.

3. Height of the web

$$h \, x \, \frac{h}{n} + (A - \kappa A) = A$$

$$h = \sqrt{n\kappa A} \qquad (4.14)$$

4. Moment of inertia I_x with respect to the axis x-x

$$A_1 = \frac{(1 - \kappa)A}{2}$$

$$I_{xx} = \frac{2\,(1 - \kappa)A}{2} \, x \left(\frac{h}{2}\right)^2 + \frac{dh^3}{12} = \frac{h^2 \left[3n(1 - \kappa)A + h^2 \right]}{12n} \qquad (4.15)$$

5. Section modulus

$$S = \frac{2I}{h} = \frac{\sqrt{A^3}\,\sqrt{n\kappa}\,(3 - 2\kappa)}{6} \qquad (4.16)$$

or

$$S = \frac{\sqrt{A^3}h}{6\sqrt{A}}(3 - 2\kappa) = \frac{Ah}{6}(3 - 2\kappa) \qquad (4.17)$$

6. The maximum section modulus of the beam will be at n = constant if

$$\kappa = \frac{A_w}{A} = 0.5$$

$$S_{max} = \frac{Ah}{6}(3 - 1) = \frac{Ah}{3} \qquad (4.18)$$

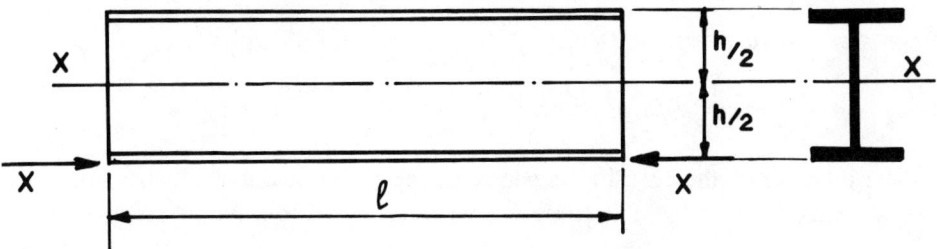

Fig. 4.12   Prestressed beam tendon at the bottom flange.

4.5.2 Stresses Under Prestress

If the tendon is placed at the bottom flange, stresses under prestress are (Fig. 4.12)

$$M_p = X\frac{h}{2} = \text{moment due to prestressing}$$

Stress at the bottom flange is

$$\sigma_b = -\frac{X}{A} - \frac{M_p}{S} = \frac{X}{A} - \frac{X(h/2)}{(Ah/6)\,(3-2\kappa)} = -\frac{2X(3-\kappa)}{A\,(3-2\kappa)} \quad (4.19)$$

Stress at the top flange is

$$\sigma_t = -\frac{X}{A} + \frac{3X}{A(3-2\kappa)} = \frac{2X\kappa}{A(3-2\kappa)} \quad (4.20)$$

The ratio of the stresses at flanges is

$$\frac{\sigma_b}{\sigma_t} = -\frac{2X(3-\kappa)}{A(3-2\kappa)} \div \frac{2X\kappa}{A(3-2\kappa)} = -\frac{3-\kappa}{\kappa} \quad (4.21)$$

The effectiveness of the prestressing increases with the increase of absolute value of the ratio

$$\frac{\sigma_b}{\sigma_t}$$

The maximum value of this ratio is at $\kappa = 1$. Therefore, prestressing is essentially effective for beams with small cross-sectional areas of the flanges. At $\kappa = 0.5$ the tensile stress at the upper flange is

$$\frac{\sigma_b}{\sigma_t} = \frac{3-0.5}{0.5} = 5.0$$

$$\sigma_t = \frac{\sigma_b}{5} = 0.2\sigma_b$$

σ_t = 20% of compressive stress in bottom flange

The increase of the carrying capacity of the beam at initial prestressing may be estimated as follows. The force X creates at the bottom flange stress

$$\sigma = \alpha f \tag{4.22}$$

where α is the coefficient and f is the allowable stress.
The tensile stress at the upper flange is

$$\sigma_t = \alpha f \frac{\kappa}{3 - \kappa} \tag{4.23}$$

The increase of the carrying capacity of the beam at a given value of κ is

$$\frac{f + \sigma_t}{f} = \frac{f + \alpha f \dfrac{\kappa}{3 - \kappa}}{f} = 1 + \alpha \frac{\kappa}{3 - \kappa} \tag{4.24}$$

Table 4.1 gives numerical values of increasing of the carrying capacity of the beam at different values of κ and at $\alpha = 1$.
During the determination of prestressing force X it is necessary to consider that the beam with the tendons represents a statically indeterminate system, and it is during the bending of the beam at the tendon arises the increment of the force, determined from the following equation:

$$\Delta X = \frac{\displaystyle\int_0^1 \frac{M_x M_p}{EI} dl + \sum \frac{N_x N_p\, l}{EA}}{\displaystyle\int_0^e \frac{M_x^2}{EI} dl + \sum \frac{N_x^2}{EA}} \tag{4.25}$$

where

M_x = bending moment due to unit load in the tendon with respect to neutral axis

Table 4.1 Increase of the Carrying Capacity of the Beam at Different Values of κ and at $\alpha = 1$

κ	0.3	0.4	0.5	0.6	0.7	0.8
$1 + \dfrac{\kappa}{3 - \kappa}$	1.111	1.154	1.2	1.25	1.304	1.364

M_p = bending moment due to external loading
N_x = axial force due to unit load in the tendon
N_p = axial force due to external load
I = moment of inertia of the beam
A = cross-sectional area of the beam
E = module of elasticity of the beam

4.5.3 Optimum Section of an I Beam

The optimum section of an I beam is determined from the following conditions.
At $\alpha = 1$ bending moment of the prestressed beam is

$$M_p = Sxf = f \frac{\sqrt{A^3} \sqrt{nk}}{6} (3 - 2\kappa) \frac{3}{3 - \kappa} \qquad (4.26)$$

The maximum value of M_p we obtain at κ-maximum

$$\frac{dM_p}{d\kappa} = \left(\frac{3f \sqrt{A^3 n}}{6} \right) \frac{d}{d\kappa} \left[\frac{\sqrt{\kappa} (3 - 2\kappa)}{3 - \kappa} \right] = 0$$

The maximum value of κ will be at

$$\frac{d}{d\kappa} \left[\frac{\sqrt{\kappa}(3 - 2\kappa)}{3 - \kappa} \right] = \frac{9 - 3\kappa}{2 \sqrt{\kappa}} + (\kappa - 6) \sqrt{\kappa} = 0$$

$$\kappa^2 - 7.5\kappa + 4.5 = 0$$

$$\kappa = 0.658$$

By substituting this value into formula (4.26), we obtain

$$\max M_p = 1.75 \frac{\sqrt{A^3} \sqrt{n}}{6} f \qquad (4.27)$$

Therefore, under prestressing the symmetrical I beam has optimum section and
cross-sectional area of the flanges equal to $1 - \kappa = 0.342$ of the total cross
section of the beam.

The optimum cross section of an I beam under bending moment and nonprestressed from formula (4.16) at constant n and $\kappa = 0.5$ is

$$S_{\max} = \frac{\sqrt{A^3} \sqrt{n} \sqrt{0.5}}{6} (3 - 1) = 1.414 \frac{\sqrt{A^3} \sqrt{n}}{6} \qquad (4.28)$$

The bending capacity of a prestressed beam having an optimum section, in comparison with an optimal nonprestressed beam, when both have the same cross-sectional areas, is

$$\frac{_{max}M_p}{M_{max}} = \frac{1.75}{1.414} = 1.24$$

or 24% greater. The cross-sectional area of the tendon is (from equation 4.19)

$$A_t = \frac{X}{f_t} \qquad \alpha f = \frac{2X}{A} = \frac{2P}{A} \cdot \frac{3 - \kappa}{3 - 2\kappa}$$

and

$$\frac{A_t}{A} = \alpha \frac{f}{f_t} \cdot \frac{3 - 2\kappa}{2\,(3 - \kappa)} \tag{4.29}$$

Table 4.2 gives the ratio, in percent, of the cross-sectional area of the tendon, A_t, to the cross-section A of the beam at $\alpha = 1$. The ratio is

$$\frac{f_t}{f} = \mu$$

The cross-sectional area of the tendon is 4–15% of the beam cross section.

While the beam is under dead load, an interchange of prestressing force is possible. In this case it is possible to obtain over the whole cross section a rectangular diagram of stress, or compression over the I member and tension in the tendon.

The possible limiting bending moment at a prestressed beam is

$$M_p = fA\,\frac{h}{2} \tag{4.30}$$

The ratio of this moment to the limiting moment of the same beam without prestressing is

Table 4.2 The ratio A_t/A

κ	$A_t/A\%$ at μ Equal to					
	3	4	5	6	7	8
0.3	14.8	11.1	8.9	7.4	6.3	5.5
0.5	13.3	10	8	6.7	5.7	5
0.7	11.6	8.7	7	5.8	5	4.3

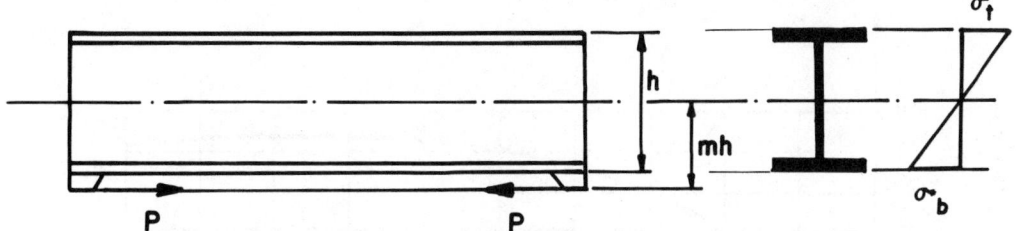

Fig. 4.13 Prestressed beam tendon under bottom flange.

$$\frac{M_p}{M} = \frac{3}{3 - 2\kappa} \tag{4.31}$$

At $\kappa = 0.5$, moment M_p is increased by 50%.

4.5.4 Beams Having a Tendon Under a Bottom Flange

At the location of the tendon at distance mh from the neutral axis of the beam (Fig. 4.13), stresses at the flanges under prestressing by force P will equal

$$\sigma_b = -\frac{P}{A} - \frac{M}{S} = -\frac{P}{A} - \frac{Pmh}{(Ah/6)(3 - 2\kappa)} = -\frac{P}{A} \cdot \frac{3(2m + 1) - 2\kappa}{3 - 2\kappa} \tag{4.32}$$

$$\sigma_t = -\frac{P}{A} + \frac{6Pmh}{Ah(3 - 2\kappa)} = \frac{P}{A} \cdot \frac{3(2m - 1) + 2\kappa}{3 - 2\kappa} \tag{4.33}$$

The increase in carrying capacity of such a prestressed beam is

$$\frac{f + \sigma_t}{f} = 1 + \alpha \cdot \frac{3(2m - 1) + 2\kappa}{3(2m + 1) - 2\kappa} \tag{4.34}$$

Figure 4.14 shows the increase in carrying capacity plotted against the change of m for the case $\alpha = 1$ and $\kappa = 0.5$. At $m = \infty$, carrying capacity doubles.

The cross-sectional area of the tendon is determined from the equation

$$\frac{A_t}{A} = \alpha \frac{f}{f_t} \cdot \frac{3 - 2\kappa}{3(2m + 1) - 2\kappa} \tag{4.35}$$

4.6 OPTIMUM PARAMETERS OF AN ASYMMETRICAL GIRDER

4.6.1 Geometric Cross-Sectional Characteristics

For the determination of the optimum geometric parameters of a girder, it is convenient to express them in dimensionless parameters.[7,8,9] In the case of a

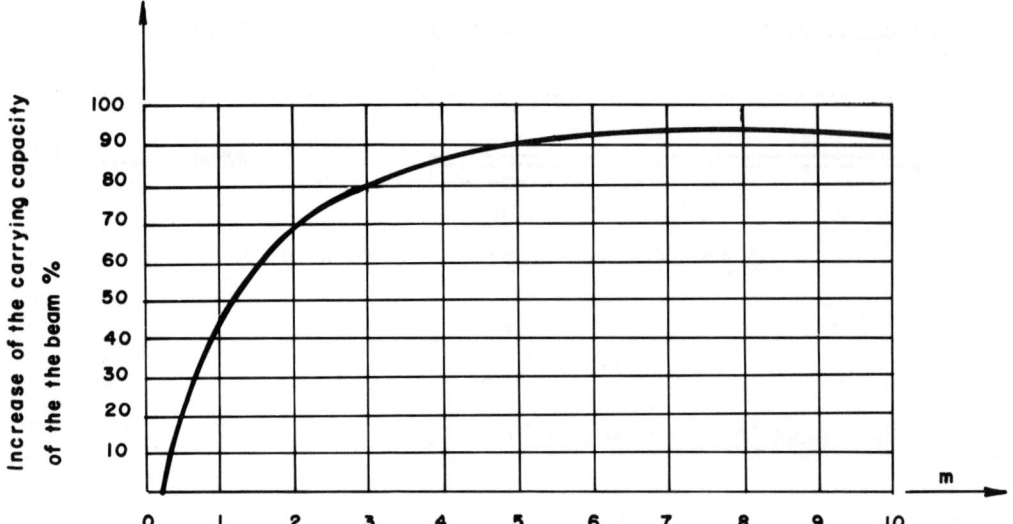

Fig. 4.14 The percent increase in the carrying capacity of a prestressed beam with the tendon placed under the bottom flange versus the distance of the tendon to the axis of the beam.

single-span girder having an asymmetric cross section prestressed by a straight tendon located at the bottom chord of a mixed span section, the main geometric cross-sectional characteristics are shown in Fig. 4.15.

Assume that the centers of gravity of the tendon's cross-sectional area A_t and the girder bottom chord are located at the same level and that the web height is equal to that of the girder: $h_w = h = h_1 + h_2$. The following dimensionless parameters are thus introduced:

1. *Parameter characterizing the asymmetry of an I girder:*

$$a = \frac{h_2}{h_1} = \frac{S_1}{S_2} \tag{4.36}$$

2. *Parameter characterizing the flexibility of a web:*

$$k = \frac{h}{t_w} \tag{4.37}$$

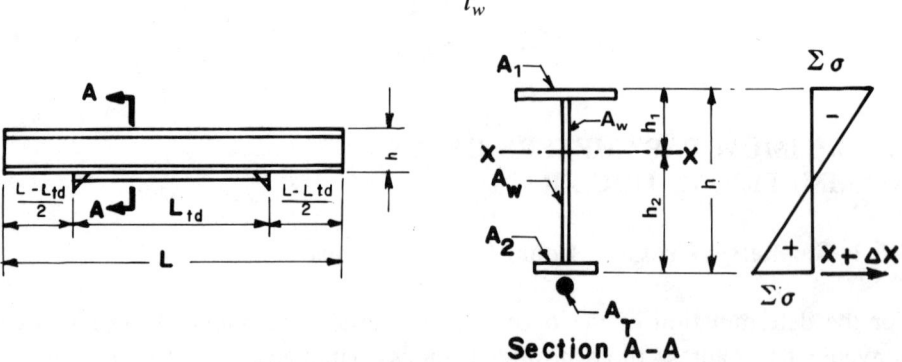

Fig. 4.15 Determination of the optimum parameters of a girder.

3. *Parameter characterizing the distribution of material in a cross section:*

$$m = \frac{A_w}{A} \qquad (4.38)$$

where $A = A_1 + A_2 + A_w =$ the total cross-sectional area of the I girder.

4. *Cross-sectional area of the flanges:* Taking the first moment, with respect to the axis at the bottom flange

$$h_2 A = A_1 h + Am \left(\frac{h_1 + h_2}{2} \right)$$

and introducing a ratio $a = h_2/h_1$, we obtain

$$A_1 = A \left(\frac{a}{a + 1} - \frac{m}{2} \right) \qquad (4.39)$$

and similarly

$$A_2 = A \left(\frac{1}{a + 1} - \frac{m}{2} \right) \qquad (4.40)$$

5. *Heights h, h_1, and h_2:* From basic parameters, we have

$$A_w = Am = ht_w = \frac{h^2}{k}$$

and

$$h = \sqrt{Akm} \qquad (4.41)$$

From

$$h = \sqrt{Akm} = h_1 + h_2 = h_1 + ah_1$$

Hence

$$h_1 = \frac{\sqrt{Akm}}{1 + a} \qquad (4.42)$$

and

$$h_2 = h - h_1 = \frac{a\sqrt{Akm}}{1 + a} \qquad (4.43)$$

6. *Moment of inertia I_x with respect to the axis x-x:*

$$I_x = \frac{1}{12} t_w h^3 + t_w h \left(h_2 - \frac{h}{2} \right)^2 + A_1 h_1^2 + A_2 h_2^2$$

After substituting the corresponding values from expressions (4.36), (4.37), (4.38), (4.39), (4.40), and (4.16), we obtain

$$I_x = A^2 km \frac{6a - (a + 1)^2 m}{6(a + 1)^2} \qquad (4.44)$$

7. *Sectional moduli S_1 and S_2:*

$$S_1 = \frac{I_x}{h_1} = \sqrt{A^3 km} \frac{6a - (a + 1)^2 m}{6(a + 1)} \qquad (4.45)$$

and

$$S_2 = \frac{I_x}{h_2} = \sqrt{A^3 km} \frac{6a - (a + 1)^2 m}{6a(a + 1)} \qquad (4.46)$$

4.6.2 Determination of a Bending Moment

Considering those parameters for an optimum girder cross section, which may take a maximum bending moment, the following equations may be used:

$$f_t = -\frac{X + \Delta X}{A} - \frac{M}{S_1} + \frac{(X + \Delta X) h_2}{S_1} = F \qquad (4.47)$$

$$f_b = -\frac{X + \Delta X}{A} + \frac{M}{S_2} - \frac{(X + \Delta X) h_2}{S_2} = F \qquad (4.48)$$

$$f_{bf} = \frac{X}{A} + \frac{X h_2}{S_2} = F \qquad (4.49)$$

Equations (4.47) and (4.48) indicate that from a bending moment M due to an external load, it follows that the force due to prestressing X and increment of stresses at the top edge and bottom tendon are equal to an allowable stress F. Equation (4.49) provides an additional requirement that the stresses at the bottom edge of a girder, after the tendon had been tensioned, should also be equal to the allowable stress F.

By introducing into equations (4.47) through (4.49) a factor β equal to

$$\beta = \frac{X + \Delta X}{X}$$

we obtain

$$-\frac{M}{S_1} - \frac{\beta X}{A} + \frac{\beta X h_2}{S_1} = F \tag{4.50}$$

$$\frac{M}{S_2} - \frac{\beta X}{A} - \frac{\beta X h_2}{S_2} = F \tag{4.51}$$

$$\frac{X}{A} + \frac{X h_2}{S_2} = F \tag{4.52}$$

We solve equations (4.50) and (4.51) with respect to the design bending moment M after eliminating in them the value X by using equation (4.52) and expressing their geometric characteristics through dimensionless parameters with the aid of formulas (4.37) to (4.44).

From equation (4.50) we have

$$M = \frac{FA\sqrt{Akm}}{6} \cdot \frac{[6a - (a + 1)^2 m][6a - (a + 1)(1 - \beta)m]}{(a + 1)[6a - (a + 1)m]} \tag{4.53}$$

and from equation (4.51)

$$M = \frac{FA\sqrt{Akm}}{6} \cdot \frac{[6a - (a + 1)^2 m](1 + \beta)}{a(a + 1)} \tag{4.54}$$

Equalizing expressions (4.53) and (4.54) we obtain an equation containing the parameters a, m, and β for the optimum stressed state

$$\frac{6a - (1 - \beta)(a + 1)m}{6a - (a + 1)m} = \frac{1 + \beta}{a} \tag{4.55}$$

Hence, the values of parameters m and a are

$$m = \frac{6a[a - (1 + \beta)]}{(a + 1)[a(1 - \beta) - (1 + \beta)]} \tag{4.56}$$

and parameter a in terms of m and β is

$$a = \frac{m\beta - 3(1 + \beta) - \sqrt{m^2 - 6m(1 + \beta)^2 + 9(1 + \beta)^2}}{m(1 - \beta) - 6} \tag{4.57}$$

After substituting m from equation (4.56) into equation (4.53), we obtain the expression for the design bending moment

$$M = fC \sqrt{A^3 k} \tag{4.58}$$

and

$$A = \sqrt[3]{\frac{M^2}{C^2 f^2 k}} \tag{4.59}$$

where

$$C = (1 + \beta) \sqrt{\frac{6a^3(1 - a)^2[a - (1 + \beta)]}{(a + 1)^3[a(1 - \beta) - (1 + \beta)]^3}} \tag{4.60}$$

An analysis shows that by assuming $m = 0.55$ (for $\beta = 1$), the value of parameter C for all the values of a and β remains practically constant.

The above assumptions produce an error in the maximum value of the bending moment of less than 1%. If the assumed value of m is satisfied in equation (4.57), and the expression is obtained for the optimum asymmetry of the cross section a and accordingly for C as a function of parameter β

$$a_{opt} = \frac{3 + 245\beta + \sqrt{0.303 + 5.7(1 + \beta)^2}}{5.45 - 0.55\beta} \tag{4.61}$$

Therefore, if the coefficient β is known, it is possible, after the value of C is obtained and the flexibility of the web k is specified, to calculate with the aid of equation (4.59) the required area A of the girder cross section. Also, after the value of the optimum asymmetry a of the cross section is found from equation (4.57), all other girder parameters can thus be determined from equations (4.37) and (4.44).

4.6.3 Effect of Various Loadings on Optimum Parameters

We now consider three types of girder loadings as follows: the moments acting at supports, a uniformly distributed load, and a concentrated load at mid-span (Fig. 4.16).

The general expression for a prestressing force is obtained from equation (4.49) after substituting values from equations (4.37) to (4.44).

$$X = \frac{FA[6a - (a + 1)^2 m]}{(a + 1)[6a - (a + 1)m]} \tag{4.62}$$

The value of an increment of prestressing force after formula (2.14) is

$$\Delta X = -\frac{\Delta_{11}}{\delta_{11}} = -\frac{\displaystyle\int \frac{M_1 M}{EI_x} dx}{\displaystyle\int \frac{M_1^2 dx}{EI_x} + \frac{l_t}{E_t A_t} + \frac{l_t}{EA}} \tag{4.63}$$

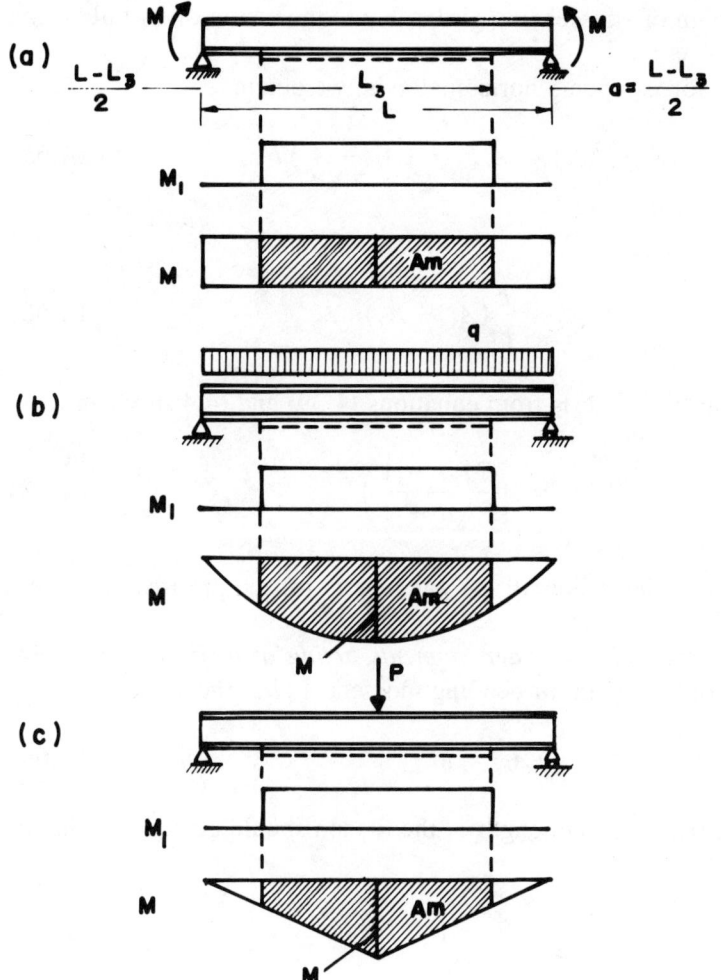

Fig. 4.16 The determination of tendon forces for various types of loading.

For the types of loadings and girders having straight tendons at their bottom levels (Fig. 4.16), the expression for ΔX may be simplified as follows

$$\Delta X = \frac{\dfrac{M_1}{I_x} A_m}{\left[\dfrac{M_1^2}{I_x} + \dfrac{l}{lA_t} + \dfrac{l}{A} \right] l_t} \tag{4.64}$$

where

A_m = an area of the diagram of a bending moment acting along the length of the tendon, $e = E_t/E$

$M_1 = 1 \times h_2$ = a bending moment due to $X = 1$, acting in the tendon

The cross-sectional area A_t of a tendon may be found by considering the equilibrium of the diagram of stresses in a girder due to the action of its full design load (Fig. 4.15).

Projecting all the forces upon a horizontal axis, we obtain

$$(X + \Delta X) = A_t F_t = (A_1 - A_2) F \tag{4.65}$$

from which we get

$$A_t = \frac{F}{F_t}(A_1 - A_2) \tag{4.66}$$

Substituting the values A_1 and A_2 from equations (4.39) and (4.40), we have

$$A_t = A\frac{F}{F_t}\left(\frac{a - 1}{a + 1}\right) \tag{4.67}$$

In the following, we determine all the parameters for the prestressed girders shown in Fig. 4.16.

Case 1—A prestressed girder under moments acting at their supports (Fig. 4.16(a)): The area of a diagram of bending moment A_m has the value

$$A_m = Ml_t \tag{4.68}$$

In this case of pure bending, the length of the tendon should be equal to that of its whole span or

$$l_t = 1$$

Case 2—A girder under a uniformly distributed load (Fig. 4.16(b)): The area of a diagram of bending moment A_m has the value

$$A_m = \left(\frac{2M + S_2 F}{3}\right)l_t \tag{4.69}$$

where $M_1 = S_2 F =$ the bending moment value, which is taken only by the girder cross section, the prestressing of the latter then being unnecessary.

Under a uniformly distributed load, the length of a tendon is determined by the conditions of total capacity of the girder cross section at the location of its anchorages

$$M_a = S_2 F = FA\sqrt{Akm}\,\frac{6a - (a + 1)^2 m}{6a(a + 1)} \tag{4.70}$$

$$M_a = \frac{4M}{l^2}\left(lc - c^2\right)$$

and

$$c = \frac{1}{2}\left(l - \sqrt{l - \frac{M_a}{M}}\right)$$
(4.71)

$$l_t = l - 2c = \sqrt{1 - \frac{M_a}{M}}$$
(4.72)

By substituting $M_a = S_2 F$ and using equation (4.58), $M = FCA\sqrt{Ak}$, we obtain

$$l_t = l^1 \sqrt{1 - \frac{\sqrt{m}}{C} \cdot \frac{6a - m(a + 1)^2}{6a(a + 1)}} = l\sqrt{\alpha}$$
(4.73)

Case 3—A girder under a concentrated load at mid-span (Fig. 4.16(c)): The span of a diagram of bending moment A_m has the value

$$A_m = \frac{(M + S_2 F)}{2} l_t$$
(4.74)

Under a concentrated load, the length of the tendon is this determined as follows:

$$M_a = \frac{2cM}{l} \qquad c = \frac{M_a}{M} \cdot \frac{1}{2}$$

By substituting $M = FCA\sqrt{Ak}$

$$c = \frac{1}{2} \cdot \frac{\sqrt{m}}{C} \cdot \frac{6a - (a + 1)^2 m}{6a(a + 1)}$$
(4.75)

and

$$l_t = l - 2c = l\left[1 - \frac{\sqrt{m}}{C} \cdot \frac{6 - (a + 1)^2 m}{6a(a + 1)}\right] = \alpha l$$
(4.76)

Substituting into equation (4.39) the value of A_m for respective loading patterns and expressing other parameters through the values A, a, k, m, and μ, we obtain formulas that define increment prestressing forces for all loading cases.

The value of μ is expressed as follows:

$$\mu = \frac{E_t}{E} \cdot \frac{f}{f_t}$$
(4.77)

The resulting value of ΔX is substituted into equations for the optimum stressed state (4.23), and having solved them for M, we obtain formulas for bending moments expressed in terms of A, a, k, m, and μ.

Equating the magnitudes of the bending moments in formulas (4.22) and (4.23), we obtain expressions that set up the relationship between a and m and determine the optimum stress state of the beam in bending.

1. For pure bending

$$\frac{6a}{[6a - (a + 1)m][\mu(a - 1) + 1]} = \frac{2}{a} \qquad (4.78)$$

2. For a uniformly distributed load

$$\frac{6a - (a + 1)^2 m}{(a - 1)[6a - (a + 1)m]} = \mu \cdot \frac{m(a + 1)(5a + 7) + 6a(A - 7)}{18.4(2 - \alpha) - 6\mu(\alpha + 1)}$$

$$(4.79)$$

3. For a concentrated load at mid-span

$$\frac{6a - (a + 1)^2 m}{(a - 1)[6a - (a + 1)m]} = \mu \cdot \frac{m(a + 1)(3a + 5) + 6a(a - 5)}{12a(2 - a) - 4\mu(a + 1)}$$

$$(4.80)$$

In the following we develop Table 4.3, showing three different cases of a loaded beam (Figs. 4.3(a), (b), and (c). In all three cases, assuming values for $\mu = 0.1$, 0.2, 0.3, and 0.4, we calculated the values of a and C as follows.

Stage 1: Using $\mu = 0.55$ and substituting into formula (4.78)

$$\frac{6a}{[6a - (a + 1)m][\mu(a - 1) + 1]} = \frac{2}{a}$$

we obtain the corresponding value of a.

Stage 2: Considering formula (4.35)

$$C = (1 + \beta) \sqrt{\frac{6a^3(1 - a^2)[a - (1 + \beta)]}{(a + 1)^3[a(1 - \beta) - (1 + \beta)]^3}}$$

and from the condition to obtain a maximum value of C

$$\frac{dC}{da} = 0$$

we obtain the relation showing β in function of value a, as follows:

Table 4.3 Values of μ, *a*, and *C*

Type of Loading	μ	a	C	Length of Tendon
	0.1	1.87	0.348	
	0.2	2.11	0.369	
	0.3	2.56	0.399	$l_t = l$
	0.4	3.60	0.446	
	0.1	1.83	0.344	
	0.2	1.98	0.357	
	0.3	2.16	0.371	$l_t = l\sqrt{\alpha}$
	0.4	2.36	0.384	
	0.1	1.82	0.342	
	0.2	1.94	0.353	
	0.3	2.06	0.363	$l_t = \alpha l$
	0.4	2.19	0.373	

$$\frac{1}{a(a+1)} - \frac{1-\beta}{a(1-\beta)-(1-\beta)} + \frac{3a-2(1+\beta)-1}{3(a-1)[a-(1+\beta)]} = 0 \quad (4.81)$$

By substituting into this relation the known value of a, we obtain β.

Stage 3: After substituting into formula (4.35) known values of a and β, we obtain C.

Table 4.3 shows, for each value of μ, the corresponding values of a and C.

4.6.4 Optimum Design of Plate Girders

Optimum design of prestressed plate girders was discussed in detail by Vedenikov.[10] He analyzed different configurations of the tendons and structural shapes of prestressed girders and confirmed his finding by numerical examples. Final results of this investigation may be summarized as follows:

1. The choice of girder cross section should be made considering its maximum carrying capacity under total loading.

2. The tensioning of the tendon should be maximum to realize the complete carrying capacity of the girder.

3. Considering the effective prestressing, the magnitude of the additional prestressing ΔX under live load is very important. Keeping the same cross section of girder will increase the capacity of the whole system.

4. The increase of ΔX depends on the tendon configuration and its length; ΔX increases with the straight and shorter tendons installed along bottom flange of the girder.

5. An increase in girder height and reduction of web thickness leads to the reduction of girder weight.

6. It is advantageous to design prestressed a girder as an asymmetric section, using a larger top flange. This is because at the prestressed girder the top flange participates more intensively and the resulting displacement of the neutral axis upward increases the eccentricity of the tendon.

7. The load-carrying capacity and the rigidity of the structure may be increased by the use of multistage prestressing, in which the prestressing and the loading of a structure are carried out in several steps.

4.6.5 Deflection

Prestressed girders are more prone to deformation in their elastic range of behavior than are conventional girders as they generally have a smaller sectional area and therefore a lesser moment of inertia. However, the positive characteristics of prestressing are as follows:

- The stiffness of the girder is increased.
- Prestressed girders usually have substantially smaller deflections.
- A girder may have a smaller construction depth, although its stiffness will be the same as that for a girder without prestressing.

The design deflection of a prestressed girder in bending is calculated as follows:

$$\Delta = \Delta_{D+L} - \Delta_x - \Delta_{xl} \tag{4.82}$$

where

$$\Delta_{D+L} = \text{a nonprestressed girder deflection under dead and live loads}$$
$$\Delta_x \text{ and } \Delta_{xl} = \text{the reverse deflection of a girder due to prestressing of the tendon}$$

1. The deflection due to dead and live loads will be determined by conventional formulas from statics considering a nonprestressed girder.

2. When a tendon is placed at the bottom cross-sectional area of a simple-span girder and is prestressed, a deflection originates in an upward vertical deflection. A deflection due to dead and live loads is then produced in the opposite direction.

The deflection due to prestressing is calculated by applying the virtual work method using the general equation

$$\Delta = \int_0^1 \frac{M_i m_k}{E_s I} dl + \int_0^1 \frac{N_i n_k}{E_s A} dl \tag{4.83}$$

where M_i and N_i are the moments and axial loads, respectively, produced by the corresponding loads on the statically determinate structure. The term m_k represents the moment under the virtual unit force applied in the direction in which the deflection is sought. The second term in equation (4.83) is applied when parabolics of polygonal tendons are used. However, it is usually neglected because its order of magnitude is relatively small.

The determination of a prestressed girder deflection is shown for a simple-span girder, having its tendon shorter than its span,[11] in Fig. 4.17.

The bending moment due to force X_t, as shown in Fig. 4.17(a), is

$$M_a = -X_t e \qquad (4.84)$$

where

$$X_t = X + \Delta X, \text{ the total tendon force}$$

X and ΔX = the prestressing and increment of prestressing force in a tendon, respectively

e = the eccentricity of a tendon force with respect to beam centroidal axis x-x

The deflection of a girder is expressed by the first term of Equation (4.83).

In Fig. 4.17(a), the moment diagram for a given tendon's force is shown, and in Fig. 4.17(b), the same is shown for a unit load at the mid-span of a girder, where the maximum deflection occurs.

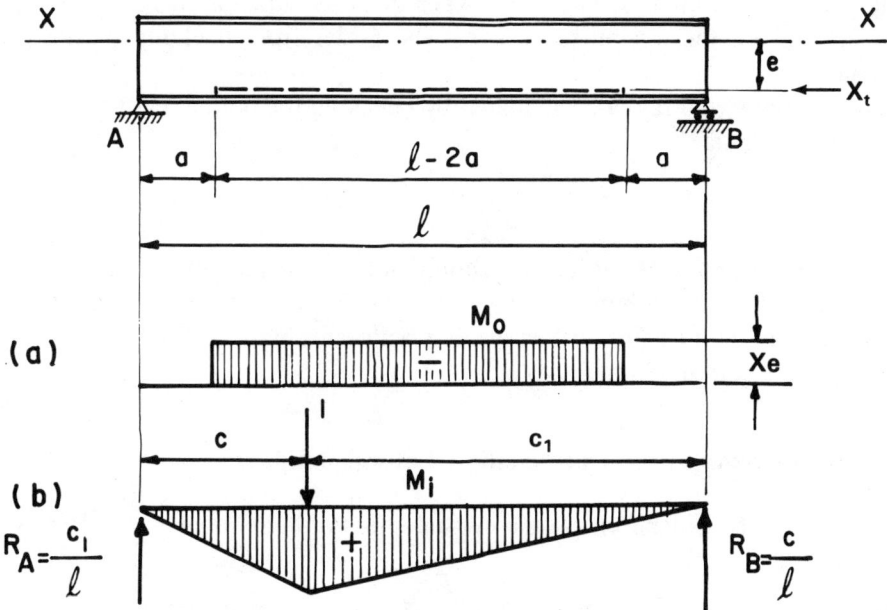

Fig. 4.17   The deflection of a prestressed girder.

For the case under consideration,

$$M_1 = \frac{1}{2}x \quad \text{(for } 0 \le x \le 1\text{)} \tag{4.85}$$

where M_1 is a bending moment due to a unit force acting at the point and direction of the required displacement.

After introducing expressions (4.84) and (4.85) into equation (4.83), we have

$$\Delta_{max} = -\frac{2}{EI} \int_a^{1/2} \frac{1}{2} X_t xe\, dx = -\frac{X_t e}{2EI}\left(\frac{l^2}{4} - a^2\right)$$
$$= -\frac{el^2}{8EI}\left[1 - 4\left(\frac{a}{l}\right)^2\right] \tag{4.86}$$

The total deflection will be determined as the summary of those deflections for a nonprestressed loaded girder and a girder under the action of a prestressed tendon.

4.7 BUCKLING STRENGTH OF PLATE GIRDERS

4.7.1 Stability of Plate Girder Bottom Chord During Prestressing

Analysis of the behavior of a steel member prestressed by a tendon installed along the center of gravity of the member and connected to it at separate points proved that the member may lose its stability only between tendon connections.[12] The tendons are usually connected to the girder at certain intervals by means of diaphragms, ribs, clamps, and other types of grips which allow a longitudinal displacement of the tendon but prevent the girder from buckling during prestressing.

The safety margin may be calculated by checking the bottom chord for buckling using formula

$$\sigma_x \le \varphi\sigma \tag{4.87}$$

where φ is the coefficient of lateral bending or buckling determined based on the flexibility of the girder chord with respect to the vertical axis with the free length of the bottom chord equal to the spacing between points of connection of the tendon to the bottom chord.

1. Determination of the Coefficient of Lateral Bending

According to the Euler formula the critical stress of buckling is

$$\sigma_{cr} = \frac{\pi^2 E}{\left(\dfrac{1}{r}\right)^2} = \frac{\pi^2 E}{\lambda^2} \tag{4.88}$$

The ratio of critical stress to the allowable stress of given steel may be expressed as follows

$$\varphi = \frac{\sigma_{cr}}{\sigma_{all}} \quad \text{and} \quad \sigma_{cr} = \varphi \sigma_{all} \tag{4.89}$$

However, the Euler formula is valid only until stress σ_{cr} remains within the proportionality limit. For structural steel with a proportionality limit of 30×10^3 psi and $E = 30 \times 10^6$ psi, we find from equation (4.88)

$$\frac{1}{r} = \sqrt{1000\pi^2} = 100 \tag{4.90}$$

Hence, for $1/r < 100$ Euler's formula is not valid.

In this case the values of coefficient φ are determined on the basis of empirical data given by the Navier formula

$$\varphi = \frac{1}{1 + 0.00008(1/i)^2} \tag{4.91}$$

The diagram in Fig. 4.18 indicates the values of coefficient φ in function of slenderness ratio.[13]

2. Determination of Critical Prestressing Buckling Force

Compressive stress in the bottom chord due to prestressing force in the tendon is

$$\sigma_x = \frac{X}{A} + \frac{Xe}{S} \tag{4.92}$$

Therefore, a minimum possible tendon force is, after substitution of (4.92) into (4.87)

$$\sigma\varphi = \frac{X(S + Ae)}{AS} \quad \text{and} \quad X = \frac{\varphi\sigma AS}{S + eA} \tag{4.93}$$

Let us find the coefficient φ for a prestressed girder having a tendon connected at spacings a, shown in Fig. 4.19.

$$I_y = \frac{hb^3}{12} \qquad A = bh$$

$$r_y = \sqrt{\frac{I_y}{A}} = \sqrt{\frac{hb^3}{12bh}} = \frac{b}{\sqrt{12}}$$

$$\text{Slenderness ratio } \lambda_y = \frac{a}{r_y} = \frac{a\sqrt{12}}{b} \tag{4.94}$$

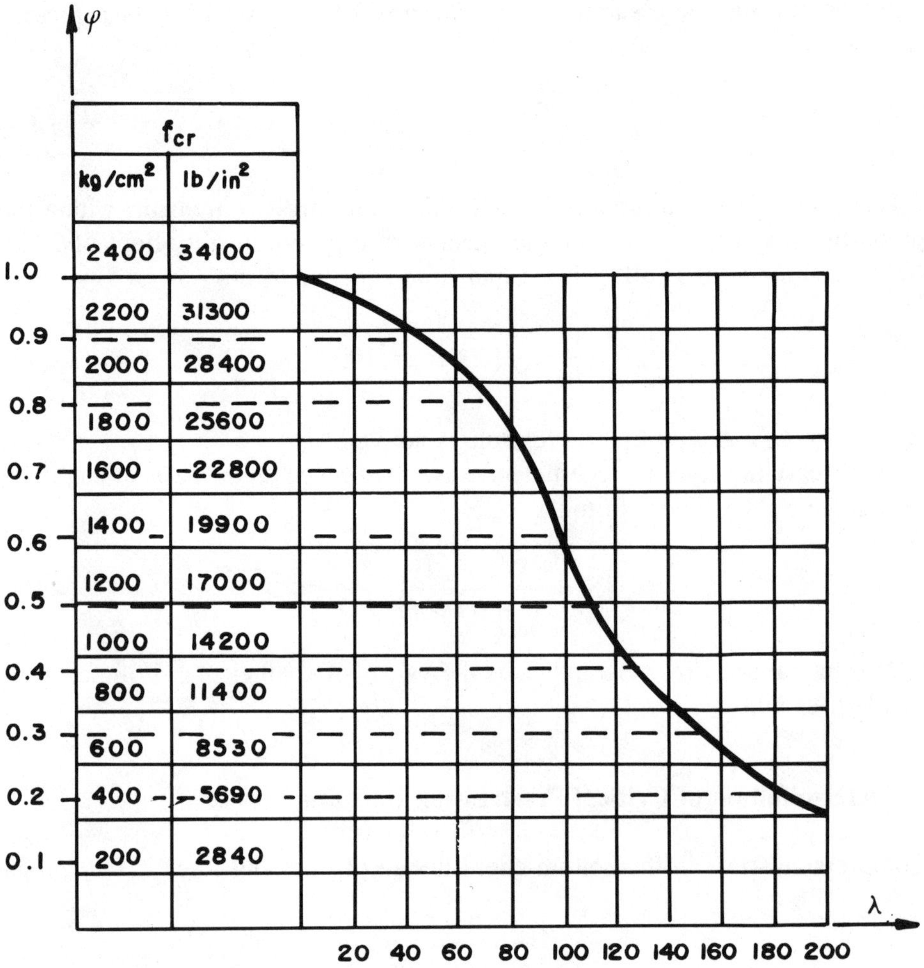

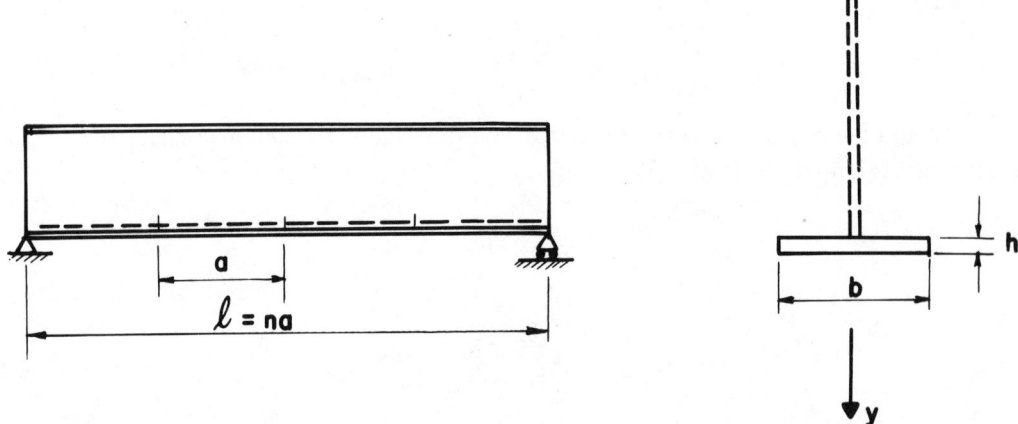

φ
f_cr
kg/cm²
lb/in²
2400 34100
2200 31300
2000 28400
1800 25600
1600 22800
1400 19900
1200 17000
1000 14200
800 11400
600 8530
400 5690
200 2840
1.0
0.9
0.8
0.7
0.6
0.5
0.4
0.3
0.2
0.1
λ
20 40 60 80 100 120 140 160 180 200

Fig. 4.18 Diagram of coefficient φ.

a
ℓ = na
h
b
y

Fig. 4.19 Tendon connected to the bottom chord.

Therefore, for given λ_y, we may find the corresponding coefficient φ from Fig. 4.18 and calculate the permissible prestressing force X, from formula (4.93).

In the following, the design of a prestressed plate girder against buckling is based on the abbreviated provisions of the American Association of State Highway and Transportation Officials (AASHTO) Specifications.[14]

4.7.2 Flanges

1. *Welded girders*: The ratio of compressive flange plate width to thickness shall not exceed the value determined by the formula

$$\frac{b}{t} = \frac{3250}{\sqrt{f_b}} \qquad (4.95)$$

F_y = yield point of the steel
b = flange plate width
t = flange plate thickness

In no case shall b/t exceed 24, and $f_b = 0.55\,F_y$ where f_b = compressive bending stress in pounds per square inch.

2. *Riveted or bolted girder*: The width of outstanding legs of flange angles in compression, except those reinforced by plates, shall not exceed the value determined by the formula

$$\frac{b'}{t} = \frac{1625}{\sqrt{f_b}} \qquad (4.96)$$

but in no case shall b'/t exceed 12.

4.7.3 Thickness of Web Plates

The web plate thickness of plate girders without longitudinal stiffeners shall not be less than determined by the formula

$$t_w = \frac{D\sqrt{f_b}}{23,000} \qquad (4.97)$$

where D is the depth of the web, but in no case shall the thickness be less than $D/170$.

4.7.4 Transverse Intermediate Stiffeners

Transverse intermediate stiffeners may be omitted if the web thickness is not less than $D/150$ and the average calculated unit shearing stress in the gross section of the web plate at the point considered, f_v, is less than the value given by the following equation

$$F_v = \frac{5.625 \times 10^7}{(D/t_w)^2} \leq \frac{4F_y}{3} \tag{4.98}$$

where

D = unsupported depth of web plate between flanges in inches
t_w = thickness of the web plate in inches
F_v = allowable shear stress in psi

Where transverse intermediate stiffeners are required, their spacing shall be such that the actual shearing stress will not exceed the value given by the following equation (the maximum spacing is limited to $1.5D$) (Figs. 4.20 and 4.21)

$$F_x = \frac{F_y}{3} \left[C + \frac{0.87(1 - C)}{\sqrt{1 + \left(\dfrac{d_0}{D}\right)^2}} \right] \tag{4.99}$$

where

$$C = \frac{2.2 \times 10^8 \left[1 + \left(\dfrac{D}{d_0}\right)^2 \right]}{F_y \left(\dfrac{D}{t_w}\right)^2} \leq 1 \tag{4.100}$$

and where d_0 is the spacing of the intermediate stiffener.

The spacing of the first intermediate stiffener at the simple support end of a girder shall be such that the shearing stress in the end panel shall not exceed the value given by the following equation (the maximum spacing is limited to $D/2$)

$$F_v = \frac{7 \times 10^7 \left[1 + \left(\dfrac{D}{d_0}\right)^2 \right]}{(D/t_w)^2} \leq \frac{F_y}{3} \tag{4.101}$$

If a girder panel is subjected to simultaneous action of shear and bending moment with the magnitude of the shear stress higher than $0.6F_v$, the bending stress, F_s, shall be limited to

$$F_s = \left(0.754 - 0.34 \frac{f_v}{F_v} \right) F_y \tag{4.102}$$

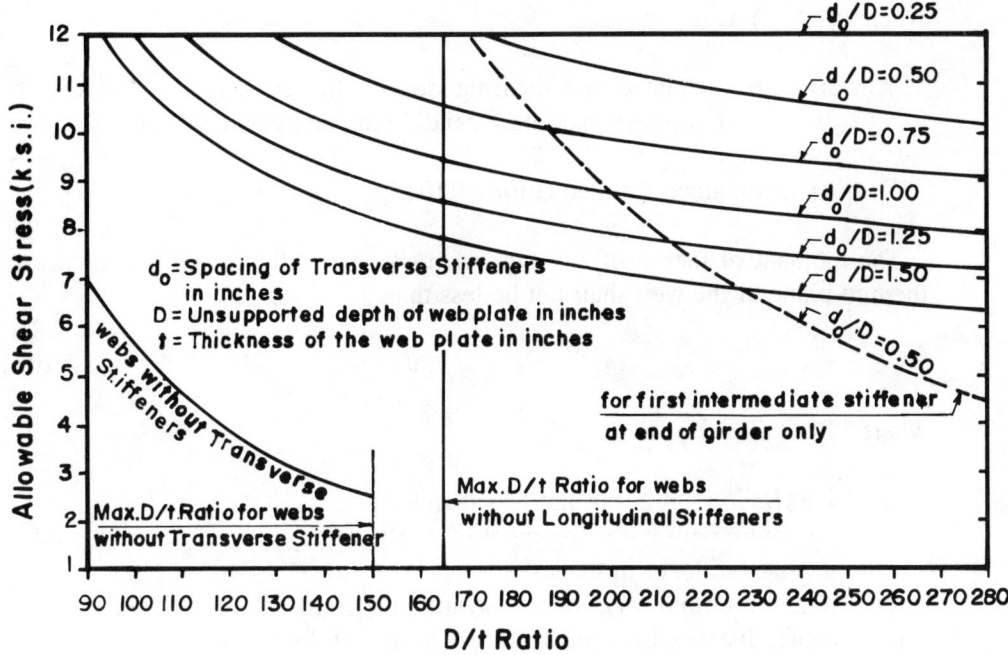

Fig. 4.20   Allowable shear stress vs. D/t ratio (M183 steel, $F_y = 36$ ksi).

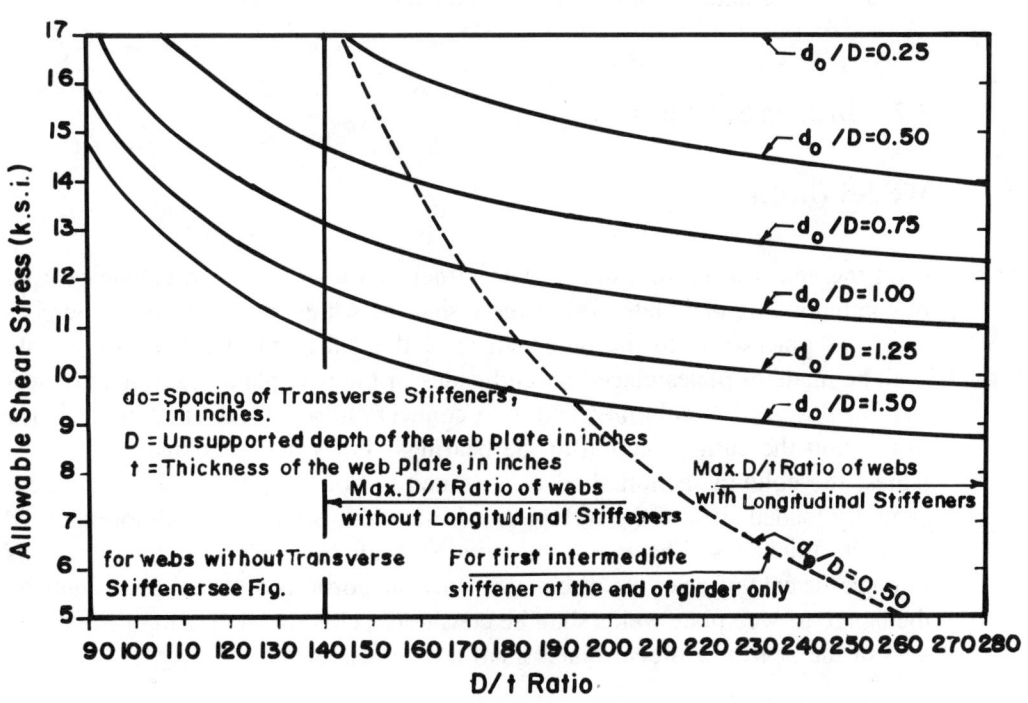

Fig. 4.21   Allowable shear stress vs. D/t ratio (M222 steel, $F_y = 50$ ksi).

where

f_v = average calculated unit shearing stress at the section, live load shall be the load to produce maximum bending moment at the section under consideration

F_v = value obtained from equation (4.101)

The moment of inertia of any type of transverse stiffeners with reference to the mid-plane of the web shall not be less than

$$I = d_0 t_w^3 J \tag{4.103}$$

where

J = $2.5(D/d_0)^2 - 2$, but not less than 0.5

I = minimum permissible moment of inertia of any type of transverse intermediate stiffener in inches

J = required ratio of one transverse stiffener to that of the web plate

d_0 = actual distance between stiffeners in inches

t_w = thickness of the web plate in inches

When stiffeners are in pairs, the moment of inertia shall be taken about the centerline of the web plate. When single stiffeners are used, the moment of inertia shall be taken about the face in contact with the web plate.

4.7.5 Bearing Stiffeners

Welded Girders

Over the end bearing of welded plate girder and over the intermediate bearing of continuous welded plate girders there shall be stiffeners. They shall extend as nearly as practicable to the outer edges of the flange plates. They preferably shall be made of plates placed on both sides of the web plate. Bearing stiffeners shall be designed as columns, and their connection to the web shall be designed to transmit the entire reaction to the bearings. For stiffeners consisting of two plates, the column section shall be assumed to comprise the two plates and a centrally loaded strip of the web plate whose width is equal to not more than 18 times its thickness. The radius of gyration shall be computed about the axis through the centerline of the web plate. Only the portions of the stiffeners outside the flange-to-web plate welds shall be considered effective in bearing. The thickness of the bearing stiffener plates shall not be less than

$$\frac{b'}{12} \sqrt{\frac{F_y}{33,000}} \tag{4.104}$$

The allowable compressive stress and the bearing pressure on the stiffeners shall not exceed the values given in AASHTO Specifications.

4.8 GIRDER UNDER LIVE LOAD

4.8.1 Prestressing Due to Concentrated Load

The increment of the tendon force under concentrated load (Fig. 4.22) is

$$\Delta X = \frac{\displaystyle\int_a^{1-a} MM_1e\, dx}{\displaystyle\int_a^{1-a} M_1^2\, dx + \frac{I}{A}l_t + \frac{EI}{E_tA_t}l_t}$$

$$M_1 = 1 \times e \qquad M_1^2 = \int_a^{1-a} e^2\, dx = e^2(1 - 2a) = e^2l_t$$

The area of the moment diagram is

$$\int_a^{1-a} Mdx = P\int_0^{x_1} \frac{(1-1)x}{1}dx + P\int_0^{1-x_1} \frac{x_1x}{1}dx - P\int_0^a \frac{(1-x_1)}{1}x\, dx$$

$$-P\int_0^a \frac{x_1x}{1}dx = \frac{P}{2}(lx_1 - x_1^2 - a^2)$$

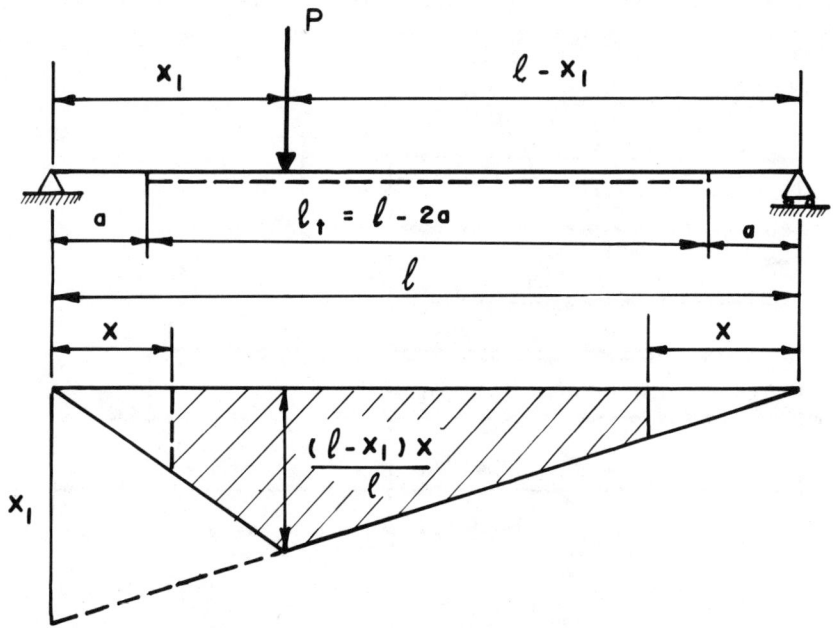

Fig. 4.22 Prestressed girder under single concentrated load.

Therefore,

$$\Delta X = \frac{Pe(lx_1 - x_1^2 - a^2)}{2(1 - 2a)\left(e^2 + \dfrac{I}{A} + \dfrac{EI}{E_t A_t}\right)} \tag{4.105}$$

4.8.2 Girder Under Truck Load

The absolute maximum bending moment due to the action of truck load on a simple-span girder occurs when the middle of the span is halfway between the load closest to the resultant of all the loads on the span. The position of the truck to determine maximum moment is shown in Fig. 4.23.

In this case the intensity of the increment of prestressing force is calculated as summary for each concentrated load after formula (4.105), or

$$\Delta X = \Delta X_1 + \Delta X_2 + \Delta X_3 \tag{4.106}$$

where dX_1, dX_2, and dX_3 are componental increments under action of concentrated truck loads, respectively.

4.8.3 Length of the Tendon in Girders Stressed by a Live Load

The length of the tendon in girders stressed by a live load is established as a function of the distance a from the bearing to the beginning of the tendon. This distance depends on the two requirements as follows.[15]

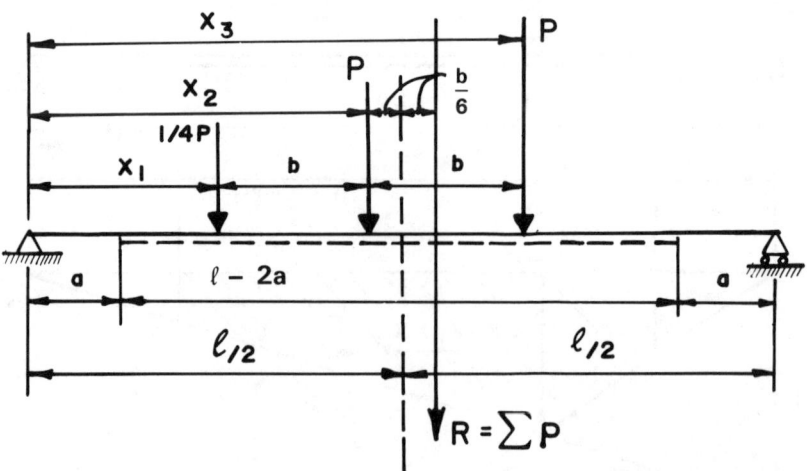

Fig. 4.23   Position of truck for maximum moment.

1. The Maximum Distance, a_{max}

The maximum distance a_{max} is found from the condition that the cross section of the girder at the point of tendon anchoring is required to withstand the acting bending moment. At this point the girder cross section is considered without the tendon.

Before a_{max} is attained, the strength of an asymmetric cross section is checked on the value of stresses in the bottom chord

$$M_a \leq S_2 f \qquad (4.107)$$

The admissible value of moment M_a may be calculated and the maximum value of a_{max} determined graphically, provided that the envelope of the diagram of moments is available, if the cross section as given by formula (4.107) is known. If the cross section has not yet been selected, the value of the moment M_a at the anchor location may be found in terms of maximum design moment M as follows.

By substitution of the value from formula (4.59)

$$A^3 = \frac{M^2}{C^2 f^2 k}$$

into the formula for section modulus (4.46)

$$S_2 = \sqrt{A^3 km} \cdot \frac{6a - (a + 1)^2 m}{6a(a + 1)}$$

we have

$$S_2 = \frac{M}{Cf} \sqrt{m} \cdot \frac{6a - (a + 1)^2 m}{6a(a + 1)}$$

By using the value of C from formula (4.60)

$$C = (1 + \beta)\frac{a}{a + 1} \sqrt{\frac{6a(1 - a)^2[a - (1 + \beta)]}{(a + 1)[a(1 - \beta) - (1 + \beta)]^3}}$$

and substituting $\beta = 1$ and $a = 1.71$, we obtain $C = 0.33$, and for $m = 0.55$

$$M_a = S_2 f = \frac{M\sqrt{0.55}}{0.33} \cdot \frac{6 \times 1.71 - 2.71^2 \times 0.55}{6 \times 1.71 \times 2.71} = 0.5\ M \quad (4.108)$$

Therefore, M_a and a_{max} may be found, if the maximum design bending moment M in the girder and the envelope of moments are known.

By substituting into formula (4.59)

$$A = \sqrt[3]{\frac{M^2}{C^2 f^2 k}}$$

the value of C from formula (4.60) for $\beta = 1$, we obtain

$$A = \sqrt[3]{\frac{M^2}{f^2 k} \cdot \frac{a + 1}{a^3 \sqrt{3(2 - a)(1 - a)^2}}} \qquad (4.109)$$

2. The Minimum Distance, $a_{\min}$

The minimum distance $a_{\min}$ is found considering that the increment of stresses in the bottom chord at the end point of tendon should be tensile for any position of the live load. The stresses are then checked for safe values.

If the tendon is anchored near the bearing, the live load may be positioned on the girder so as to produce a greater increment of stress σ_x in the bottom chord of the girder due to the compression increment of the tendon than the increment of tensile stresses σ_t due to live load.

In this case the compressive stresses due to prestress X (generally equal to σ) add up with the resultant compressive stresses due to the action of external load, with the effect that the bottom chord is overstressed.

The closer the tendon to the bearing, the greater the probability of overstressing the bottom chord in the tendon anchoring zone where the bending moment from the external load will be small.

The initial equation for establishing the minimum distance from the girder support to the tendon anchoring is

$$\sigma_b' = \sigma_x = \sigma_t \geq 0 \qquad (4.110)$$

where

  σ_x = compressive stress due to increment of the compressive force in the tendon

  σ_t = tensile stress due to bending moment from external load in a section of girder without tension, calculated from the following formulas

$$\sigma_x = -\frac{\Delta X e}{S_2} - \frac{\Delta X}{A} \quad \text{and} \quad \sigma_t = \frac{M_t}{S_2} \qquad (4.111)$$

For a load located within the tendon length at a distance x from the support and a cross section located to the left of the load, we find the stress σ_b' from equation (4.110) after substituting the following values.

From formula (4.105)

$$\Delta X = \frac{Pe(x^2 - xl + a^2)}{2(1 - 2a)\left(e^2 + \dfrac{I}{A} + \dfrac{EI}{E_t A_t}\right)}$$

we designate by μ the expression

$$\mu = \frac{e}{2\left(e^2 + \dfrac{I}{A} + \dfrac{EI}{E_t A_t}\right)} \tag{4.112}$$

and from (4.111)

$$\sigma_x = -\Delta X\left(\frac{e}{S_2} + \frac{1}{A}\right) = -\Delta X \alpha \tag{4.113}$$

where

$$\alpha = \frac{e}{S_2} + \frac{1}{A} \tag{4.114}$$

By designating the distance from the left support to the cross-section considered as η, we have

$$M_t = \frac{P(l - x)}{l} \eta \tag{4.115}$$

By substituting into expression (4.110) the values from equations (4.111), (4.112), (4.113), and (4.114), we obtain

$$\sigma_b' = \frac{P\mu\alpha}{1 - 2a}(x^2 - xl - a^2) + \frac{P\eta(l - x)}{lS_2} \tag{4.116}$$

Equating the derivative of σ_b' with respect to x to zero, we obtain the location of the load for which the compressive stresses in the cross section of coordinate η attain a maximum value.

$$\bar{x} = \frac{1}{2} + \frac{\eta}{2lS_2} \frac{(l - 2a)}{\mu\alpha} \tag{4.117}$$

We designate by γ the value

$$\gamma = \mu\alpha S_2 = \frac{e\left(e + \dfrac{S_2}{A}\right)}{2\left(e^2 + \dfrac{I}{A} + \dfrac{EI}{E_t A_t}\right)} \tag{4.118}$$

where, from (4.114)

$$\alpha S_2 = \left(e + \frac{S_2}{A}\right)$$

and

$$\bar{x} = \frac{1}{2} + \frac{\eta(l - 2a)}{2\lambda l} \tag{4.119}$$

To determine the minimum distance a_{min} from the support, substitute $\bar{x}$ in formula (4.116), assuming $\eta = a$, and equate σ'_b to zero.

By solving the resultant equation for a, we obtain

$$a_{min} = \gamma l \tag{4.120}$$

4.8.4 Calculation of a Prestressed Girder Under a Movable Concentrated Load

1. We are considering a prestressed girder under the influence of a movable unit load (Fig. 4.24).

In intervals

$$M_x = \frac{x(l - x_1)}{l} \qquad \ldots (0 \text{ to } x_1)$$

$$m_1 = -e \qquad \ldots \left(\frac{l - l_t}{2} \text{ to } x_1\right)$$

$$M_x = \frac{x(l - x_1)}{l} - 1(x - x_1) \qquad \ldots (x_1 \text{ to } l)$$

$$m_1 = -e \qquad \ldots \left(x_1 \text{ to } \frac{l + l_t}{2}\right)$$

$$N_1 = 1 \qquad \ldots \left(\frac{l - l_t}{2} \text{ to } \frac{l + l_t}{2}\right)$$

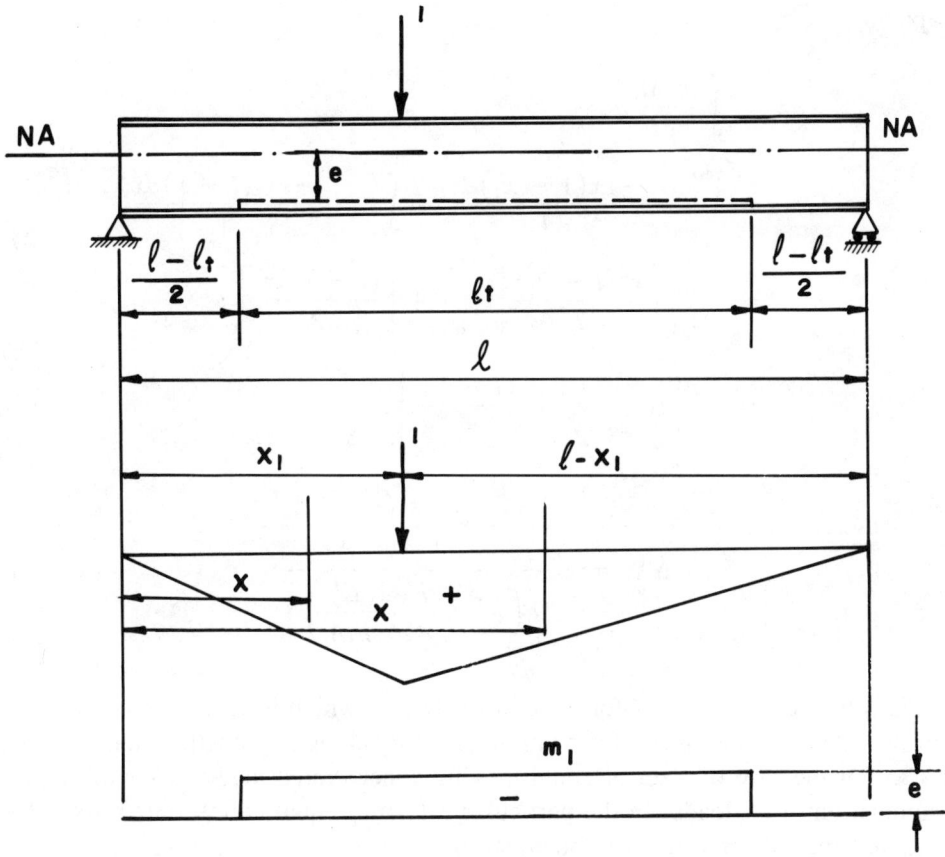

Fig. 4.24 Diagrams of bending moments due to unit load.

2. An increment of prestressing under a unit load acting along the tendon length is

$$\delta_{11}\, \Delta X_1 + \delta_{1L} = 0 \qquad \Delta X = -\frac{\delta_{1L}}{\delta_{11}}$$

$$\delta_{11} = \int_0^1 \frac{m_1^2}{EI}\, dx + \int_0^1 \frac{N_1^2\, dx}{EA} + \int_0^1 \frac{N_1^2\, dx}{E_t A_t}$$

But the integration over the interval 0 to $(l - l_t)/2$ and $(l + l_t)/2$ to 1 is zero; therefore in interval $(l - l_t)/2$ to $(l + l_t)/2$, we have

$$\delta_{11} = + \int_{l-l_t/2}^{l+l_t/2} \frac{l^2 dx}{EI} + \int_{l-l_t/2}^{l+l_t/2} \frac{dx}{EA} + \int_{l-l_t/2}^{l+l_t/2} \frac{dx}{E_t A_t} = \frac{l_t}{EI}\left(e^2 + \frac{I}{A} + \frac{EI}{E_t A_t} \right)$$

$$(4.121)$$

and

$$\delta_{1L} = \int \frac{m_1 M_x}{EI} dx$$

$$EI\delta_{1L} = \int_{l - l_t/2}^{x_1} \frac{-ex(1 - x_1)dx}{1} + \int_{x_1}^{l + l_t/2} \frac{-ex_1(l - x)dx}{1}$$

$$= -\frac{e}{1} \left\{ \left[\frac{x^2(l - x_1)}{2} \right]_{l - l_t/2}^{x_1} + \left[\frac{(1 - x)^2 x_1}{2} \right]_{x_1}^{l + l_t/2} \right\}$$

$$= -\frac{e}{2} \left[lx_1 - x_1^2 - \frac{(l - l_t)^2}{4} \right]$$

(4.122)

$$\Delta X_1 = \frac{e \left[lx_1 - x_1^2 - \dfrac{(l - l_t)^2}{4} \right]}{2l_t \left(e^2 + \dfrac{I}{A} + \dfrac{EI}{E_t A_t} \right)}$$

(4.123)

This is the equation of ΔX for the unit load at x_1, which is actually the equation of the influence line of ΔX for a unit load acting along the length of the tendon.

3. For the case of a set of equal moving concentrated loads, ΔX will be the summation of all loads. In the particular case of n equal concentrated loads P, the incremental force ΔX will be equal to

$$\Delta X_2 = \Sigma \frac{Pe \left\{ \left[lx_1 - x_1^2 - \left(\dfrac{(l - l_t)^2}{2} \right) \right] + \cdots \left[lx_n - x_n^2 - \left(\dfrac{l - l_t}{2} \right)^2 \right] \right\}}{2l_t \left(e^2 + \dfrac{I}{A} + \dfrac{EI}{E_t A_t} \right)}$$

(4.124)

or

$$\Delta X_3 = \Sigma \frac{Pe \left\{ \left[1(x_1 + x_2 + \cdots + x_n) - (x_1^2 + x_2^2 + \cdots + x_n^2) \right] - n \left(\dfrac{l - l_t}{2} \right)^2 \right\}}{2l_t \left(e^2 + \dfrac{I}{A} + \dfrac{EI}{E_t A_t} \right)}$$

(4.125)

4. For the particular case when unit loading is acting before the point of a short tendon on the left side, or $x_1 < (l - l_t)/2$ (Fig. 4.25)

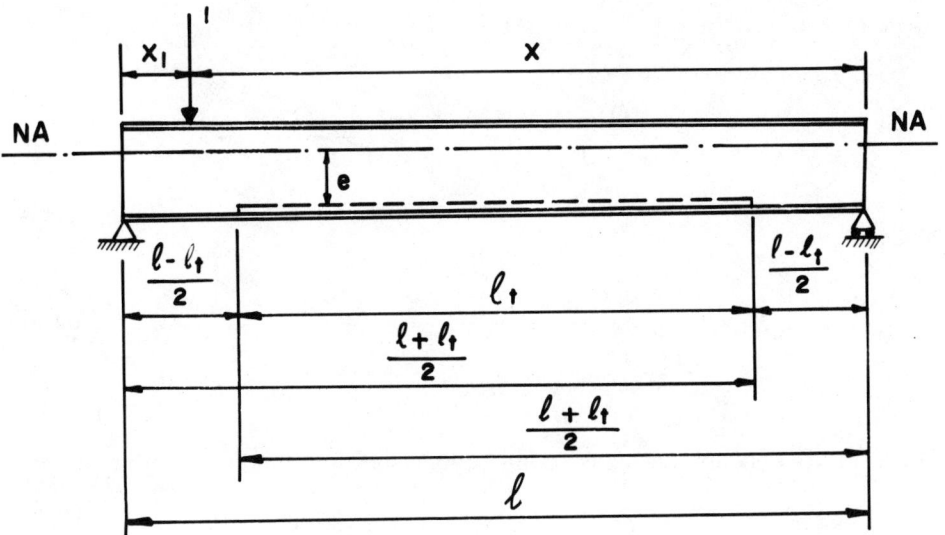

Fig. 4.25 Intervals along prestressed girders for action of unit load.

In interval

$$M_x = \frac{x_1(l - x)}{1} \qquad \cdots (x_1 \text{ to } 1)$$

$$EI\delta_{1L} = \int_{l-l_t/2}^{l+l_t/2} \frac{ex_1(1 - x)\,dx}{1} = -\frac{ex_1}{1}\left[\frac{(1 - x)^2}{-2}\right]_{l-l_t/2}^{l+l_t/2} = -\frac{ex_1 l_t}{2} \quad (4.126)$$

and

$$\Delta X = \frac{ex_1 l_t}{2l_t\left(e^2 + \dfrac{I}{A} + \dfrac{EI}{E_t A_t}\right)} = \frac{ex_1}{2\left(e^2 + \dfrac{I}{A} + \dfrac{EI}{E_t A_t}\right)} \qquad (4.127)$$

For a set of equal concentrated forces P

$$\Delta X_4 = \sum \frac{Pe(x_1 + x_2 + \cdots + x_n)}{2\left(c^2 + \dfrac{I}{A} + \dfrac{EI}{E_t A_t}\right)} \qquad (4.128)$$

5. For the particular case when $x_1 > l + l_t/2$

$$M_x = \frac{x}{1}(l - x_1) \qquad (0 \text{ to } 1)$$

$$m_1 = -e \qquad \left(\frac{l - l_t}{2} \text{ to } \frac{l + l_t}{2}\right)$$

$$\delta_{1L} = \int_{l-l_t/2}^{l+l_t/2} \frac{-ex(1-x_1)}{1} \, dx$$

$$= -\frac{1(1-x_1)}{1}\left(\frac{x^2}{2}\right)_{l-l_t/2}^{l+l_t/2} = -\frac{el_t(l-x_1)}{2} \qquad (4.129)$$

and

$$\Delta X_5 = \frac{Pel_t(1-x_1)}{2l_t\left(e^2 + \dfrac{I}{A} + \dfrac{EI}{E_tA_t}\right)} = \frac{Pe(1-x_1)}{2\left(e^2 + \dfrac{I}{A} + \dfrac{EI}{E_tA_t}\right)} \qquad (4.130)$$

For a set of concentrated loads acting beyond the right tendon anchorage

$$\Delta X_6 = \sum \frac{Pe\left[(l-x_1) + (l-x_2) + \cdots + (l-x_n)\right]}{2\left(e^2 + \dfrac{I}{A} + \dfrac{EI}{E_tA_t}\right)} \qquad (4.131)$$

6. Transformation of the denominator in expression (4.125) can be found similar to previous calculation for δ_{11}, or after formula (4.121)

$$\delta_{11} = \frac{l_t}{EI}\left(e^2 + \frac{I}{A} + \frac{EI}{E_tA_t}\right)$$

For the calculation of δ_1, the maximum ordinate at the mid-span may be found from expression (4.122), after the substitution $x_1 = 1/2$, namely,

$$\frac{1}{2}\left[lx_1 - x_1^2 - \frac{(l-l_t)^2}{4}\right] = \frac{1}{2}\left[\frac{l^2}{2} - \frac{l^2}{4} - \frac{(l-l_t)^2}{4}\right] = \frac{l_t(2l-l_t)}{8} \qquad (4.132)$$

The envelope curve to the maximum value of the moment diagram due to moving loads is of a second-degree curve equation, having boundary conditions of zero at the two ends and the ordinate $[l_t(2l-l_t)]/8$ at mid-span. We assume such a parabolic equation expressed as

$$y = Ax^2 + Bx + C \qquad (4.133)$$

Applying boundary conditions

at $x = 0$ $y = 0$ and $C = 0$

at $x = \dfrac{1}{2}$ $y = \dfrac{l_t}{8}(2l - l_t)$ and $\dfrac{l_t}{8}(2l - l_t) = \dfrac{Al^2}{4} + \dfrac{Bl}{2}$

at $y = 0$ $Al + B = 0$

Solutions of the above equations yield

$$A = -\frac{l_t}{2l^2}(2l - l_t) \qquad B = +\frac{l_t l}{2l^2}(2l - l_t) \qquad (4.134)$$

After substitution of (4.134) into (4.133), we obtain

$$y = \frac{l_t(2l - l_t)}{2l^2}(-x^2 + lx) \qquad (4.135)$$

For a set of unit concentrated loads

$$y = \frac{l_t(2l - l_t)}{2l^2}[l(x_1 + x_2 + \cdots + x_n) - (x_1{}^2 + x_2{}^2 + \cdots + x_n{}^2)]$$

and considering that $\delta_{1L} = \Sigma(Mm/EI)$

$$EI\delta_{1L} = -\frac{el_t(2l - l_t)}{2l}[l(x_1 + x_2 + \cdots + x_n) - (x_1{}^2 + x_2{}^2 + \cdots + x_n{}^2)]$$

Considering that $\delta X = -(\delta_{1L}/\delta_{11})$, for a set of concentrated loads P, we have

$$\Delta X_6 = \Sigma \frac{Pe(2l - l_t)[l(x_1 + x_2 + \cdots + x_n) - (x_1{}^2 + x_2{}^2 + \cdots + x_n^2)]}{2l^2\left(e^2 + \dfrac{I}{A} + \dfrac{EI}{E_t A_t}\right)}$$

$$(4.136)$$

7. For the case of different concentrated moving loads (Fig. 4.26), we consider the summary of expressions (4.128), (4.125), and (4.131), or

$$\Delta X_4 = \Sigma \frac{e(P_1 x_1 + P_2 x_2 + \cdots)}{2\left(e^2 + \dfrac{I}{A} + \dfrac{EI}{E_t A_t}\right)}$$

$$\Delta X_3 =$$

$$\Sigma \frac{e\left[l(P_3 x_3 + P_4 x_4 + \cdots + P_n x_n) - (P_3 x_3{}^2 + \cdots + P_n x_n{}^2) - (n - 2)\left(\dfrac{l - l_t}{2}\right)^2\right]}{2l_t\left(e^2 + \dfrac{I}{A} + \dfrac{EI}{E_t A_t}\right)}$$

$$\Delta X_6 = \Sigma \frac{e[P_{n+1}(l - x_{n+1}) + P_{n+2}(1 - x_{n+2})]}{2\left(e^2 + \dfrac{I}{A} + \dfrac{EI}{E_t A_t}\right)}$$

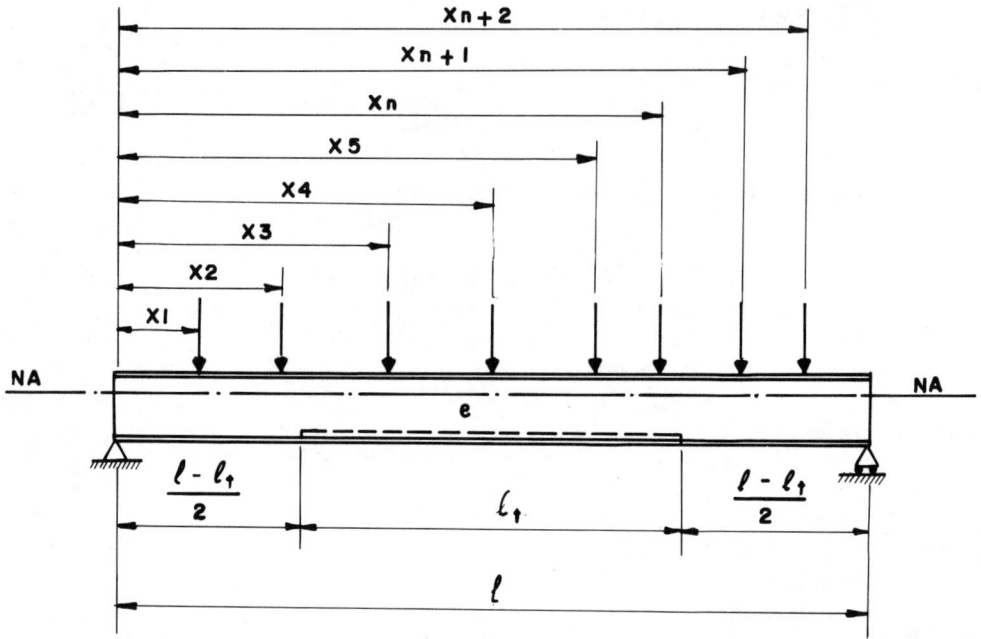

Fig. 4.26 Different concentrated moving loads.

4.9 NUMERICAL EXAMPLE

Design a prestressed steel plate girder to support a uniformly distributed dead load of $g = 2.45$ k/ft and a live load of $p = 1.63$ k/ft. The girder is simply supported with a span of $l = 60$ ft 0 in. Materials: steel having $F = 30$ ksi and a tendon of high-strength wire, $d = 0.196$ in., $F_t = 138$ ksi. Moduli of elasticity for the steel and tendon $E = E_t = 30 \times 10^3$ ksi. Limit deflection $f = 1/250$.

Solution

1. Maximum Bending Moment

$$M_{max} = \frac{(2.45 + 1.63)}{8} \times 60^2 = 1836 \text{ k/ft}$$

Reaction

$$R = \frac{(2.45 + 1.63)60}{2} = 122.4 \text{ k}$$

2. For the Design of a Cross Section, the Ratio μ

$$\mu = \frac{E_t \times F}{E \times F_t} = \frac{30 \times 10^3 \times 30}{30 \times 10^3 \times 138} = 0.217$$

For a uniformly distributed load from Table 4.3, by interpolation, the values of coefficients a and C are shown.

$$a = 2.01 \qquad C = 0.36$$

Assume the web flexibility as $K = 100$, and determine from formulas (4.30) to (4.40) and (4.53)

$$A = \sqrt[3]{\frac{M^2}{C^2 f^2 K}} = \sqrt[3]{\frac{1836^2}{0.36^2 \times 30^2 \times 144^2 \times 100}} \qquad (4.59)$$

$$= \sqrt[3]{0.0139} = 0.24 \text{ ft}^2 = 34.65 \text{ in.}^2$$

$$A_w = mA = 0.55 \times 34.65 = 19.05 \text{ in.}^2 \qquad (4.38)$$

where $m = 0.55$ (see Section 4.8.2).

$$A_1 = A\left(\frac{a}{a+1} - \frac{m}{2}\right) = 34.65\left(\frac{2.01}{2.01 + 1} - 0.275\right) = 13.55 \text{ in.}^2 \quad (4.39)$$

$$A_2 = A\left(\frac{1}{a+1} - \frac{m}{2}\right) = 34.65\left(\frac{1}{2.01 + 1} - 0.275\right) = 1.98 \text{ in.}^2 \quad (4.40)$$

$$h = \sqrt{mAk} = \sqrt{0.55 \times 34.65 \times 100} = 43.65 \text{ in.} \qquad (4.41)$$

$$A_t = A\frac{F}{F_t} \cdot \frac{a-1}{a+1} = 34.65 \times \frac{30}{138} \times \frac{2.01 - 1}{2.01 + 1}$$

$$= 2.52 \text{ in.}^2 \ (0.018 \text{ ft}^2) \qquad (4.67)$$

On the basis of the calculated values of the parameters, choose the girder cross section, Fig. 4.27.

$$A_1 = 20 \times 0.75 = 15 \text{ in.}^2$$

$$A_2 = 5 \times 0.5 = 2.5 \text{ in.}^2$$

$$A_w = 43 \times 0.5 = 21.5 \text{ in.}^2$$

$$A = 15 + 2.50 + 21.5 = 39.0 \text{ in.}^2 = 0.27 \text{ ft}^2$$

3. *Tendon Cross Section*

$$a_w = \frac{\pi d^2}{4} = \frac{3.14 \times 0.196^2}{4} = 0.03 \text{ in.}^2$$

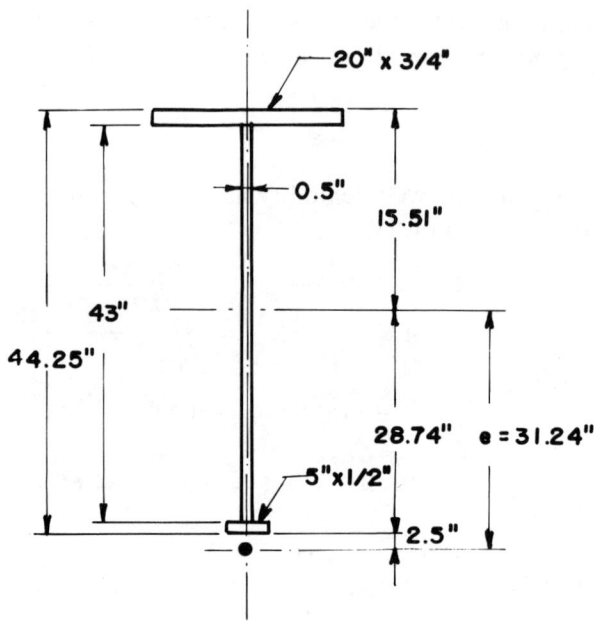

Fig. 4.27 A plate girder cross section.

Number of wires:

$$n = \frac{A_t}{A_w} = \frac{2.52}{0.03} = 84$$

Use 95 wires at $A_1 = 95 \times 0.03 = 2.85$ in.^{2} = 0.0147 ft^2.

4. Length of the Tendon

$$\alpha = l - \frac{\sqrt{m}}{C} \cdot \frac{6a - m(a + 1)^2}{6a(a + 1)}$$

$$= l - \frac{\sqrt{0.55}}{0.36} \cdot \frac{6 \times 2.01 - 0.55 (2.01 + 1)^2}{6 \times 2.01(2.01 + 1)} \qquad (4.73)$$

$$= 0.6$$

$$l_t = l\sqrt{\alpha} = 60\sqrt{0.6} = 46.47 \text{ ft} \qquad (4.73)$$

5. Location of the Center of Gravity of a Girder Cross Section

$$y_b = \frac{20 \times 0.75 \times 43.875 + 43 \times 0.5 \times 21.5 + 5 \times 0.5 \times 0.25}{39}$$

$$= 28.74 \text{ in.}$$

$$y_t = 44.25 - 28.74 = 15.51 \text{ in.}$$

Eccentricity of a tendon

$$e = 28.74 + 2.5 = 31.24 \text{ in.} = 2.6 \text{ ft}$$

6. Moment of Inertia and Section Moduli of a Girder

$$I_x = \frac{1}{12} \times 20 \times 0.75^3 + 20 \times 0.75 \times 15.135^2 + 5 \times 0.5 \times 28.49^2$$

$$+ \frac{1}{12} \times 5 \times 0.5^3$$

$$+ \frac{1}{12} \times 0.5 \times 43^3 + 43 \times 0.5 \times 6.75^2 = 9758.36 \text{ in.}^4 \ (0.471 \text{ ft}^4)$$

$$S_1 = \frac{0.471}{1.292} = 0.365 \text{ ft}^3$$

$$S_2 = \frac{0.471}{2.395} = 0.197 \text{ ft}^3$$

7. The Value of the Prestressing Force

$$X = \frac{\Psi F_b S_b A_b}{S_b + e A_b} = \frac{30 \times 144 \times 0.96 \times 0.197 \times 0.27}{0.197 + 0.27 \times 2.60} = 243.42\text{k} \quad (4.3)$$

The value of an increment of prestressing force ΔX

$$\Delta X = -\frac{\displaystyle\int \frac{M_1 M}{EI_x} dx}{\displaystyle\int \frac{M_1^2 dx}{EI} + \frac{l_t}{E_t A_t} + \frac{l_t}{EA}} \quad (4.63)$$

where $M_1 = l \times e = l \times 2.60 = 2.60 \text{ k/ft}$.

$$\Delta X = \frac{2 \times 1836 \times 2.60}{3\left(2.60^2 + \dfrac{0.467}{0.27} + \dfrac{0.467}{0.018}\right)} \left(2 \times \frac{46.67}{60}\right) = 143.668 \text{ k} \quad (4.64)$$

The value of the controllable prestressing force during the fabrication of girder is

$$X_c = \frac{X}{0.95} + \Delta_a \frac{E_t A_t}{l_t}$$

$$= \frac{243.42}{0.95} + \frac{0.039}{12} \times \frac{30 \times 10^3 \times 12^2 \times 0.0197}{46.67} \quad (3.19)$$

$$= 262.18\text{k}$$

8. Checking of Stresses in a Girder

a. In the Course of Prestressing

$$f_t = -\frac{X}{A} + \frac{Xe}{S_1} = -\frac{243.42}{0.27} + \frac{243.42 \times 2.60}{0.362} = 846.77 \text{ k/ft}^2$$

$$f_b = -\frac{X}{A} - \frac{Xe}{S_2} = -\frac{243.42}{0.27} - \frac{243.42 \times 2.60}{0.195} = -4147.15 \text{ k/ft}^2$$

b. Under a Load at Mid-Span

$$f_t = -\frac{X + \Delta X}{A} - \frac{M - (X + \Delta X)\,e}{S_1}$$

$$f_t = -\frac{243.42 + 143.69}{0.27} - \frac{1836 - (243.42 + 143.69) \times 2.60}{0.365}$$

$$= -3725.26 \text{ k/ft}^2$$

$$f_b = -\frac{X + \Delta X}{A} + \frac{M - (X + \Delta X)\,e}{S_2}$$

$$f_b = -1433.74 + 4254.00 = 2820.26 \text{ k/ft}^2$$

c. Under a Load at the End Point of a Tendon

$$M = 122.4 \times 6.67 - \frac{4.08 \times 6.67^2}{2} = 725.65 \text{ k/ft}$$

$$f = \frac{M}{S_2} = \frac{725.65}{0.195} = 3721.28 \text{ k/ft}^2$$

9. Stresses in a Tendon

$$\frac{X + \Delta X}{A_t} = \frac{243.42 + 143.69}{0.0197} = 19650 \text{ k/ft}^2 < 19872.00 \text{ k/ft}^2$$

where permissible stress in tendon $= 138 \text{ ksi} \times 144 = 19872.00 \text{ k/ft}^2$.

10. Girder Deflection

$$\Delta = \frac{5}{384} \times \frac{9l^4}{EI} - \frac{(X + \Delta X)el^2}{8EI}\left[l - 4\left(\frac{a}{l}\right)^2\right]$$

$$= \frac{5}{384} \times \frac{4.08 \times 60^4}{4320 \times 10^3 \times 0.467}$$

$$- \frac{(243.42 + 143.69) \times 2.60 \times 60^2}{8 \times 4320 \times 10^3 \times 0.467}\left[l - 4\left(\frac{6.67}{60}\right)^2\right]$$

$$= 0.341 - 0.213 = 0.128 \text{ ft} = 1.536 \text{ in.}$$

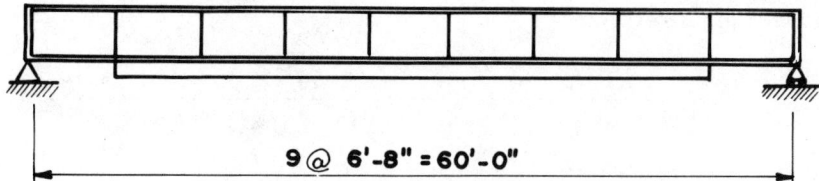

Fig. 4.28 Elevation of a girder.

$$\Delta_{\text{all}} = \frac{l}{250} = \frac{60 \times 12}{250} = 2.88 \text{ in.}$$

$$1.536 \text{ in.} < 2.88 \text{ in.}$$

11. Tangential Stress

The static moment of the half-section

$$Q_x = 20 \times 0.75 \times 15.135 + 14.76 \times 0.5 \times \frac{14.76}{2}$$

$$= 281.48 \text{ in.}^3 \ (0.163 \text{ ft}^3)$$

$$\tau = \frac{RQ_x}{It_w} = \frac{122.4 \times 0.163}{0.467 \times 0.0416} = 1026.97 \text{ k/ft}^2 \qquad (2.37)$$

12. Shear Stress in a Web

The shear load at the fourth panel is (see Fig. 4.28)

$$\frac{122.4}{30 - 3.33} = \frac{Q_1}{3.33} \qquad Q_1 = \frac{122.4 \times 3.33}{26.67} = 15.28 \text{ k}$$

$$\tau = \frac{15.28}{3.58 \times 0.0416} = 102.60 \text{ k/ft}^2$$

13. Bending Stress in a Web, Fig. 4.29

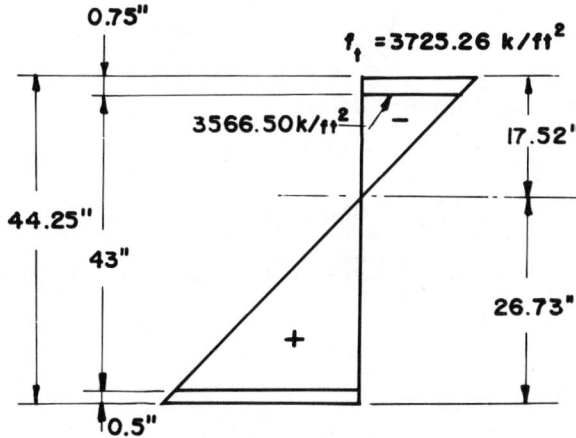

Fig. 4.29 Bending stress in a web.

14. Buckling Stresses

$$f_{cr} = \frac{K\pi^2 E}{12(1 - \mu^2)(h/t)^2} = \frac{23.9 \times 3.14^2 \times 30 \times 10^3 \times 144}{12(1 - 0.3^2)(43/0.5)^2}$$

$$= 12,604 \text{ k/ft}^2$$

$$\tau_{cr} = \left[5.34 + \frac{4}{(a/b)^2} \right] \frac{\pi^2 E}{12(1 - \mu^2)(h/t)^2}$$

$$= \left[5.34 + \frac{4}{(80/43)^2} \right] \frac{3.14^2 \times 30 \times 10^3 \times 144}{12(1 - 0.3^2)(43/0.5)^2}$$

$$= 3428.73 \text{ k/ft}^2$$

$$\sqrt{\left(\frac{3566.5}{12604} \right)^2 + \left(\frac{102.60}{3428.73} \right)^2} = 0.28 < 1$$

4.10 DESIGN EXAMPLE

Design an interior prestressed girder for a two-lane highway bridge having a span of $l = 60$ ft for HS20 traffic. Girders are of A36 steel; $\sigma_{all} = 30$ ksi. High-strength steel wires $\sigma_{all} = 200 \times 10^3$ psi. Moduli of elasticity for steel and wires are $E = E_t = 30 \times 10^3$ ksi. An allowance for future paving is 25 psf. A cross section of the bridge is shown in Fig. 4.30.

Solution

1. Dead Load; no Composite Action

Slab $7/12 \times 6.75 \times 0.150 = 0.590$ k/ft
Hunch $1/12 \times 2.10 \times 0.15 = 0.025$ k/ft
Weight of steel beam $\qquad = 0.100$ k/ft

$$g_{DL} = 0.715 \text{ k/ft}$$

2. Superimposed Dead Load

Asphalt paving $\qquad 0.025 \times 26.0 = 0.650$ k/ft
Parapet $\qquad 2 \times 1.0 \times 0.75 \times 0.15 = 0.225$ k/ft
Railing $\qquad 2 \times 0.15 = 0.300$ k/ft
Sidewalk $\qquad 2 \times 3.05 \times 0.75 \times 0.15 = 0.686$ k/ft

$$g_{SD} = 1.861 \text{ k/ft}$$

per beam

$$g_{SD} = \frac{1.861}{5} = 0.372 \text{ k/ft}$$

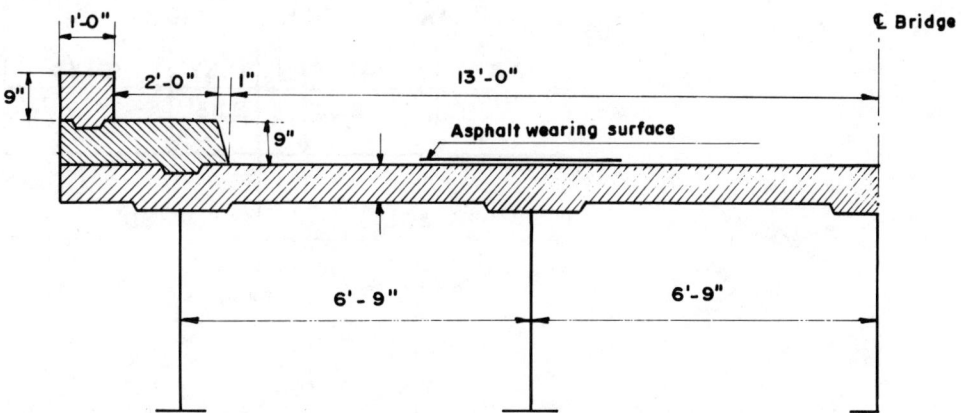

Fig. 4.30 Noncomposite steel beam bridge.

3. Concentrated Dead Load (Diaphragm)

Assume diaphragm weight at 0.04 k/ft.

$$0.04 \times 6.75 = 0.27 \text{ k}$$

4. Dead-Load Moment

$$M_{DL} = \frac{0.715 \times 60^2}{8} + \frac{0.27 \times 60}{4} = 321.75 + 4.05 = 325.8 \text{ kft}$$

5. Superimposed Dead-Load Moment

$$M_{SD} = \frac{0.372 \times 60^2}{8} = 167.4 \text{ kft}$$

6. Live Load (Fig. 4.31)

$$x = \frac{16 \times 14 + 4 \times 28}{36} = \frac{336}{36} = 9.33 \text{ ft}$$

$$R = \frac{16(18.33 + 32.33) + 4 \times 46.33}{60} = \frac{995.88}{60} = 16.598 \text{ k}$$

$$M_{LL} = 16.598 \times 27.67 - 4 \times 14 = 403.267 \text{ kft}$$

$$\text{Fraction} = \frac{6.75}{5.5} = 1.22$$

$$\text{Impact} = I = \frac{50}{L + 125} = \frac{50}{60 + 125} = \frac{50}{185} = 0.27$$

Use $I = 1.30$.

$$\text{Total } M_{LL} = 403.267 \times 1.22 \times 1.30 = 639.58 \text{ kft}$$

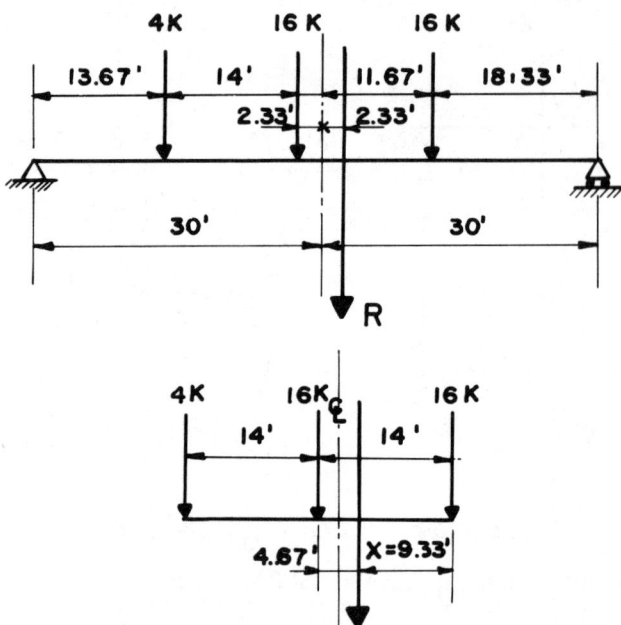

Fig. 4.31 Position of live load for M_{max}.

7. Total moment

$$M = M_{DL} + M_{SDL} + M_{LL} = 325.8 + 167.4 + 639.58 = 1132.78 \text{ kft}$$

8. Determination of Geometric Characteristics of the Cross Section

After formula (4.109)

$$A = \sqrt[3]{\frac{M^2}{f^2 k}} \cdot \frac{a + 1}{a \sqrt[3]{3(1 - a)^2 (2 - a)}}$$

$$A = \sqrt[3]{\frac{1132.78^2}{4320^2 \times 120}} \cdot \frac{(1.71 + 1)}{1.71 \sqrt[3]{3(1 - 1.71)^2 (2 - 1.71)}} = \frac{0.649}{3.746}$$

$$= 0.173 \text{ ft}^2 = 24.965 \text{ in.}^2$$

$$h_2 = \frac{a \sqrt{Akm}}{a + 1} = \frac{1.71 \sqrt{24.965 \times 120 \times 0.55}}{2.71}$$

$$= 25.6 \text{ in.} \cong 26 \text{ in.}$$

Assume center of prestressing tendon is 4 in. below bottom flange.
    Eccentricity $e = 26 + 4 = 30$ in.

$$h_1 = \frac{h_2}{a} = \frac{26}{1.71} = 15.2 \text{ in.} \cong 15 \text{ in.}$$

$$h = h_1 + h_2 = 15 + 26 = 41 \text{ in.}$$

$$(4.36)$$

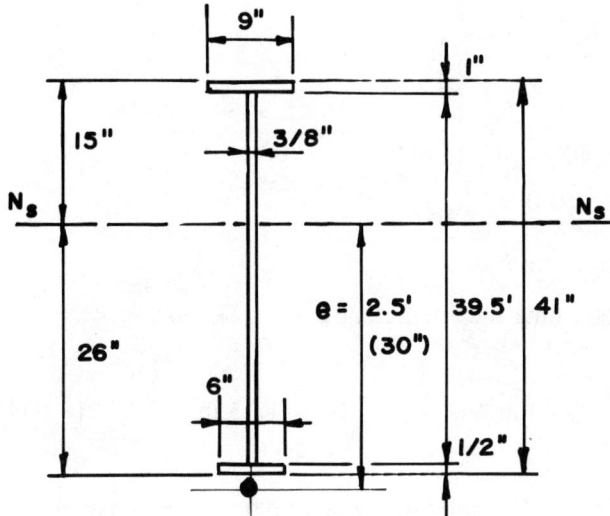

Fig. 4.32 Cross section of the beam.

$$t_w = \frac{h}{k} = \frac{41}{120} = 0.341 \text{ in.} \qquad \text{use } t_w = \frac{3}{8} \text{ in. } (0.375 \text{ in.}) \qquad (4.37)$$

$$A_1 = A\left[\frac{a}{a+1} - \frac{m}{2}\right] = 24.965\left[\frac{1.71}{1.71+1} - \frac{0.55}{2}\right] = 9 \text{ in.}^2 \quad (4.39)$$

$$A_2 = A\left[\frac{1}{a+1} - \frac{m}{2}\right] = 24.965\left[\frac{1}{1.71+1} - \frac{0.55}{2}\right] = 2.35 \text{ in.}^2 \quad (4.40)$$

Arrange the cross section on the basis of values above (Fig. 4.32)

$$A_1 = 9 \times 1 = 9 \text{ in.}^2 \qquad A_2 = 6 \times \frac{1}{2} = 3 \text{ in.}^2$$

$$A_w = A - (A_1 + A_2) = 24.965 - (9 + 3) = 12.965 \text{ in.}^2$$

$$A_w = 39.5 \times \frac{3}{8} = 14.81 \text{ in.}^2$$

$$A = 9 + 3 + 14.81 = 26.81 \text{ in.}^2 \ (0.186 \text{ ft}^2)$$

$$I_b = 9 \times (15 - 0.5)^2 + 6 \times \frac{1}{2} \times (26 - 0.25)^2$$

$$+ \frac{0.375 \times 39.5^3}{12} + 39.5 \times \frac{1}{2} \times 5.75^2$$

$$= 1892.25 + 1989.18 + 1925.93 + 652.98 = 6460.34 \text{ in.}^4$$

$$S_1 = \frac{6460.34}{15} = 430.69 \text{ in.}^3 \ (0.249 \text{ ft}^3)$$

$$S_2 = \frac{6460.34}{26} = 248.47 \text{ in.}^3 \ (0.144 \text{ ft}^3)$$

$$A\frac{F}{F_t}\left(\frac{a-1}{a+1}\right) = 26.81 \times \frac{30}{200} \times \frac{1.71 - 1}{1.71 + 1} = \frac{26.81 \times 30 \times 0.71}{200 \times 2.71} \tag{4.67}$$

$$= 1.054 \text{ in.}^2 \ (0.0073 \text{ ft.}^2)$$

9. Determination of Prestressing Force

For the bottom chord, we determine the coefficient φ.

$$I_y = \frac{0.5 \times 6^3}{12} = 9 \text{ in.}^4 \qquad r_y = \sqrt{\frac{9}{3}} = 1.73 \text{ in.} \tag{4.94}$$

Assuming the spacing of points of fastening of the tendon to the beam equals 48 in.

$$\lambda_y = \frac{48}{1.73} = 27.74 \tag{4.94}$$

From Fig. 4.18, we find $\varphi = 0.98$.

The value of prestressing force after formula (4.93) is

$$X = \frac{\sigma\varphi AS_b}{(S_b + eA)} = \frac{4320 \times 0.98 \times 0.186 \times 0.14}{0.14 + 2.5 \times 0.186} = \frac{110.24}{0.605} = 182.2 \text{ k}$$

10. Determination of the tendon length

The maximum distance of the tendon from the bearing $a_{\max}$ is a function of the load-carrying capacity of the main cross section σW_2. To determine $a_{\max}$ we calculate moments cross-section spaced at 2-ft intervals (Table 4.4).

Live loads:

$$4.0 \times 1.3 \times 1.22 = 6.34 \text{ k}$$

$$16 \times 1.3 \times 1.22 = 25.38 \text{ k}$$

where

4 k and 16 k = wheel loads
1.3 = impact
$S/5.5$ = 1.22 distribution factor

Dead load:

Weight of steel beam	0.715 k/ft
Superimposed dead load	0.372 k/ft
	1.087 k/ft

Table 4.4 Girder Under Dead and Moving Live Load. Moments Are Calculated at 2-ft Intervals.

Distance a (ft)	Live Load Reaction R_A (kips)	Moments $M_a = M_{LL} + M_g$
2	$R_A = \dfrac{6.34 \times 30 + 25.38(44 + 58)}{60}$ $= 46.32$	$M_{LL} = 46.32 \times 2 = 92.64$ $M_g = \dfrac{1.087 \times 2}{2}(60 - 2) = 63.05$ $M_2 = 155.69$
4	$R_A = \dfrac{6.34 \times 28 + 25.38(42 + 56)}{60}$ $= 44.41$	$M_{LL} = 44.41 \times 4 = 177.64$ $M_g = \dfrac{1.087 \times 4}{2}(60 - 4) = 121.74$ $M_4 = 299.38$
6	$R_A = \dfrac{6.34 \times 26 + 25.38(40 + 54)}{60}$ $= 42.51$	$M_{LL} = 42.51 \times 6 = 255.06$ $M_g = \dfrac{1.087 \times 6}{2}(60 - 6) = 176.09$ $M_6 = 431.15$
8	$R_A = \dfrac{6.34 \times 24 + 25.38(38 + 52)}{60}$ $= 40.61$	$M_{LL} = 40.61 \times 8 = 324.88$ $M_g = \dfrac{1.087 \times 8}{2}(60 - 8) = 226.09$ $M_g = 550.97$
10	$R_A = \dfrac{6.34 \times 22 + 25.38(36 + 50)}{60}$ $= 38.70$	$M_{LL} = 38.70 \times 10 = 387.00$ $M_g = \dfrac{1.087 \times 10}{2}(60 - 10) = 271.75$ $M_{10} = 658.75$

Envelope of the moment (Fig. 4.34)

$$S_2 = \frac{0.5M_{\max}}{\sigma} = \frac{0.5 \times 1132.78}{30 \times 144} = 0.131 \text{ ft}^3$$

$$M = \sigma S_2 = 30 \times 144 \times 0.131 = 565.92 \text{ kft}$$

Graphically, we obtain $a_{\max} = 8.15$ ft.

11. Determination of $a_{\min}$

Determine the minimum distance, $a_{\min}$, between the tendon and the bearing for the specific truck load with the aid of formula (4.118) $a_{\min} = \gamma l$.

$$\gamma = \frac{e\left[e + \dfrac{S_2}{A}\right]}{2\left[e^2 + \dfrac{I}{A} + \dfrac{EI}{E_t A_t}\right]} = \frac{30\left[30 + \dfrac{248.47}{26.81}\right]}{2\left[30^2 + \dfrac{6460.34}{26.81} + \dfrac{6460.34}{1.059}\right]} = 0.0899$$

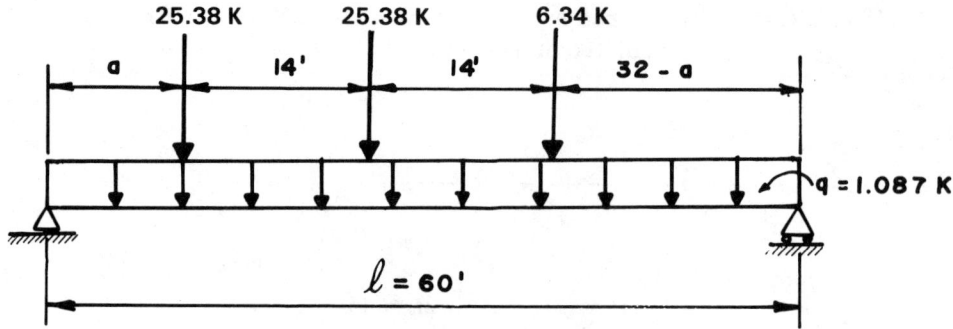

Fig. 4.33   Girder under dead and moving live load.

The minimum distance from the bearing is

$$a_{min} = \gamma l = 0.0899 \times 60 = 5.396 \text{ ft}$$

Assume $a_{min} = 5.5$ ft.

12. Determination of the Increment of Force in the Tendon Due to Live Load
Considering the position of the truck for the maximum moment (Fig. 4.35)

$$\Delta X = \frac{p_e (cl - a^2 - c^2)}{2(l - 2a)\left[e^2 + \dfrac{I}{A} + \dfrac{EI}{A_t A_t}\right]}$$

The numerator for three loads is calculated in Table 4.5.

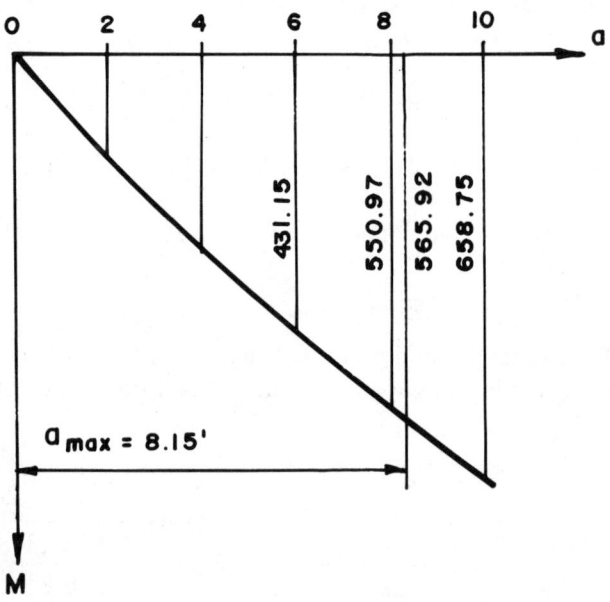

Fig. 4.34   Diagram of moment envelope.

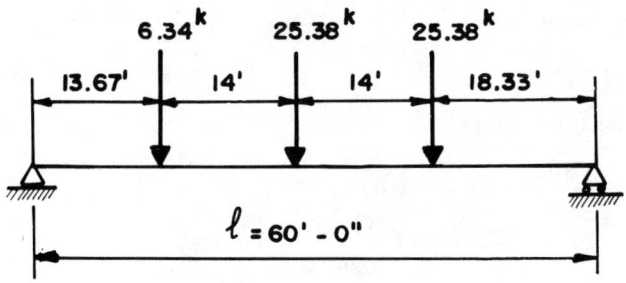

Fig. 4.35 Position of truck for maximum moment after formula (4.105).

Denominator

$$2(l - 2a)\left[e^2 + \frac{I}{A} + \frac{EI}{E_t A_t}\right] = 2(60 - 2 \times 5.5)\left[2.5^2 + \frac{0.266}{0.186} + \frac{0.266}{0.0073}\right]$$

$$= 4322.78$$

$$\Delta X = \frac{111018.54}{4322.78} = 25.68 \text{ k}$$

13. Check for Tendon Strength

$$\sigma_t = \frac{X + \Delta X}{A_t} = \frac{164.72 + 25.68}{0.0073} = \frac{190.40}{0.0073} = 26{,}082.19 \text{ k/ft}^2$$

$$= 181{,}126.32 < 200{,}000 \text{ psi}$$

14. Check of Beam for Stresses During Prestressing

In the course of prestressing, normal stresses in the top chord

$$\sigma_t = -\frac{X}{A} + \frac{Xe}{S_1} = -\frac{182.20}{0.186} + \frac{182.20 \times 2.5}{0.249} = -979.57 + 1829.32$$

$$= 849.75 \text{ k/ft}^2 = 5901.04 \text{ psi}$$

Normal stresses in the bottom chord

$$\sigma_b = -\frac{X}{A} - \frac{Xe}{S_1} = -\frac{182.20}{0.186} - \frac{182.20 \times 2.5}{0.249} = -979.57 - 1829.32$$

$$= -2868.89 \text{ k/ft}^2 = -19922.85 \text{ psi}$$

Table 4.5

p	c	c^2	a	a^2	cl	$cl - a^2 - c^2$	p_e	$p_e(cl - a^2 - c^2)$
6.34	13.67	186.87	5.5	30.25	820.2	603.08	15.85	9558.82
25.38	27.67	765.63	5.5	30.25	1660.20	864.32	63.45	54841.10
25.38	41.62	1732.22	5.5	30.25	2497.20	734.73	63.45	46618.62
							Σ	111018.54

15. Check of Beam During Loading

a. Normal stresses in the top chord

$$\sigma_t = \frac{X + \Delta X}{A} - \frac{M + (X - \Delta X)e}{S_1}$$

$$= -\frac{182.20 + 25.68}{0.186} - \frac{1{,}132.78 - (182.20 + 25.68)2.5}{0.249}$$

$$= -\frac{207.88}{0.186} - \frac{1{,}132.78 - 519.70}{0.249} = -1{,}117.63 - \frac{613.08}{0.249}$$

$$= -1{,}117.63 - 2{,}462.17 = -3{,}579.80 \text{ k}/\text{ft}^2 = -24{,}859.72 \text{ psi}$$

b. Normal stresses in the bottom chord

$$\sigma_b = -\frac{X + \Delta X}{A} + \frac{M - (X + \Delta X)e}{S_2}$$

$$= -\frac{182.20 + 25.68}{0.186} + \frac{1{,}132.78 - (182.20 + 25.68)25}{0.144}$$

$$= -1{,}117.63 + \frac{613.08}{0.144} = 3{,}139.87 \text{ k}/\text{ft}^2 = 21{,}804.65 \text{ psi}$$

16. Check of Tendon Anchors for Stresses

The bending moment at a distance of $a = 5.5$ ft from the bearing should be calculated (Fig. 4.36).

Dead load

$$g_{DL+SD} = 0.175 + 0.372 = 1.087 \text{ kft}$$

Diaphragm $= 0.27$ k.
  Moment due to dead load

$$\text{Reaction } R_{DL} = \frac{1.087 \times 60}{2} + \frac{0.27 \times 60}{4} = 36.66 \text{ k}$$

$$M_{DL} = 36.66 \times 5.5 = 201.63 \text{ kft}$$

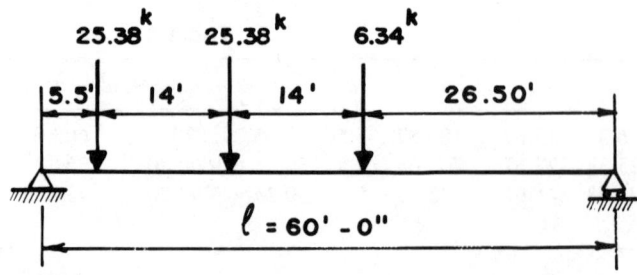

Fig. 4.36 Bending moment for tendon anchor.

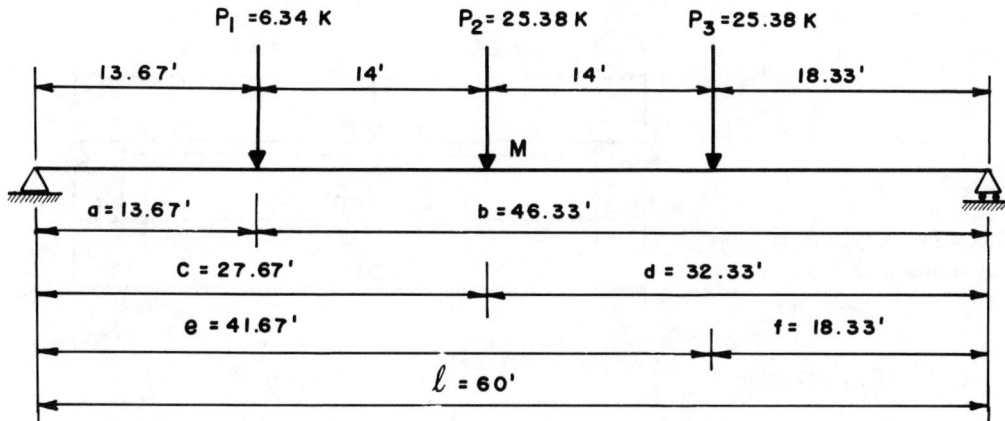

Fig. 4.37 Position of truck for maximum moment at section M.

Reaction due to live load

$$R_{LL} = \frac{6.34 \times 26.5 + 25.38\,(40.5 + 54.5)}{60} = 42.99 \text{ k}$$

$$M_{LL} = 42.99 \times 5.5 = 236.45 \text{ kft}$$

$$M_{\text{tot}} = M_{DL} + M_{LL} = 201.63 + 236.45 = 438.08 \text{ kft}$$

Normal stresses

$$\sigma_t = \frac{M}{S_1} = \frac{438.08}{0.294} = 1490.07 \text{ k/ft}^2 = 10{,}347.69 \text{ psi}$$

$$\sigma_b = \frac{M}{S_2} = \frac{438.08}{0.144} = 3042.22 \text{ k/ft}^2 = 21{,}126.54 \text{ psi}$$

Table 4.6 Deflection of the Girder Under Truck Loads.

Load	Deflection at Section M
	$\Delta_1 = \dfrac{P_1 a^2 b^2}{6EIl}\left[2\dfrac{x}{b} + \dfrac{x}{a} - \dfrac{\lambda 3}{ab^2}\right]$
P_1	$= \dfrac{6.34 \times 46.33^2 \times 13.67^2}{6EI \times 60}\left[2 \times \dfrac{32.33}{46.33} + \dfrac{32.33}{13.67} - \dfrac{32.33^2}{46.33^2 \times 13.67}\right]$
	$= \dfrac{18{,}365.35}{EI}$
P_2	$\Delta_2 = \dfrac{P_2 c^2 d^2}{3EIl} = \dfrac{25.38 \times 27.67^2 \times 32.33^2}{3EI \times 60} = \dfrac{112{,}834.02}{EI}$
P_3	$\Delta_3 = \dfrac{P_3 fx}{6EIl}(l^2 - f^2 - c^2) = \dfrac{25.38 \times 18.33 \times 27.67}{6EIl \times 60} = \dfrac{89{,}332.79}{EI}$
	Total $\sum_1^3 \Delta = \dfrac{220{,}536.16}{4{,}320 \times 10^3 \times 0.266} = 0.19 \text{ ft} = 2.30 \text{ in.}$

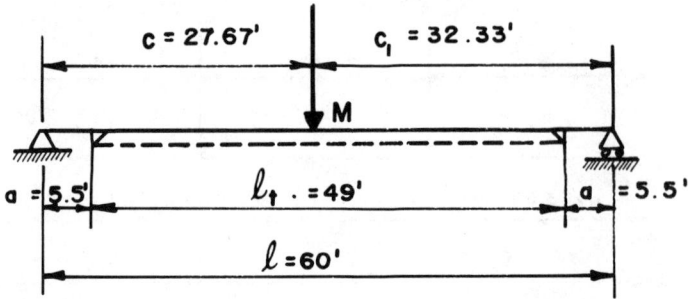

Fig. 4.38 Deflection at arbitrary section M due to prestressing.

17. Deflection

Deflection due to live load plus impact shall not exceed $1/800$ of the span.
Case I. Deflection at section M under truck loads (Fig. 4.37 and Table 4.6).
Case II. Negative deflection at section M due to prestressing (Fig. 4.38).
After formula (4.86)

$$\Delta_{\max} = -\frac{X_t e l^2}{8EI}\left[1 - 4\left(\frac{a}{1}\right)^2\right]$$

$$\Delta_{\max} = -\frac{(182.20 + 25.68) \times 2.5 \times 60^2}{8 \times 4320 \times 10^3 \times 0.312}\left[1 - 4\left(\frac{5.5}{60}\right)^2\right]$$

$$= -0.167 \text{ ft.} = -2.148 \text{ in.}$$

Resulting deflection

$$\Delta = \Delta_{LL} + \Delta_{\max} = 2.30 - 2.00 = 0.30 \text{ in.}$$

$$\frac{0.30}{60 \times 12} = \frac{1}{2400} < \frac{1}{800}$$

REFERENCES

1. Streletskii, N. S., *Metal Constructions*, Edition of Literature on Civil Engineering, Moscow, 1965, pp. 29–31 (in Russian).
2. Brodka, J., Jerka-Kulawinska, K., and Kwasniewski, M., "Prestressed Steel Girder, Statical Calculation," Verlags Gesellschaft Rudolf Muller, Cologne-Braunsfeld, 1968, pp. 10–14, (in German).
3. Brodka, J., and Klobukowski, J., "Prestressed Steel Constructions," Edition Wilhelm Ernst und Sohn, Berlin, 1969, pp. 96–103 (in German).
4. Belenya, E., *Prestressed Load-Bearing Metal Structures*, State Edition, Moscow, 1963, pp. 119–130 (in Russian).
5. Troitsky, M. S., and Rabbani, N. F., "Tendon Configuration of Prestressed Steel Girder Bridges," CSCE Centennial Conference, May 19–22, 1987, Montreal, *Proceedings*, vol. I, pp. 171–182.
6. Meljnikov, N. P., *Design Handbook—Metal Constructions*, 2nd ed., vol. III, 1980, Stroiizdat, Moscow, pp. 637–638 (in Russian).
7. Belenya, op. cit., pp. 115–125.

8. Ferjencik, P., and Tochacek, M., "Prestressing of Steel Structures, Theory and Construction Practice Edition Wilhelm Ernst und Sohn, Berlin, 1975, pp. 307–313, (in German).

9. Troitsky, M. S., Zielinski, Z. A., and Pimprikar, M. S., "Optimum Design of Prestressed Steel Girders," CSCE Annual Conference, May 12–16, 1986, Toronto, *Proceedings*, **4,** pp. 1–18.

10. Vedenikov, G. S., *Some Considerations on Optimum Shape of Steel Prestressed Beams*, Nauchniye Doklady Visshey Shkoly, Stroiteljstvo, 1958, pp. 126–134 (in Russian).

11. Brodka, J., Jerka-Kulawinska, J., and Kwasniewski, op. cit., pp. 76–79, (in German).

12. Belenya, E., *Prestressed Load-Bearing Metal Structures*, Mir Publishers, Moscow, 1977, pp. 92–94.

13. Moukhanov, K., "Metal Constructions," Edition Mir, Moscow, 1980, pp. 68–73, (in French).

14. American Association of State Highway and Transportation Officials, *Standard Specifications for Highway Bridges*, 13th ed., 1983.

15. Belenya, E., 1977, op. cit., pp. 193–196.

Chapter 5
Composite Plate Girders

5.1 INTRODUCTION

The prestressing of composite plate girders was first proposed by Dischinger in 1949.[1] In his two papers Dischinger developed a basic analysis for prestressed composite systems for simple and continuous spans plate girders, suspension bridges, and arch-type bridges in highway and railway bridge applications. He proposed that tendons of high-strength steel cables be used as tendons. In addition, he proposed reinforcement of the bottom chord of continuous plate girders by using concrete.

In 1950 Coff was granted a U.S. patent for a composite steel beam and concrete slab system prestressed by means of draped cables.[2] In 1959 Szilard discussed the analysis of prestressed composite beams using both elastic assumptions and approximate ultimate methods.[3] Hoadley in 1963 investigated a simply supported prestressed composite steel beam with a constant eccentric prestressing force applied along the entire length of the beam.[4]

Reagan and Krahl discussed the use of cover plates to increase the moment of inertia of the cross section in conventional composite design for working stresses.[5] Prestressed composite beams of this type have been used for bridge construction.

Knowles in his book[6] edited in 1973 discussed simple span plate girders prestressed by cables. He also investigated the method of prestressing by cambering. Roik discussed methods of prestressing continuous composite girders,[7] and Gibshman investigated methods of increasing the work of reinforced concrete slabs in composite beams by prestressing.[8]

Anand studied prestressed composite designs for simply supported longitudinal stringers for short span deck bridges using welded high-strength cover plates on the flange of initially stressed rolled beams.[9]

The strengthening of simple-span steel beam concrete bridges was extensively investigated at Iowa State University, and the results were published in a number of publications in the period 1981–1985.[10, 11, 12, 13] The analytical and experimental studies on composite prestressed and poststressed steel girder bridges were performed at Concordia University.[14]

5.1.1 Effective Modulus Methods

In the design of bridges the loads are considered to be of either short or long duration. In composite girders, loads of long duration, such as dead loads of concrete, cause creep. It is assumed that curbs, parapets, and railings are cast after the slab has cured and that this superimposed dead load is distributed equally between the longitudinal stringers. In flexural members, creep reduces the intensity of the compressive stresses in concrete. Thus, under sustained loads the concrete deck is less effective than for temporary loads. Stresses due to long-time dead loads on the composite section are usually computed with section properties based on a modular ratio $n = 10$.

The effect of creep under live loads is accounted for in composite construction by increasing the modular ratio n by a factor of 3, that is, $3n$.[15]

5.1.2 Construction Methods

The full economy of composite construction can only be achieved if consideration is given to the method by which the structure is to be erected. Because the composite action is more efficient than the steel beam alone, it would seem sensible to achieve composite action as early as possible—ideally for all types of loading. In shored construction, the steel beam is supported temporarily until the slab has become composite with it. The object of shoring is to let the composite girder take all the loading as compared to nonshored construction where the dead load acting before composite action has been achieved is taken by the steel beam alone.

Different construction procedures affect the stress analysis. Basically, there are three methods of construction with prestressed composite plate girders, according to their span.

Construction Method I—Small Spans

In this case the dead load is carried by the steel section alone, and the live load, impact, and superimposed dead loads are carried by the composite section in connection with the prestressing force. Figure 5.1 shows the corresponding moment diagrams, where the subscripts "s" and "cp" respectively indicate in the further analysis whether the moment is carried by the steel or composite section, following designations used by Szilard.[16]

Construction Method II—Medium Spans

Temporary supports are used generally in the form of wooden piles. These temporary supports must be carefully positioned to maintain the beam flange horizontal and rigid enough to resist settlement. The supports may be removed as

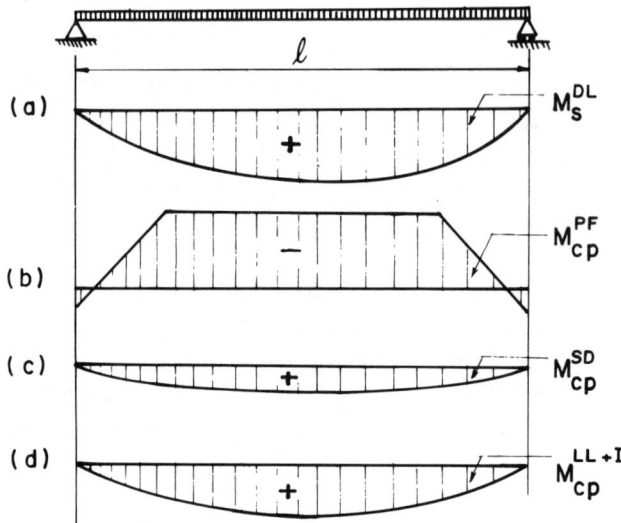

Fig. 5.1 Construction method I—moment diagrams.

soon as the composite section has developed its action and has been prestressed. It can be assumed that this stage is reached when the slab concrete has attained 75% of its 28-day cube strength. The action of various loads is shown in Fig. 5.2.

Construction Method III—Long Spans

For extremely long spans, a continuous temporary support of steel girder is used so that all the loads are carried by the composite section. An equally effective method is to place the beam on the ground, cast the slab on it, and then erect both beam and slab. This technique has advantages in bridge construction where temporary supports may be difficult or impossible to position. Diagrams of moments for construction method III are shown in Fig. 5.3.

5.1.3 Basic Data and Notation

In further computations we will use construction method II from which other construction methods might be derived by substituting zero into the formulas for the corresponding moments which do not occur in these construction methods.

The structure is statically indeterminate in its prestressed form in spite of determinate external supports and even though the structure performs like a tied arch. Consequently, the three basic equations of equilibrium of statics

$$\Sigma V = 0 \quad \Sigma H = 0 \quad \Sigma M = 0$$

are not sufficient in determining all the internal forces. Therefore, an additional equation of elasticity must be applied.

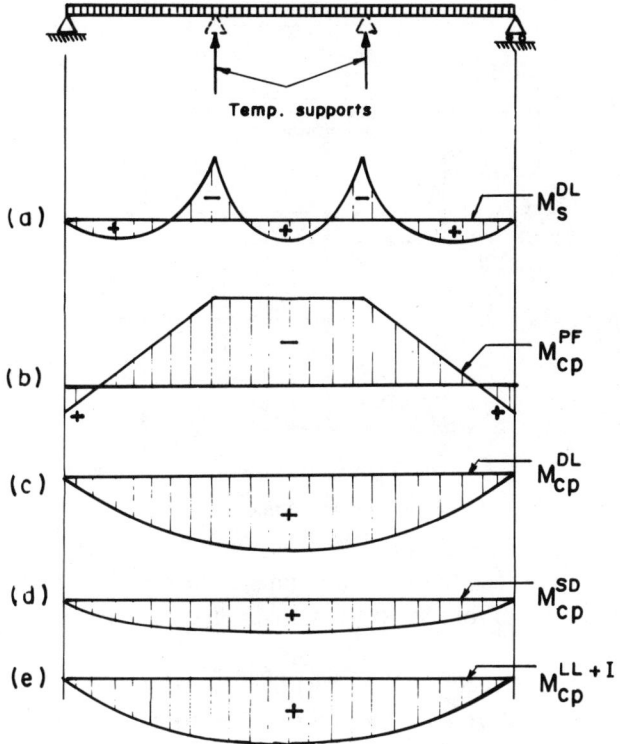

Fig. 5.2 Construction method II—moment diagrams.

The basic assumptions used in the analysis are as follows; the composite section is achieved due to the effective connection between the steel and the concrete, provided by means of shear connectors.

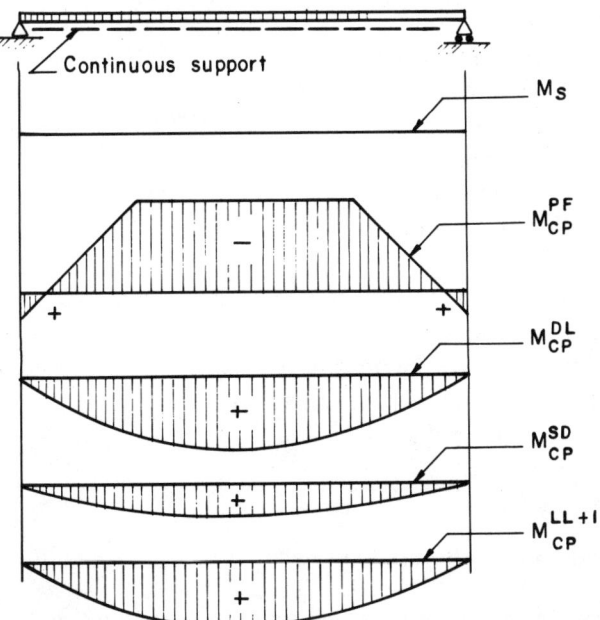

Fig. 5.3 Construction method III—moment diagrams.

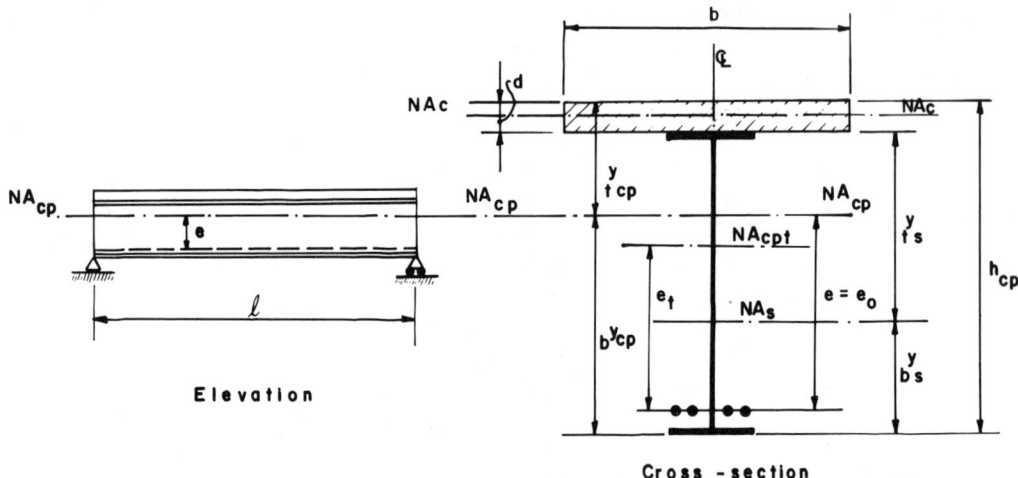

Fig. 5.4   Elevation and cross section of the typical prestressed composite section.

The sectional properties are determined based on the transformed steel section using $n = 30$ for dead loads and $n = 10$ for live loads, impact, and superimposed dead loads.

For a typical prestressed composite section (Fig. 5.4) the sectional properties can be determined based on the transformed steel section. The following notations are used:

$$A_s = \text{net cross-sectional area of steel}$$
$$A_t = \text{cross-sectional area of tendons}$$
$$A_c = \text{cross-sectional area of concrete}$$
$$NA_s = \text{neutral axis of steel}$$
$$NA_c = \text{neutral axis of concrete}$$
$$_ty_s,\ _by_s = \text{extreme fiber distances of steel}$$
$$I_s = \text{moment of inertia of steel with respect to the neutral axis of steel}$$
$$b = \text{effective width of concrete slab}$$
$$d = \text{thickness of concrete}$$
$$E_s = \text{modulus of elasticity of steel}$$
$$E_c = \text{modulus of elasticity of concrete}$$
$$E_t = \text{modulus of elasticity of the tendon}$$
$$n = E_s/E_c = \text{modular ratio}$$
$$A_{cp} = \text{transformed cross-sectional area of the composite section}$$
$$NA_{cp} = \text{neutral axis of the composite section}$$
$$_ty_c,\ _by_{cp} = \text{extreme fiber distances of the composite section}$$
$$I_{cp} = \text{moment of inertia of the composite section}$$
$$e = \text{distance from the tendon to } NA_{cp}$$
$$h_{cp} = \text{depth of the composite girder}$$

5.1.4 Analysis of Prestressed Composite Girder

Knowing all the sectional properties, it is thus necessary to investigate the stresses at the critical section, at the mid-span, in order to determine the required prestressing force. For the other sections the same procedure should be followed, after a prestressing force has already been determined.

Generally, the analysis of a prestressed composite girder is divided into the determination of the following data:

1. Stresses due to the bending moment in the steel section
2. Determination of the prestressing force and incremental prestressing
3. Stresses due to the bending moments in the composite section
4. Stresses due to shear
5. Loss of prestress
6. Deflection
7. Checking of the web buckling

Stage I. Stresses in the Steel Girder Due to the Dead Load

$$t\sigma_s^{\mathrm{I}} = -\frac{M_s^{DL}}{I_s} {}_t y_s \qquad (5.1)$$

$$b\sigma_s^{\mathrm{I}} = +\frac{M_s^{DL}}{I_s} {}_b y_s \qquad (5.2)$$

where

$\quad {}_t\sigma_s^{\mathrm{I}}$ = stress at the top fiber of the steel beam due to DL
$\quad {}_b\sigma_s^{\mathrm{I}}$ = stress at the bottom fiber of the steel beam due to DL
$\quad M_s^{DL}$ = dead load moment in the steel beam

Stage II. Prestressing of the Composite Section

After the concrete has reached the prescribed cylinder strength, the composite section will be prestressed by the force X.

$$t\sigma_{cp}^{\mathrm{II}} = -\frac{X}{A_{cp}} + \frac{Xe}{I_{cp}} {}_t y_{cp} \qquad (5.3)$$

$$b\sigma_{cp}^{\mathrm{II}} = -\frac{X}{A_{cp}} - \frac{Xe}{I_{cp}} {}_b y_{cp} \qquad (5.4)$$

where

$\quad {}_t\sigma_{cp}^{\mathrm{II}}$ = stress at the top fiber of the composite section due to prestress
$\quad {}_b\sigma_{cp}^{\mathrm{II}}$ = stress at the bottom fiber of the composite section due to prestress
$\quad X$ = prestressing force

Stage III. The Prestressing Force Lifts the Girder from the Temporary Supports and Produces the Following Stresses

The composite section will carry a moment of M_{cp}^{DL+SD} moment, producing the following stresses

$$_t\sigma_{cp}^{III} = -\frac{M_{cp}^{DL+SD}}{I_{cp}} \,_t y_{cp} \qquad (5.5)$$

$$_b\sigma_{cp}^{III} = +\frac{M_{cp}^{DL+SD}}{I_{cp}} \,_b y_{cp} \qquad (5.6)$$

where

$_t\sigma_{cp}^{III}$ = stresses at the top fiber due to dead load and superimposed dead load
$_b\sigma_{cp}^{III}$ = stresses at the bottom fiber due to dead load and superimposed dead load

Stage IV. The Live Load and Impact Load Will Produce the Following Stresses

$$_t\sigma_{cp}^{IV} = -\frac{M_{cp}^{LL+I}}{I_{cp}} \,_t y_{cp} \qquad (5.7)$$

$$_b\sigma_{cp}^{IV} = +\frac{M_{cp}^{LL+I}}{I_{cp}} \,_b y_{cp} \qquad (5.8)$$

where

$_t\sigma_{cp}^{IV}$ = stresses at the top fiber due to live load and impact
$_b\sigma_{cp}^{IV}$ = stresses at the bottom fiber due to live load and impact

Stage V. Action of External LL + I and Superimposed Dead Load SD

The external $LL + I$ and SD loads produce a statically indeterminate prestressing force increase, ΔX, which can be determined by the virtual work method, where

$$\Delta X = -\frac{\delta_{01}^{LL+I+SD}}{\delta_{11}} \qquad (5.9)$$

where

$$\delta_{01}^{LL+I+SD} = \int_0^1 \frac{(M_{cp}^{LL+I+SD})edl}{E_s I_{cp}} \qquad (5.10)$$

$$\delta_{11} = \int_0^1 \frac{e^2 dl}{E_s I_{cp}} + \int_0^1 \frac{dl}{E_s A_{cp}} + \int_0^1 \frac{dl}{E_t A_t} \qquad (5.11)$$

and

$$\Delta X = \frac{\displaystyle\int_0^1 \frac{M_{cp}^{LL+I+SD}}{E_s I_{cp}}}{\displaystyle\int_0^1 \frac{e^2 dl}{E_s I_{cp}} + \int_0^1 \frac{dl}{E_s A_{cp}} + \int_0^1 \frac{dl}{E_t A_t}} \qquad (5.12)$$

The prestressing force increment produces the following stresses

$$_t\sigma_{cp}^V = -\frac{\Delta X}{A_{cp}} + \frac{\Delta X e}{I_{cp}} \, _t y_{cp} \qquad (5.13)$$

$$_b\sigma_{cp}^V = -\frac{\Delta X}{A_{cp}} - \frac{\Delta X e}{I_{cp}} \, _b y_{cp} \qquad (5.14)$$

where

$_t\sigma_{cp}^V$ = stresses at the top fiber due to live load, impact, and superimposed dead load

$_b\sigma_{cp}^V$ = stresses at the bottom fiber due to live load, impact, and superimposed dead load

Stage VI. Final Stresses

The sum of all stresses is shown in Fig. 5.5. The actual stresses should be less than the allowable,

$$\sum_I^V \, _t\sigma \leq f_c \qquad (5.15)$$

$$\sum_I^V \, _b\sigma \leq f_s \qquad (5.16)$$

where

f_c = allowable stress in the concrete (compression)
f_s = allowable stress of the steel

Stage VII. Loss of Prestress

The magnitude of the required prestress for X is reduced by the number of factors, and the total loss of prestress may be expressed as follows

$$\sum \Delta X_{loss} = -\Delta X_{ca} - \Delta X_{fr} - \Delta X_{cr} - \Delta X_{sr} - \Delta X_t \qquad (5.17)$$

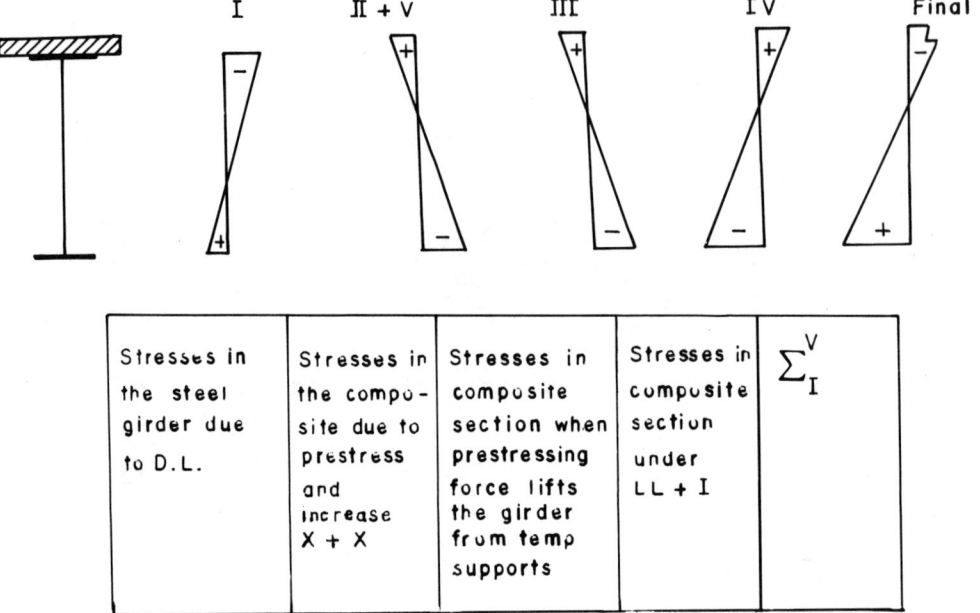

Stresses in the steel girder due to D.L.	Stresses in the composite due to prestress and increase X + X	Stresses in composite section when prestressing force lifts the girder from temp supports	Stresses in composite section under LL + I	$\sum_{I}^{V}$

Fig. 5.5   Diagram of stresses.

where

ΔX_{ca} = loss due to the creep of tendons and anchorages
ΔX_{fr} = frictional loss
ΔX_{cr} = loss due to creep of concrete
ΔX_{sr} = loss due to shrinkage of concrete
ΔX_{t} = loss due to unequal temperature changes

Consequently, the required initial and final prestressing forces

$$X_0 = X_t - \Delta X_0 - \Sigma \Delta X_{\text{loss}}$$

$$X_t = X_0 + \Delta X_0 - \Sigma X_{\text{loss}}$$

For quick estimation purposes

$$X_t = 0.85 \, X_0 \text{ may be used}$$

Loss due to creep of tendons and anchorages

Stress relaxation in steel, also called creep, is the loss of stress when it is prestressed and maintained at a constant strain for a period of time. Relaxation varies with different steels. For most kinds of steel available in the market that is stressed to the usual allowable values, the percentage of creep varies from 1% to 5% and an average of 3% of the initial stress is used.

When a tendon is tensioned to its fullest value, the jack is released and the prestress is transferred to the anchorage. The anchorage will tend to deform, and the friction wedges employed to hold the wires will slip a little distance before the wires can be firmly gripped.

A general formula for computing the loss of prestress due to anchorage deformation Δ_a is

$$\Delta_a = \frac{Pl}{AE_s} = \frac{\Delta_s l}{E_s} \tag{5.18}$$

where

Δ_a = anchorage slip loss, in.
Δ_s = prestressing of the tendon, psi
l = tendon length

Frictional Loss

Suppose that at one saddle the directional change of the cable is (Fig. 5.6)

$$\Delta\alpha = \alpha_1 - \alpha_2 \tag{5.19}$$

The frictional loss at a saddle can be expressed as follows

$$\Delta X_{fr} = X_1 - X_2 = \mu \frac{1}{2} \text{arc } \Delta\alpha \, (X_1 + X_2) \tag{5.20}$$

where μ is the coefficient of friction.

The total frictional loss for a number of saddles is

$$\Delta X_{fr} = \sum_0^n \Delta X_{fr} \tag{5.21}$$

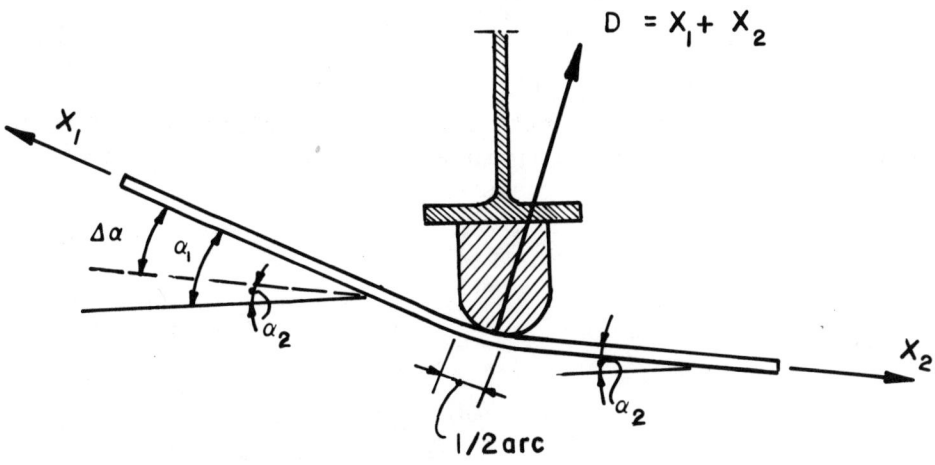

Fig. 5.6 Forces acting on the saddle.

It should be noted that the frictional loss is relatively small for two reasons: First, the directional changes of the tendons are small and they occur at saddles only; second, the coefficient of friction existing between the lubricated steel saddles and greased tendons is of small magnitude, namely

$$\mu \approx 0.05 - 0.10$$

Loss due to unequal temperature changes

Since the coefficient of expansion in steel is nearly the same as that for concrete, there is practically no loss of prestress due to the uniform change in temperature.

However, uneven temperature changes produce prestressing forces. Therefore, it is convenient to consider an uneven temperature change if $T = \pm 60°F$. We assume that the temperature change at the top fiber of the concrete, T_c, is smaller than the temperature change at the bottom fiber of the steel, T_s. Consider a part of the girder between sections M-M and N-N (Fig. 5.7). Assume that the linear change of temperature, section N-N, will occupy the N'-N' position.

The shortening of the neutral axis of the girder is caused by the temperature

$$T_{NA} = T_c + \frac{T_s - T_c}{h} \, y_{cp} \tag{5.22}$$

The horizontal displacement is

$$\delta'_{01T} = \int_0^{\ell} \epsilon_T T_{NA} \, dl \tag{5.23}$$

where $\epsilon_T = 0.0000067$ °F.

The unit angular change caused by the uneven temperature change is

$$K = \frac{\epsilon_T (T_s - T_c)}{h_{cp}} \tag{5.24}$$

The horizontal displacement produced by the angular change at the mid-span of the tendon is expressed by the virtual work as

$$\delta^{11}_{01T} = \int_0^1 Ke\,dl \tag{5.25}$$

Therefore

$$\Delta X_T = \frac{\delta^1_{01T} + \delta^{11}_{01T}}{\delta_{11}} \tag{5.26}$$

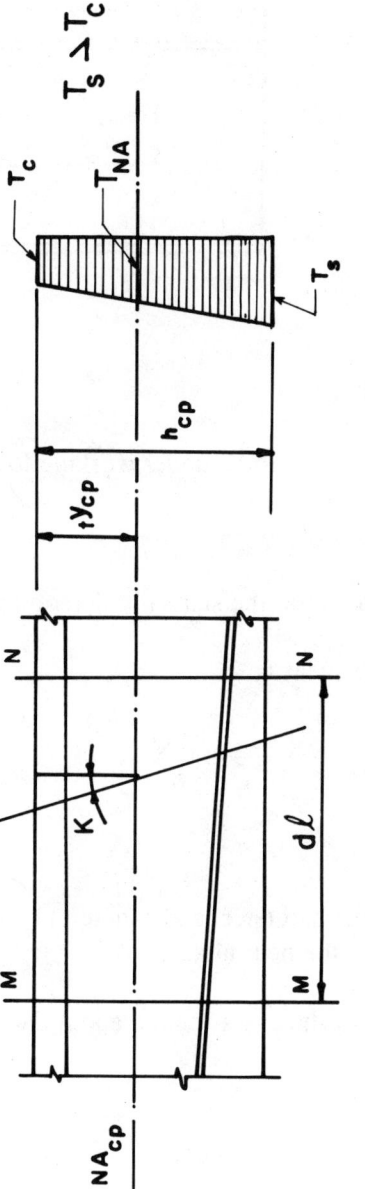

Fig. 5.7 Uneven temperature change.

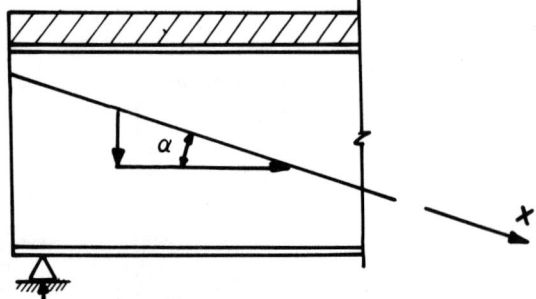

Fig. 5.8   Shear forces.

Stage VIII.   Shear Stresses

The general equation for a shear force under working loads (Fig. 5.8) is

$$V = V_0 - X \sin \alpha \qquad (5.27)$$

where V_0 is the shearing force on the statically determinate structure at the section under consideration.

The formula for the web shear is

$$\tau = \frac{VQ}{It_w} \qquad (5.28)$$

where

Q = static moment of that part (steel and concrete) lying above the fiber under consideration about the neutral axis

t_w = web thickness

I = moment of inertia of the cross-section about the neutral axis.

In the case of straight tendons

$$V = V_0 \qquad (5.29)$$

Table 5.1 indicates different cases of loads to evaluate shear stresses.

Table 5.1

Web Shear	V	$X \sin \alpha$	Q	I
τ_s^{DL}	V_s^{DL}	—	Q_s	I_s
τ_{cp}^{DL}	V_{cp}^{DL}	—	Q_{cp}	I_{sp}
τ_{cp}^{X}	—	$X \sin \alpha$	Q_{cp}	I_{sp}
τ_{cp}^{LL}	V_{cp}^{LL+I}	$\Delta X_{\sin}^{LL+I}$	Q_{cp}	I_{sp}
τ_{cp}^{SD}	V_{cp}^{SD}	$\Delta X_{\sin}^{LL+I}$	Q_{cp}	I_{sp}

Table 5.2

Deflection	Direction	M_i	E	I	N_i	A
Δ_s^{DL}	↓	M_s^{DL}	E_s	I_s	—	—
Δ_{cp}^{DL}	↓	M_{cp}^{DL}	E_s	I_{cp}	—	—
Δ_{cp}^{X}	↑	M_{cp}^{X}	E_s	I_{cp}	X	A_{cp}
Δ_{cp}^{LL+I}	↓	$M_{cp}^{LL+I} - \Delta Xe$	E_s	I_{cp}	ΔX^{LL+I}	A_{cp}
Δ_{cp}^{SD}	↓	$M_{cp}^{SD} - \Delta X^{SD}e$	E_s	I_{cp}	ΔX^{SD}	A_{cp}

Stage IX. Deflection

The deflection of the composite girder is calculated by applying the virtual work method by using the general formula (4.83)

$$\Delta_{cp} = \int_0^1 \frac{M_i m_k \, dl}{E_s I} + \int_0^1 \frac{N_i n_k}{E_s A} \, \Delta l \qquad (5.30)$$

However, the proper combination of moments, moments of inertia, and cross-sectional areas should be used, as shown in Table 5.2.

5.1.5 Preliminary Determination of Steel Cross Section for Composite Girder

The cross section of a composite girder usually has two working stages. The first stage corresponds to the work of a steel girder without a reinforced-concrete slab under the weight of a steel structure and reinforced concrete slab. The second stage involves work of a steel girder acting as a composite under loading of the deck, superimposed dead load, and live loading.

The cross section of a steel girder is usually found by the method of successive approximation.[17] As a first approximation we determine the cross-sectional areas of the steel flanges. We assume the bending moments are basically taken by the flanges and reinforced concrete slab. For the bottom and top we are applying the following equations:

$$\frac{M_1}{H_s A_b} + \frac{M_2}{H_{cp} A_b} = f_{\text{all}} \qquad (5.31)$$

$$\frac{M_1}{H_s A_t} + \frac{M_2}{H_{cp}(A_t + A_c)} = f_{\text{all}} \qquad (5.32)$$

where

$M_1 = M_{DL} + M_{SD}$ = bending moment due to dead load and superimposed dead load

$$M_2 = M_{LL+I} = \text{bending moment due to live load and impact}$$

H_s = height of steel beam between centroids of top and bottom flanges

H_{cp} = height between centroids of bottom steel flange and reinforced concrete slab

A_t, A_b = cross-sectional areas of the top and bottom steel flanges, respectively

A_c = transformed cross-sectional area of the concrete slab

Considering that

$$N_1 = \frac{M_1}{H_s} \qquad N_2 = \frac{M_2}{H_{cp}} \tag{5.33}$$

After substitution of (5.33) into (5.31), we have

$$\frac{N_1}{A_b} + \frac{N_2}{A_b} = f_{\text{all}}$$

or

$$A_b' = \frac{N_1 + N_2}{f_{\text{all}}} \tag{5.34}$$

By the introduction of a prestressed tendon, we may reduce the cross-sectional area of the bottom flange as follows

$$A_b = A_b' - a_t x \frac{f_t}{f_{\text{all}}} \tag{5.35}$$

where

a_t = assumed cross-sectional area of the tendon

f_t = allowable stress of the tendon

f_{all} = allowable stress of the beam

and, after substitution of (5.33) and (5.34) into (5.32), we have

$$\frac{N_1}{A_t} + \frac{A_b f_{\text{all}} - N_1}{(A_t + A_c)} = f_{\text{all}}$$

$$A_t^2 + A_t \left[A_c - \frac{(N_1 - N_2)}{f_{\text{all}}} \right] - \frac{N_1 A_c}{f_{\text{all}}} = 0$$

$$A_t = -\frac{1}{2} \left[A_c - \frac{(N_1 - N_2)}{f_{\text{all}}} \pm \sqrt{\frac{N_1 A_c}{f_{\text{all}}} + \frac{1}{4} \left[A_c - \frac{(N_1 - N_2)}{f_{\text{all}}} \right]^2} \right.$$

Using the designation

$$A_R = A_c - \frac{(N_1 - N_2)}{f_{\text{all}}}$$

we have

$$A_t = \sqrt{\frac{N_1 A_c}{f_{\text{all}}} + 0.25\, A_R^2} - 0.5\, A_R \qquad (5.36)$$

5.2 CREEP AND SHRINKAGE OF CONCRETE SLAB

5.2.1 Introduction

The concrete slab of the composite section under the influence of the permanent load creeps. By this phenomenon the concrete transfers more load to the steel girder, which causes a change of stresses in the cables as well as in the girders.

Assuming the cables are cut at point 1 in Fig. 5.9, and expressing the horizontal displacement of the cable at that point as zero on the basic structure which has been made statically determinate, we have

$$\sigma_{01}^{CR} + \Delta F_{CR} \cdot \sigma_{11} = 0 \qquad (5.37)$$

or the approximate loss of prestress due to plastic flow may be expressed as follows

$$\Delta F_{CR} \approx -\frac{\sigma_{01}^{CR}}{\sigma_{11}} \qquad (5.38)$$

the curve of the prestress loss due to the plastic flow of the concrete versus time is shown in Fig. 5.10.

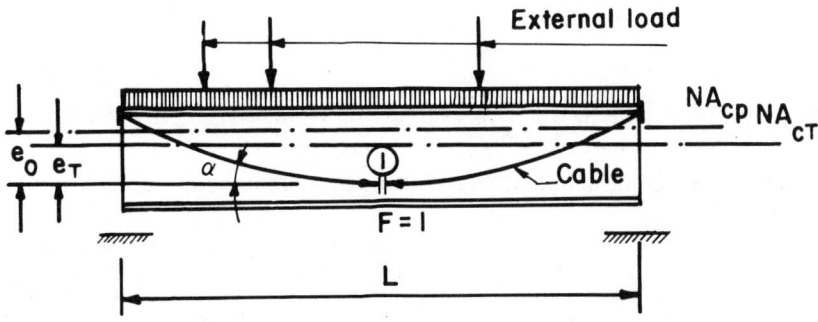

Fig. 5.9 A diagram related to virtual work.

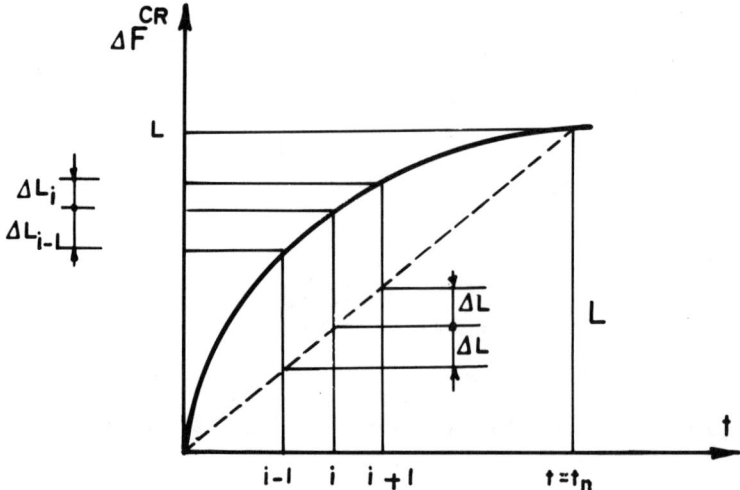

Fig. 5.10   Curve diagram of creep loss.

5.2.2 Influence of Creep

Creep of concrete is defined as its time-dependent deformation resulting from the presence of continuous stress. No data has been found regarding the experimental study of the influence of creep in prestressed composite structures. Only a limited amount of data is available concerning the creep of concrete under higher stress. Some data indicate that when the sustained stress (dead-load stress) is in excess of about one-third of the ultimate strength of the concrete, the rate of increase of strain with stress tends to get higher. As in bridges the dead loads are the long-term loads acting on the structure, and if the sustained stress is more than one-third of the ultimate strength of the concrete used for the deck slab, then the influence of creep is higher. It is possible that the increase can become quite pronounced as the stress approaches the ultimate strength of the concrete. According to Magnel, ther term *creep coefficient*, C_c, is employed to indicate the total strain δ_t (instantaneous plus creep strain) after a lengthy period of constant stress to the instantaneous strain immediately obtained upon the application of stress.[18] Thus

$$C_c = \frac{\delta_t}{\delta_i} \tag{5.39}$$

For the purpose of design it is considered safe to take $C_c = 3.0$.

The same term, *creep coefficient*, is sometimes used to denote the ratio of creep strain (excluding instantaneous strain) to instantaneous strain[19]

$$C_c = \frac{\delta_c}{\delta_1} \tag{5.40}$$

so care should be taken to find out the exact meaning of the term whenever it is employed. In the following, the creep coefficient is taken as 2.5 at the end of

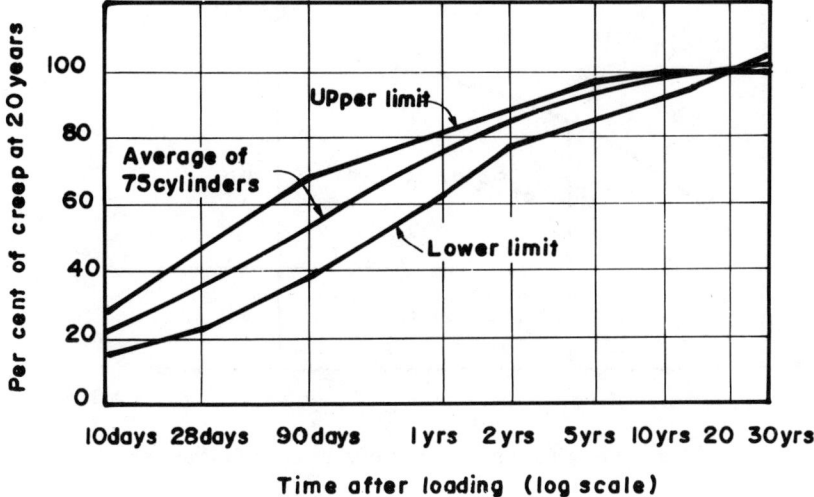

Fig. 5.11 Creep–time ratio curves.

one year; using the second definition, the same coefficient is only 1.5, as shown in Fig. 5.11. Of the total amount of creep strain, it is roughly estimated that about one-fourth takes place within the first two weeks after the application of prestress, another fourth within two to three months, another fourth within one year, and the last fourth over the course of many years.

For smaller members creep as well as shrinkage takes place faster than for larger members. To account for the inelastic shortening of concrete slabs under long-term load, the AASHTO Specifications require that the composite properties used for computing deflections and stresses from these loads be calculated using an n that is always needed for live-load stress calculations.[20]

The situation with regard to long-term composite dead loads is quite different with the subject system. These loads consist of the load from the beam and slab plus the form load, the prestress force, and any superimposed load.

In composite section, concrete under the influence of the permanent load creeps, which means that the concrete tries to transfer more load to the steel girder, causing changes in stress in the girder as well as in the prestressing cable. Therefore it is important to find the approximate loss of prestress due to creep or plastic flow and modular ratio n. The following data is based on the analysis by Szilard.[21]

1. The time-strain relation of plastic strain is shown in Fig. 5.12. According to Dischinger[22]

$$\epsilon_{pt} = \epsilon_n(1 - e^t) \tag{5.41}$$

where

ϵ_{pt} = initial plastic strain
ϵ_n = final value of plastic strain after several years
e = Naperian base
t = time

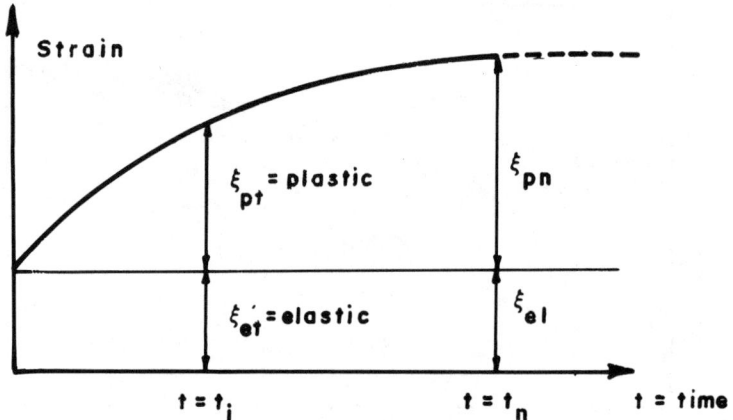

Fig. 5.12   Time-strain relation of concrete deformations.

By dividing equation (5.41) by ϵ_{el} we obtain

$$\varphi_t = \varphi_n \left(1 - e^t\right) \tag{5.42}$$

which expresses the ratio between the elastic and plastic strain as a function of time, and where

$$\varphi_t = \frac{\epsilon_{pt}}{\epsilon_{el}} = \text{creep coefficient}$$

$$\epsilon_{el} = \text{elastic strain}$$

$$\varphi_n = \frac{\epsilon_{pn}}{\epsilon_{el}} = \text{final creep coefficient}$$

2. The magnitude of φ_n is a function of the following factors: (1) climatic conditions, (b) concrete strength at time of loading, (c) water-cement ratio, and (d) dimensions of the concrete structure.

The question of what final creep coefficient should be used in practical computation can be answered only approximately. This is because the determination of the actual creep curve takes years and therefore does not yield a readily available value to the designer. The designer should first study climatic and atmospheric conditions of the region and should determine the average relative humidity (RH) to which the structure will be exposed. Consulting Fig. 5.13, one could determine creep coefficient. This value should be multiplied by C_1 and C_2 read from Figures 5.14 and 5.15. The term C_2 is given in a function of a guide number (GN),

$$GN = \frac{WM \times 0.0625}{\sqrt[3]{d} \times 0.394} \tag{5.43}$$

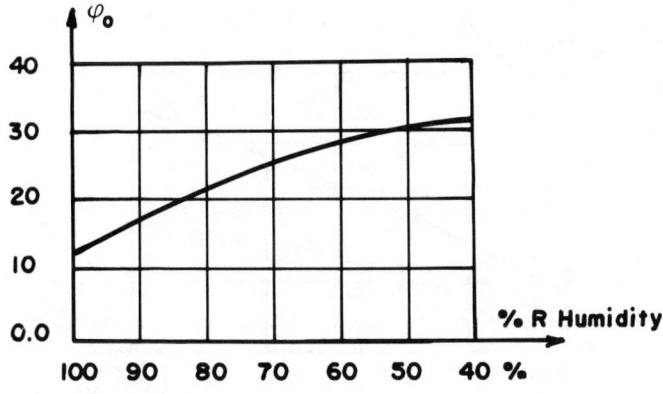

Fig. 5.13 Final creep coefficient variation.

where

    W = weight of water per cubic ft
    M = weight of fine aggregate (up to $1/4''$)/weight of the total aggregate
    d = thickness of concrete in inches

Table 5.3 gives approximate values for the final creep coefficient.
    An approximate solution of the creep effect can be obtained by means of an increased modular ratio

$$n_t = n_0(1 + \varphi_n) \qquad (5.44)$$

where

    $n_t = E_s/E_{ct}$ modular ratio at the end of the plastic flow ($t = t_n$)
    $n_0 = E_s/E_{co}$ modulus of elasticity of steel/modulus of elasticity of concrete
        at $t = 0$
    f_n = final creep coefficient

It should be noted that the present practice in designing composite girders under dead load takes $n = 30$.
    3. The change of the length of concrete at its neutral axis due to permanent loads (DL, SD, and PR) can be expressed as follows.

$$\Delta\epsilon = \epsilon_{co}^{DL}\,\varphi_n - \epsilon_{co}^{PR}\,\varphi_n + \epsilon_{co}^{PR}\,\frac{L}{2F_0}\,\varphi_n + \epsilon_{co}^{PR}\,\frac{L}{F_0} \qquad (5.45)$$

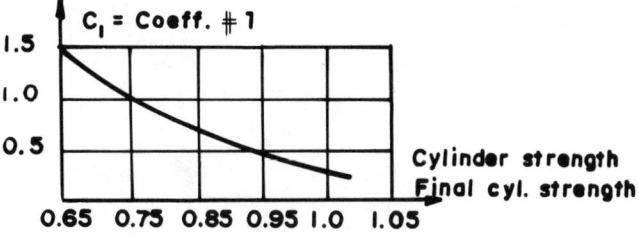

Fig. 5.14 Variation of C_1.

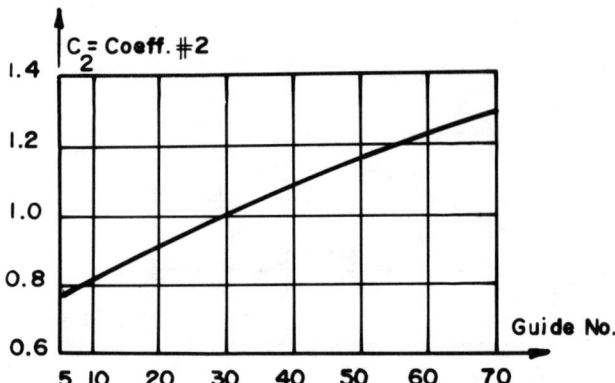

Fig. 5.15   Variation of C_2.

where

$$\epsilon_{co} = \frac{\sigma_0}{E_{co}} \tag{5.46}$$

where

$$\frac{L}{F_0} = \frac{e_0 - e_t}{e_0} \tag{5.47}$$

where

e_0 = distance to NA_{cp}
e_t = distance to NA_{cpt} at $t = t_n$

or a better value

$$\frac{L}{F_0} = \frac{\Delta F^{CR}}{F_0} = \frac{\delta_{01}^{CR}/\delta_{11}}{F_0} \tag{5.48}$$

where ΔF^{CR} = loss due to creep of concrete and approximately

$$\delta_{01}^{CR} = \int_0^1 \frac{(\epsilon_{co}^{DL} - \epsilon_{CR}^{PR})\varphi_n e_0 \, dl}{h_{cp} - d/2} \tag{5.49}$$

Table 5.3

Climatic conditions	φ_n	ϵ
Humid atmosphere	1.5–2.0	0.0001
Bridges above rivers		
Normal atmosphere	2.0–3.0	0.0002
Dry atmosphere	2.5–4.0	0.0003

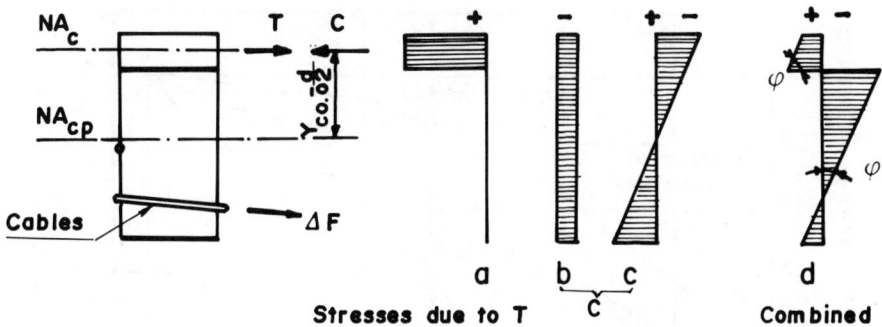

Fig. 5.16   Interpretation of plastic flow and shrinkage.

The creep may be considered as a combined effect of a tensile force acting on the concrete only (T^{CR}) and of the same force acting as a compressive force on the whole composite section (C^{CR}) (Fig. 5.16).

In this manner the concrete is relieved and more load is added to the steel girder. These forces are

$$T^{CR} = C^{CR} = E_{co} A_c \Delta\epsilon \qquad (5.50)$$

The compressive force acting eccentrically on the section creates a bending moment

$$M_{CP}^{CR} = C^{CR} (y_{cpo} - d/2) \qquad (5.51)$$

The horizontal displacement at point "1" can be expressed by virtual work (see Fig. 5.9)

$$\delta_{01}^{CR} = -\int_0^1 \frac{M_{cp}^{CR} e_0}{E_s I_{cpo}} dl + \Sigma_0^1 \frac{1}{E_s} \left(\frac{T^{CR}}{A_c/n_0} - \frac{C^{CR}}{A_{CPO}} \right) \Delta l \qquad (5.52)$$

Using equation (5.52), we can determine the prestress loss due to creep of the concrete

$$\Delta F_{CR} \approx -\frac{\delta_{01}^{CR}}{\delta_{11}}$$

5.2.3 Influence of Shrinkage

As distinguished from creep, shrinkage in concrete is its contraction due to drying and chemical changes dependent on time and moisture conditions but not on stresses. At least a portion of shrinkage resulting from drying of concrete is recoverable upon the restoration of lost water. The shrinkage time curve is shown in Fig. 5.17.[23]

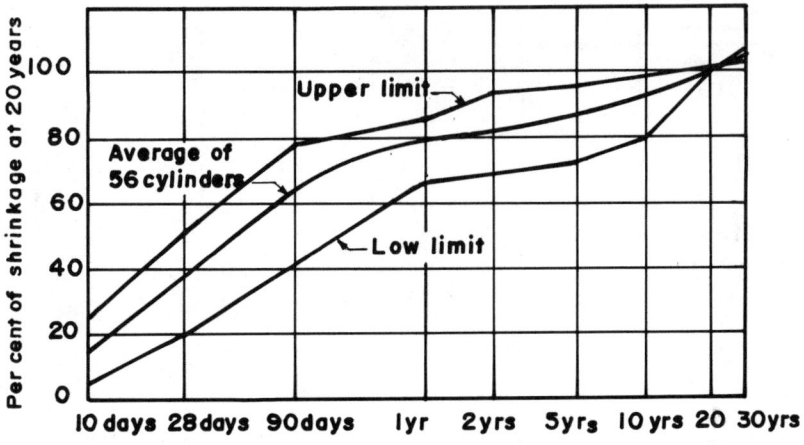

Fig. 5.17   Drying shrinkage–time ratio curves.

Shrinkage of concrete is somewhat proportional to the amount of water employed in the mix. Hence if minimum shrinkage is desired, the water-cement ratio and proportion of cement paste should be kept to a minimum. Thus aggregates of larger size, well graded for minimum void, will need a smaller amount of cement paste and shrinkage will be smaller. Harder and denser aggregates of low absorption and high modulus elasticity will exhibit smaller shrinkage. In cases of heavy reinforcing, the shrinkage is reduced.

In computing the approximate prestress loss due to shrinkage we can follow a method similar to that for creep by substituting formula (5.50)

$$T^{SR} = C^{SR} = E_{co} A_c \epsilon_{SR} \tag{5.53}$$

$\epsilon_{SR} = \epsilon_{OSR} \left(1 + n\varphi\right)$ = reduced shrinkage unit strain

where

ϵ_{OSR} = unit shrinkage strain of plain concrete

$$\varphi = \frac{A_s}{A_c} = \frac{\text{section area of reinforcing}}{\text{section area of concrete}} \tag{5.54}$$

$$n = n_0 - n_t$$

$$M^{SR} = C^{SR} \left(y_{cpo} - d/2 \right) \tag{5.55}$$

$$\delta_{01}^{SR} = - \int_0^1 \frac{M^{SR} e_0}{E_s I_{cpo}} \, dl + \sum_0^1 \left(\frac{I^{SR}}{A_c/n_0} - \frac{C^{CR}}{A_{cpo}} \right) \Delta l \tag{5.56}$$

Therefore, loss of force due to shrinkage is

$$\Delta F_{SR} = \frac{\delta_{01}^{SR}}{\delta_{11}} \tag{5.57}$$

5.2.4 Second Approach for Calculating Loss of Prestress Due to Shrinkage and Creep

In this case it is assumed by Leonhardt that tensioning force is diminishing linearly from X to X_1 and, therefore, reducing of tensioning force for F may be expressed by the middle $F/2$ (prestressing force).[24] The shortening of concrete fiber at the level of the tensioned tendon, due only to prestressing, is

$$\Delta\epsilon_x = +\left(\epsilon_x - \epsilon_x \frac{F}{2X}\right)\varphi - \epsilon_x \frac{F}{X} \tag{5.58}$$

where

$\epsilon_x = \sigma_x/E_c$ = elastic shortening of concrete under action of X

$\epsilon_x(F/X)$ = elastic elongation of concrete fiber due to reduction of prestressing

$\varphi = \epsilon_{pl}/\epsilon_{el}$ = creep coefficient

Under the action of its own weight and shrinkage, an element of concrete having length $dx = 1$ changes its length as follows

$$\Delta\epsilon_{g+s} = \epsilon_g\varphi + \epsilon_s \tag{5.59}$$

The total elongation of the concrete element is, therefore, equal to

$$\Delta\epsilon_c = \epsilon_g\varphi + \epsilon_s + \epsilon_x\varphi - \epsilon_x \frac{F}{X}\left(1 + \frac{\phi}{2}\right) \tag{5.60}$$

The change in elongation of prestressed steel is proportional to the change of the prestressing force

$$\Delta\epsilon_p = \epsilon_p \frac{F}{X} \tag{5.61}$$

Because both values $\Delta\epsilon_c$ and $\Delta\epsilon_p$ should be equal, we obtain the following equation for the determination of F

$$\frac{F}{X} = \frac{\epsilon_s + \varphi(\epsilon_g + \epsilon_x)}{\epsilon_x\left(1 + \dfrac{\varphi}{2}\right) + \epsilon_p} \tag{5.62}$$

Multiplying the numerator and denominator of equation (5.62) by $\epsilon_p = nE_0$ and considering the stresses, we obtain

$$\frac{F}{X} = \frac{\epsilon_s E_p + n\varphi \left(\sigma_g + \sigma_x\right)}{n\sigma_c \left(1 + \dfrac{\varphi}{2}\right) - \sigma_p} \tag{5.63}$$

where

σ_g = stress in concrete under dead load
σ_c = stress in concrete only under X at moment of time zero
σ_p = stress at tension at $t = 0$
ϵ_s = final value of shrinkage deformation, negative value
σ_x = stress under prestressing

All the σ's should be taken with their signs and shortening should be substituted with a negative sign.

By putting the values of the above formula with proper signs, we will get the prestress force loss in terms of tensioning force produced at the bottom of the prestressed composite girder.

In the following numerical example both methods are compared.

5.3 DESIGN EXAMPLE

This design example concerns an interior composite prestressed girder for a two-lane highway bridge having a span of 150 ft under H20-44 loading. Girders are of A36 steel, $f_{all} = 30$ ksi. High-strength steel tendons have $f_t = 200$ ksi. Moduli of elasticity for girder steel and tendons are $E = E_t = 30 \times 10^3$ ksi. Permissible deflection is $\Delta_p = 1/800$. The half cross-section of the bridge is shown in Fig. 5.18.

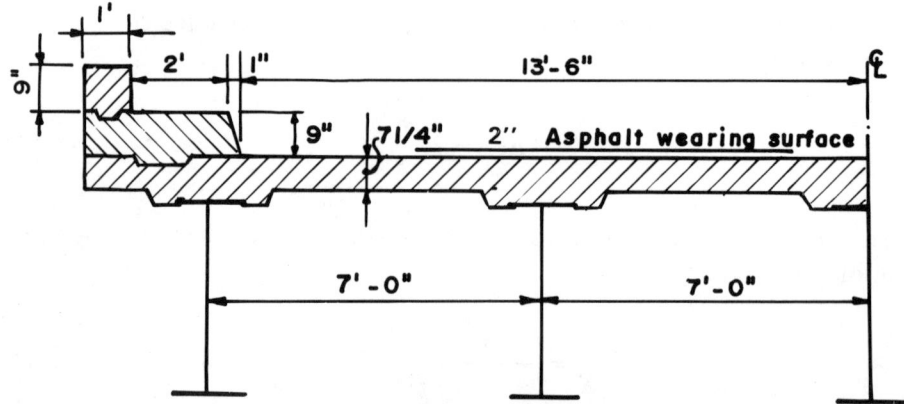

Fig. 5.18 Prestressed composite steel beam bridge.

A. Calculation of Dead and Live Loads

1. Dead Load

$$
\begin{aligned}
\text{Slab: } (7.25/12) \times 7.0 \times 0.15 &= 0.634 \text{ k/ft} \\
\text{Hunch: } (1/12) \times 2.0 \times 0.15 &= 0.025 \text{ k/ft} \\
\text{Steel beam and bracings} &= \underline{0.250 \text{ k/ft}} \\
g_{DL} &= 0.909 \text{ k/ft}
\end{aligned}
$$

2. Superimposed Dead Load

$$
\begin{aligned}
\text{Asphalt paving: } & 2/12 \times 0.14 \times 27 & = 0.680 \text{ k/ft} \\
\text{Railing: } & 2 \times 0.05 & = 0.100 \text{ k/ft} \\
\text{Parapet: } & 2 \times 1.0 \times 0.75 \times 0.15 & = 0.225 \text{ k/ft} \\
\text{Sidewalk: } & 2 \times 3.05 \times 0.75 \times 0.15 & = \underline{0.686 \text{ k/ft}} \\
& & = 1.691 \text{ k/ft}
\end{aligned}
$$

Per girder $g_{SD} = 1.691/5 = 0.338$ kft

3. Dead-Load Moment

$$
M_{DL} = \frac{0.909 \times 150^2}{8} = 2556.56 \text{ kft}
$$

4. Superimposed Dead-Load Moment

$$
M_{SD} = \frac{0.338 \times 150^2}{8} = 950.6 \text{ kft}
$$

5. Live Load—Truck (Fig. 5.19)

$$
R = \frac{16(63.33 + 77.33) + 4 \times 91.33}{150} = \frac{2615.88}{150} = 17.44 \text{ k}
$$

$$
M_{LL} = 17.44 \times 72.67 - 4 \times 14 = 1211.36 \text{ kft}
$$

6. Live Load—Lane Loading (Fig. 5.20)

Moment due to lane loading per one girder is

$$
M_{LL} = \frac{1}{2} \times \frac{0.64 \times 150^2}{8} = 900 \text{ kft}
$$

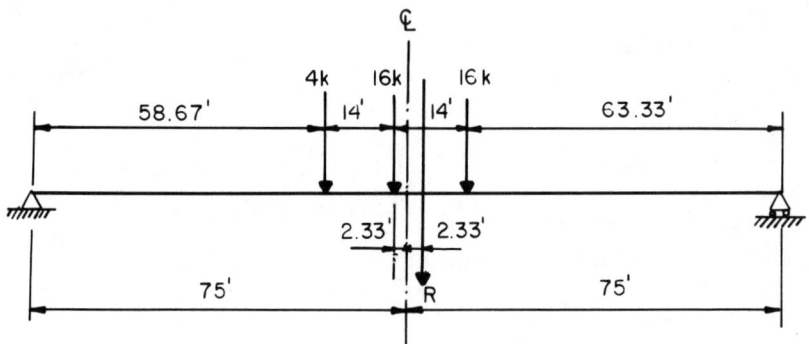

Fig. 5.19 Truck loading.

Moment due to concentrated load per one girder is

$$M_c = \frac{1}{2} \times \frac{18 \times 150}{4} = 337.5 \text{ kft}$$

Total moment: $M_{LL} + M_c = 900 + 337.5 = 1237.5$ kft. Comparing results in 5) and 6), it is found that lane loading produces the maximum bending moment in the girder, or

$$1237.50 \text{ kft} > 1211.36 \text{ kft}$$

7. *Fraction*

$$\frac{S}{5.5} = \frac{7.0}{5.5} = 1.27$$

8. *Impact*

$$I = \frac{50}{L + 125} = \frac{50}{150 + 125} = 0.18 \qquad I = 1.18$$

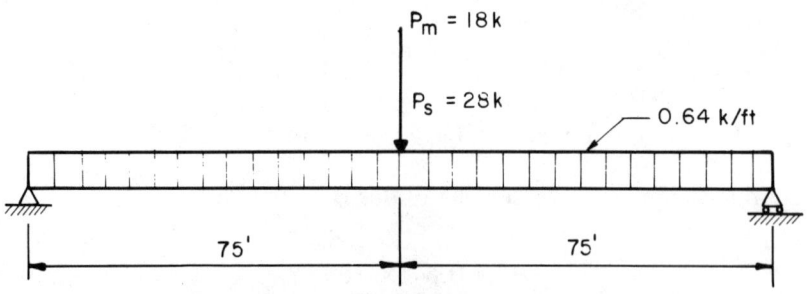

Fig. 5.20 Lane loading.

9. Maximum Bending Moment Due to Lane Load

$$M_{LL} = 900 \times 1.27 \times 1.18 = 1348.74 \text{ k ft}$$

10. Maximum Bending Moment Due to Concentrated Load

$$M_c = 337.5 \times 1.27 \times 1.18 = 505.78 \text{ k ft}$$

B. Calculation of Bending Moments—Composite Cross Section

In the design it is assumed that shoring will be used during construction. Two temporary supports are used to decrease the initial bending.

1. Temporary supports are installed and the concrete slab has been poured. Steel carries a dead load of steel and concrete.

From the formulas for continuous beams

$$R_A = R_D = 0.4 \, gl = 0.4 \times 0.909 \times 50 = 18.18 \text{ k}$$

$$R_B = R_C = 1.1 \, gl = 1.1 \times 0.909 \times 50 = 50.00 \text{ k}$$

$$M_B = M_C = -0.1 \, gl^2 = -0.1 \times 0.909 \times 50^2 = -227.25 \text{ k ft}$$

$$M_1 = M_3 = +0.08 \, gl^2 = +0.08 \times 0.909 \times 50^2 = 181.80 \text{ kft}$$

$$M_2 = +0.025 \, gl^2 = +0.025 \times 0.909 \times 50^2 = +56.81 \text{ k ft}$$

Moments under loads B and C (Fig. 5.21)

$$M_B = M_C = 50.00 \times 50 = 2500 \text{ k ft}$$

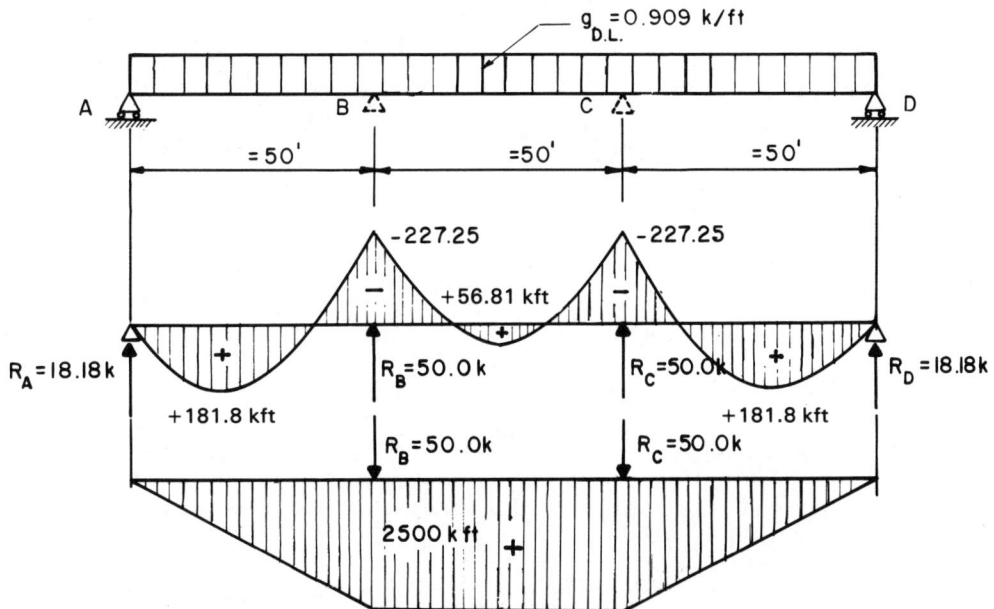

Fig. 5.21 Moments under loads B and C.

C. Approximate Determination of Steel Plate Girder Cross Section

1. Proposed Height

$$h = \frac{1}{30} = \frac{150 \times 12}{30} = 60'' \quad \text{use } 62''$$

2. Required Cross-Sectional Areas of the Flanges

a. Bottom Flange

$$M_1 = M_{DL} + M_{SD} = 2556.56 + 950.6 = 3507.16 \text{ k ft}$$

$$H_s = 62 - 1 = 61'' \ (5.08 \text{ ft}) \qquad (5.33)$$

$$N_1 = \frac{M_1}{h_s} = \frac{3507.16}{5.08} = 690.38 \text{ k}$$

$$M_2 = M_{LL+I} = 1348.74 + 505.78 = 1854.52 \text{ k ft}$$

$$H_{cp} = 61.25 + 4.625 = 66.125 \text{ in. } (5.51 \text{ ft}) \qquad (5.33)$$

$$N_2 = \frac{M_2}{h_{cp}} = \frac{1854.52}{5.51} = 336.57 \text{ k}$$

$$A_b' = \frac{N_1 + N_2}{f_{\text{all}}} = \frac{(690.38 + 336.57) \times 10^3}{30 \times 10^3} = \frac{1026.95}{30} = 34.33 \text{ in.}^2$$

$$(5.34)$$

After formula (5.35) (assume $a_t = 1.2$ in.^{2})

$$A_b = A_b' - a_t \frac{f_t}{f_{\text{all}}} = 34.23 - 1.2 \times \frac{200}{30} = 26.23 \text{ in.}^2$$

b. Top Flange

$$A_t = \sqrt{\frac{N_1 A_c}{f_{\text{all}}} + 0.25 A_R^2} - 0.5 A_R \qquad (5.36)$$

$$A_c = 2.8 \times 7.25 = 20.3 \text{ in.}^2$$

$$A_R = A_c - \frac{N_1 - N_2}{f_{\text{all}}} = 20.3 - \frac{(690.38 - 336.57) \times 10^3}{30 \times 10^3}$$

$$= 20.3 - \frac{353.81}{30} = 8.50 \text{ in.}^2$$

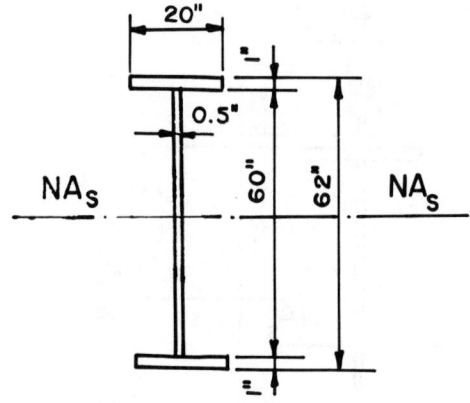

Fig. 5.22 Cross section of steel girder.

$$A_t = \sqrt{\frac{690.38 \times 10^3 \times 20.3}{30 \times 10^3} + 0.25 \times 8.50^2} - 0.5 \times 8.50$$

$$= \sqrt{467.16 + 18.06} - 4.25 = 22.03 - 4.25 = 17.78 \text{ in.}^2$$

For both top and bottom flanges we use cross-sectional areas $A_t = A_b = 20 \times 1 = 20$ in.^{2}

c. Properties of Girder Cross Section (Fig. 5.22)

$$I_s = 2 \times 20 \times 1.0 \times 30.5^2 + \frac{1}{12} \times 0.5 \times 60^3 = 37,210 + 9,000 = 46,210 \text{ in.}^4$$

$$A = 2 \times 20 \times 1.0 + 0.5 \times 60 = 40.0 + 30.0 = 70.0 \text{ in.}^2$$

$$S_1 = S_2 = \frac{46,210}{31} = 1,490.65 \text{ in.}^3$$

3. Composite Cross Section, n = 30 (Fig. 5.23; Table 5.4)

Effective flange width for composite section

    a. $150/4 = 37.5$ ft $= 450$ in.
    b. Spacing from center to center of girders $= 7$ ft $= 84$ in.
    c. 20 in. $+ 12 \times 7.25 = 107$ in.

Transformed width of flange $W = 84/30 = 2.8$ in.

$$\bar{y} = -\frac{2,076.12}{90.3} = -22.99 \text{ in.}$$

$$I_{cp} = 114,003.15 - 2,076.12 \times 22.99 = 114,003.15 - 47,729.99 = 66,273.16 \text{ in.}^4$$

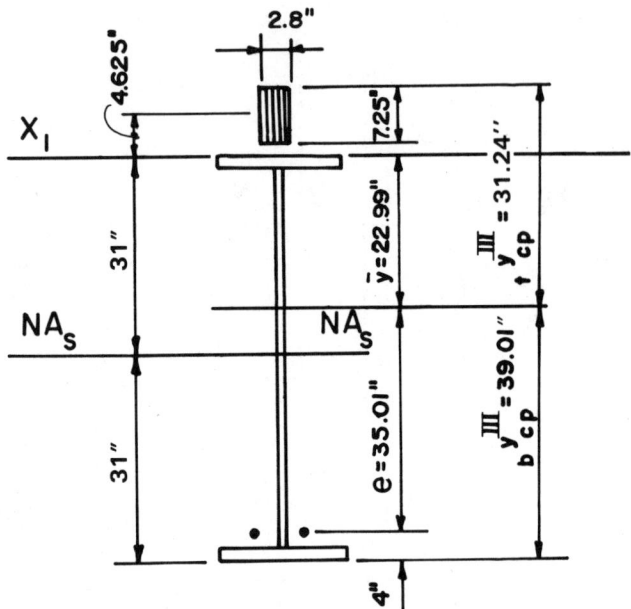

Fig. 5.23 Cross section of composite girder, $n = 30$.

Stage I. Stresses in the Steel Girder with Noncomposite Section

Check stresses at the temporary supports of the steel girder due to the weight of steel and fresh concrete. Stresses at the top and bottom flanges

$$f_t f_s^I = \pm \frac{M_{DL}}{S_1} = \pm \frac{227.25 \times 12}{1,490.65} = \pm 1.83 \text{ ksi}$$

Stage II. Stresses Due to Prestressing

When the prestressing force is applied by the high-strength cables, the composite girder lifts from the temporary supports. We use two wire strands—each one on both sides of the girder—having the following properties:

$$f_t = 200 \text{ ksi} \qquad \text{wire } d = 0.196 \text{ in.}$$

$$a_w = \frac{\pi d^2}{4} = \frac{3.14 \times 0.196^2}{4} = 0.03 \text{ in.}^2$$

Table 5.4

$n = 30$ Material	A (in.^{2})	y (in.)	Ay (in.^{3})	Ay^2 (in.^{4})	I_0 (in.^{4})	ΣI (in.^{4})
Steel	70.0	-31.00	$-2,170$	67,270	46,210	113,480
Concrete	20.3	$+4.625$	93.88	434.23	88.92	523.15
Σ	90.3	—	$-2,076.12$	—	—	114,003.15

A single strand consists of 20 wires.

$$a_t = 0.03 \times 20 = 0.6 \text{ in.}^2$$

Force in a single strand $X = 0.6 \times 200 = 120$ k.

a. Stress at the Top Due to Prestress

$$_t f_{cp}^{II} = -\frac{X}{A_{cp}} + \frac{Xe}{I_{cp}} \, _t y_{cp}^{II} = -\frac{240}{90.3} + \frac{240 \times 35.01 \times 31.24}{66,273.16}$$

$$= -2.66 + 3.96 = +1.3 \text{ ksi}$$

Actual stress in concrete:

$$_t f_{cp}^{II} = +\frac{1.3}{30} = +0.04 \text{ ksi}$$

b. Stress at Bottom Flange Due to Prestress

$$_b f_{cp}^{II} = -\frac{X}{A_{cp}} - \frac{Xe}{I_{cp}} \, x_b y_{cp}^{II} = -\frac{240}{90.3} - \frac{240 \times 35.01 \times 39.01}{66,273.16}$$

$$= -2.66 - 4.95 = -7.61 \text{ ksi}$$

Stage III. Stresses Due to Dead Load and Superimposed Dead Load, n = 30

$$M_{DL} + M_{SD} = 2,556.56 + 450.6 = 3,507.16 \text{ k ft}$$

a. Stress at the Top

$$_t f_{cp}^{III} = -\frac{M_{DL+SD}}{I_{cp}} x_t y_{cp}^{III} = -\frac{3,507.16 \times 31.24 \times 12}{66,273.16} = -19.84 \text{ ksi}$$

Actual stress in concrete

$$_t f_{cp}^{III} = -\frac{19.84}{30} = -0.66 \text{ ksi}$$

b. Stress at the Bottom

$$_b f_{cp}^{III} = +\frac{M_{DL+SD}}{I_{cp}} x_b y_{cp}^{III} = +\frac{3,507.16 \times 39.01 \times 12}{66,275.98} = +24.77 \text{ ksi}$$

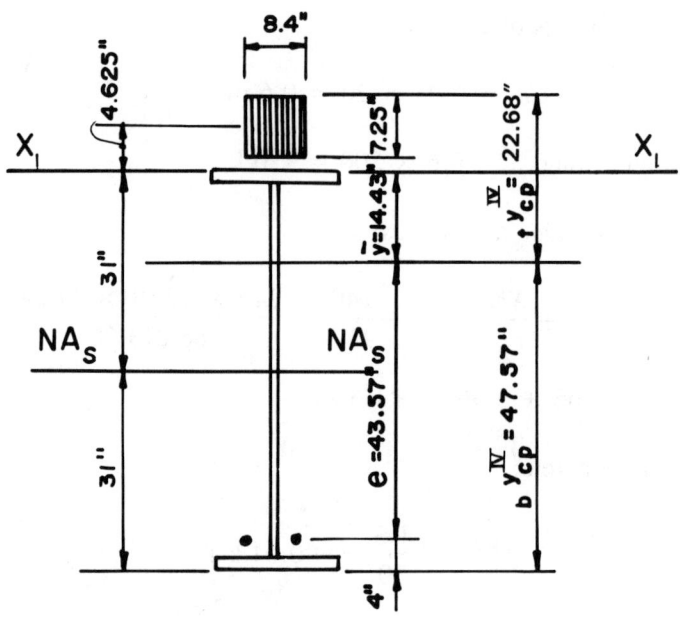

Fig. 5.24   Cross section of composite girder, $n = 10$.

Stage IV.   Stresses Due to LL + I, n = 10 (Fig. 5.24; Table 5.5)

$$\bar{y} = -\frac{1,888.34}{130.9} = -14.43 \text{ in.}$$

$$I_{cp} = 115,049.44 - 1,888.34 \times 14.43 = 87,800.74 \text{ in.}^4$$

$$M_{LL+I} = 1348.74 + 505.78 = 1854.52 \text{ k ft}$$

a. Stresses at the Top

$$_tf_{cp}^{IV} = -\frac{M_{LL+I}}{I_{cp}} x_t y_{cp}^{IV} = -\frac{1,854.52 \times 22.68 \times 12}{87,800.74} = -5.75 \text{ ksi}$$

Actual stress in concrete

$$_tf_{cp}^{IV} = -\frac{5.75}{10} = -0.58 \text{ ksi}$$

Table 5.5

$n = 10$ Material	A (in.^{2})	y (in.)	Ay (in.^{3})	Ay^2 (in.^{4})	Io (in.^{4})	ΣI (in.^{4})
Steel	70.0	−31	−2,170	67,270	46,210	113,480
Concrete	60.90	+4.625	281.66	1,302.69	266.75	1,569.44
Σ	130.90	—	−1,888.34	—	—	115,049.49

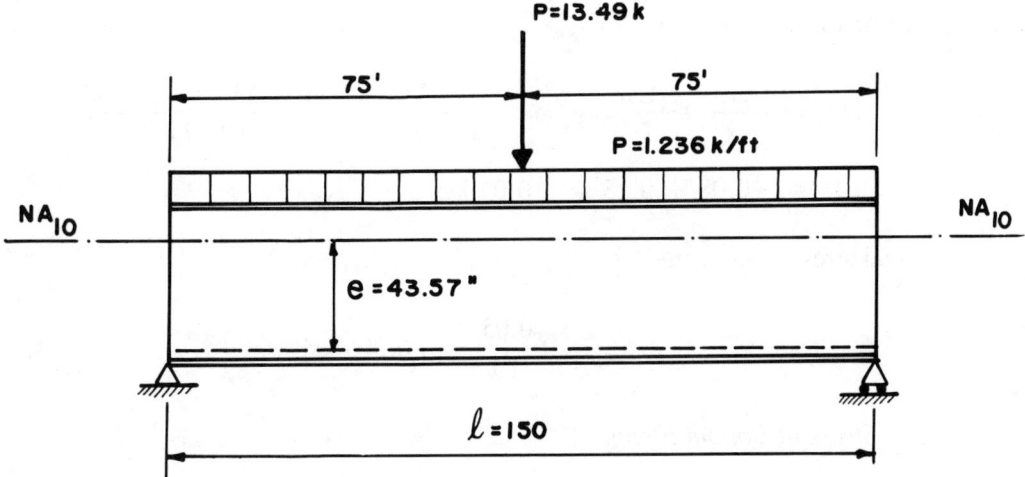

Fig. 5.25 Girder under lane load and superimposed dead load.

b. Stresses at the Bottom

$$_b f_{cp}^{IV} = +\frac{M_{LL+I}}{I_{cp}} x_b y_{cp}^{IV} = +\frac{1,854.52 \times 47.57 \times 12}{87,800.74} = +12.05 \text{ ksi}$$

Stage V. Increment of Prestressed Force in the Tendon Due to Lane Load and Superimposed Dead Load, n = 10 (Fig. 5.25)

a. Distributed Lane Load and Superimposed Dead Load

$$M_{LL+I+SD} = 1348.74 + 950.6 = 2299.34 \text{ kft}$$

$$\Delta X_1 = \frac{2(M_{LL+I+SD}) \times e}{3\left(e^2 + \frac{I_{cp}}{A_{cp}} + \frac{I_{cp}}{A_t}\right)} = \frac{2 \times 2,299.34 \times 43.57 \times 12}{3\left(43.57^2 + \frac{87,800.74}{130.9} + \frac{87,800.74}{1.2}\right)}$$

$$= \frac{80,1457.95}{75,736.37} = 10.58$$

b. ΔX_2 Under Concentrated Load

$$\Delta X_2 = \frac{Ple}{\left(e^2 + \frac{I_{cp}}{A_{cp}} + \frac{I_{cp}}{A_t}\right)} = \frac{13.49 \times 150 \times 43.57}{75,736.37} = 1.16 \text{ k}$$

Total $\Delta X = \Delta X_1 + \Delta X_2 = 10.58 + 1.16 = 11.74$ k.

c. *Stress at Top*

$$_tf_{cp}^V = -\frac{\Delta X}{A_c} + \frac{\Delta Xe}{I_{cp}}x_ty_{cp}^V = -\frac{11.74}{130.9} + \frac{11.74 \times 43.57 \times 22.68}{87,899.74}$$

$$= -0.08 + 0.13 = +0.05 \text{ ksi}$$

Actual stress in concrete

$$_tf_{cp}^V = +\frac{0.05}{10} = +0.005 \text{ ksi}$$

d. *Stress at Bottom Flange*

$$_bf_{cp}^V = -\frac{\Delta X}{A_{cp}} - \frac{\Delta Xe}{I_{cp}}x_by_{cp}^V = -\frac{11.74}{130.9} - \frac{11.74 \times 43.57 \times 47.57}{87,800.74}$$

$$= -0.08 - 0.28 = 0.36 \text{ ksi}$$

Stage VI. Final Stresses

a. *At Top*

$$\Sigma_t f = {_tf_{cp}^{II}} + {_tf_{cp}^{III}} + {_tf_{cp}^{IV}} + {_tf_{cp}^V}$$

$$= +0.04 - 0.66 - 0.58 + 0.005 = -1.195 \text{ ksi}$$

b. *At Bottom*

$$\Sigma_b f = {_bf_{cp}^{II}} + {_bf_{cp}^{III}} + {_bf_{cp}^{IV}} + {_bf_{cp}^V}$$

$$= -7.61 + 24.77 + 12.05 - 0.36 = +28.85 \text{ ksi}$$

D. *Check for Shear*

We calculate maximum web shear at the supports and at different neutral axes.

1. *Web Shear in Steel Girder (Fig. 5.26)*

a. *Shear Stress at Axis NA_{30}*

$$V = R_B = R_C = 50 \text{ k} \qquad I_s = 46210 \text{ in.}^4$$

$$Q = 20 \times 1 \times 22.49 + 21.99 \times 0.5 \times 10.99$$

$$= 449.8 + 120.83 = 570.83 \text{ in.}^3$$

$$\tau^I = \frac{VQ}{It_w} = \frac{50 \times 570.83}{46210 \times 0.5} = 1.23 \text{ ksi}$$

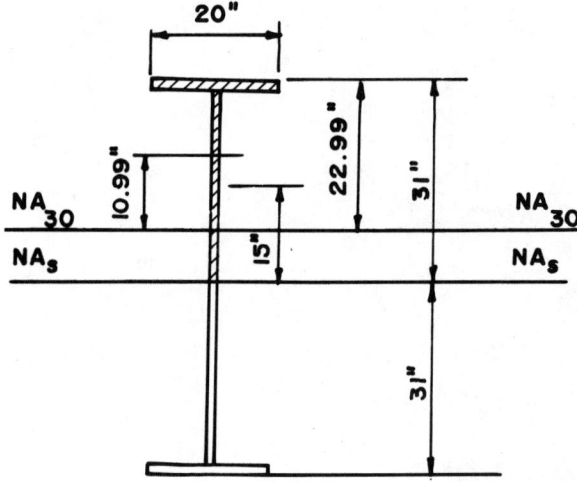

Fig. 5.26   Shear stresses.

b. Shear Stress at Axis NA_s

$$Q = 20 \times 1 \times 30.5 + 30 \times 0.5 \times 15 = 610 + 225 = 835 \text{ in.}^3$$

$$\tau = \frac{50 \times 835}{46210 \times 0.5} = 1.80 \text{ ksi}$$

2. Web Shear in Composite Girder Under Dead Load, $n = 30$ (Fig. 5.27)

a. Shear Stress at Axis NA_{30}

$$\text{Reaction } V_{DL} = \frac{0.909 \times 150}{2} = 68.175 \text{ k}$$

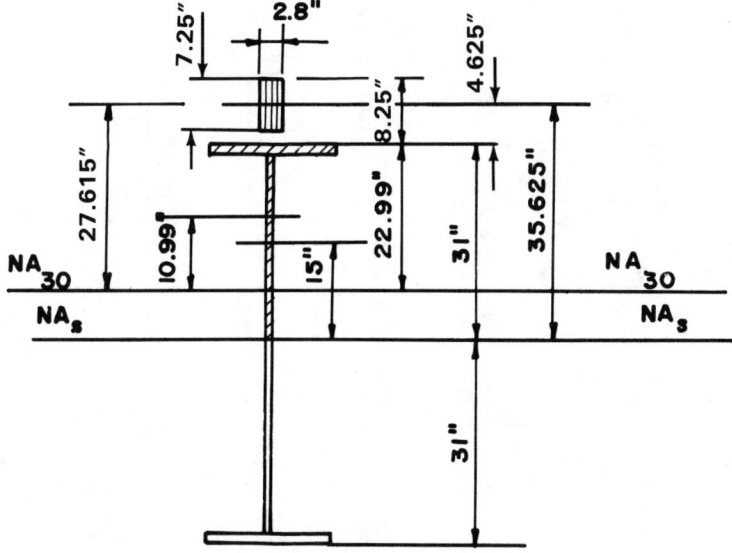

Fig. 5.27   Shear stresses at neutral axes NA_{30} and NA_s, $n = 30$.

$$Q = 2.8 \times 7.25 \times 27.615 + 20 \times 1 \times 22.49 + 21.99 \times 0.5 \times \frac{21.99}{2}$$

$$= 560.58 + 449.8 + 120.89 = 1{,}131.27 \text{ in.}^3$$

$$\tau^{\text{II}} = \frac{68.175 \times 1{,}131.48}{66{,}273.16 \times 0.5} = 2.33 \text{ ksi}$$

b. Shear Stress at Axis N_s

$$Q = 2.8 \times 7.25 \times 35.625 + 20 \times 1 \times 30.50 + 30 \times 0.5 \times 15$$

$$= 723.19 + 610 + 225 = 1{,}558.19 \text{ in.}^3$$

$$\tau = \frac{68.175 \times 1{,}558.19}{66{,}273.16 \times 0.5} = 3.20 \text{ ksi}$$

3. Web Shear Under LL + I, n = 10 (Fig. 5.28)

a. Shear Stress at Axis NA_{30}

$$\text{Reaction: } V_{LL+I} = \frac{0.64 \times 1.27 \times 1.18}{2} \times \frac{150}{2} + \frac{18 \times 1.27 \times 1.18}{2}$$

$$= 35.97 + 13.49 = 49.46 \text{ k}$$

$$Q = 8.4 \times 7.25 \times 27.615 + 20 \times 1 \times 22.49 + 21.99 \times 0.5 \times 10.99$$

$$= 1{,}681.75 + 449.80 + 120.84 = 2{,}252.39 \text{ in.}^3$$

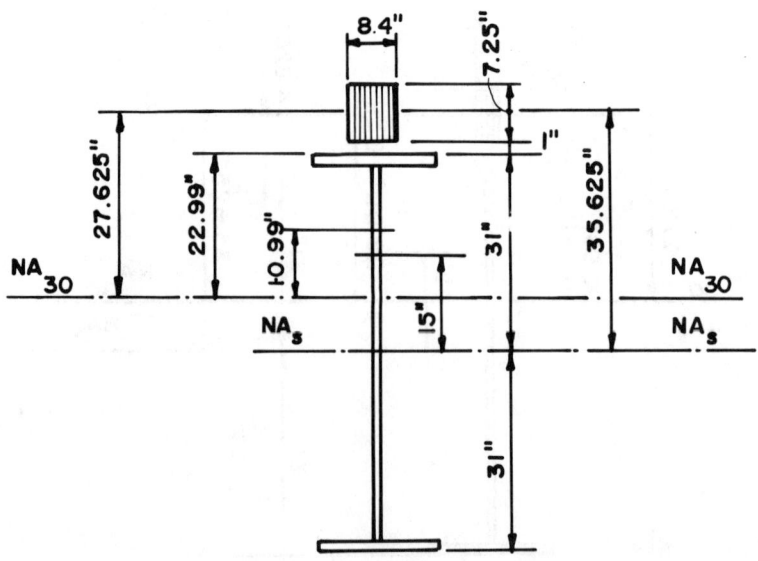

Fig. 5.28   Shear stress at neutral axes NA_{30} and NA_s, $n = 10$.

$$\tau^{III} = \frac{49.96 \times 2{,}252.39}{87{,}800 \times 0.5} = 2.56 \text{ ksi}$$

b. Shear Stresses at Axis NA_s

$$Q = 8.4 \times 7.25 \times 35.625 + 20 \times 1 \times 30.5 + 30 \times 0.5 \times 15$$
$$= 2169.56 + 610 + 225 = 3{,}004.56 \text{ in.}^3$$

$$\tau = \frac{49.46 \times 3{,}004.56}{87{,}800 \times 0.5} = 3.39 \text{ ksi}$$

4. Web Shear Under Superimposed Dead Load, n = 30

a. Shear Stress at Axis NA_{30}

$$V = \frac{0.338 \times 150}{2} = 25.35 \qquad \tau^{IV} = \frac{25.35 \times 1{,}131.27}{66{,}215.99 \times 0.5} = 0.87 \text{ ksi}$$

b. Shear Stress at Axis NA_s

$$\tau = \frac{24.6 \times 1{,}558.19}{66{,}273.16 \times 0.5} = 1.16 \text{ ksi}$$

5. Summary of Shear Stresses (Fig. 5.29)

a. Stresses at Axis NA_{30}

$$\Sigma\tau = \tau^I + \tau^{II} + \tau^{III} + \tau^{IV} = 1.23 + 2.33 + 2.56 + 0.84 = 6.96 \text{ ksi}$$

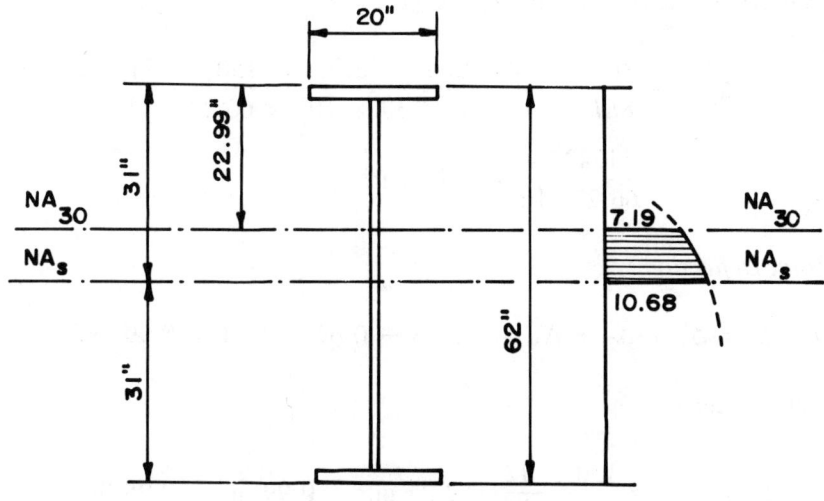

Fig. 5.29 Resulting shear stresses.

b. Stresses at Axis NA_s

$$\Sigma \tau = 1.80 + 3.20 + 3.39 + 1.16 = 9.55 \text{ ksi}$$

Allowable $\tau = 0.4 \, F_y = 0.4 \times 36 = 14.40 \text{ ksi}$

$$9.55 \text{ ksi} < 14.40 \text{ ksi}$$

E. Deflection

1. Deflection Due to Live Load and Impact, n = 10

$$P_{LL+I} = \frac{1}{2} \times 0.64 \times 1.27 \times 1.18 = 0.48 \text{ k/ft} = 40 \text{ lb/in.}$$

$$\Delta_1 = \frac{5 \, pl^4}{384 E I_{cp}} = \frac{5 \times 40 \times (150 \times 12)^4}{384 \times 30 \times 10^6 \times 87,800.74}$$

$$= \frac{182,250}{87,800.74} = 2.08 \text{ in.}$$

2. Deflection Due to Concentrated Load

$$P = \frac{1}{2} \times 18.0 \times 1.27 \times 1.18 = 13.49 \text{ k}$$

$$\Delta_2 = \frac{Pl^3}{48EI} = \frac{13,490 \times (150 \times 12)^3}{48 \times 30 \times 10^6 \times 87,800.74}$$

$$= \frac{54,634.50}{87,800.74} = 0.62 \text{ in.}$$

3. Deflection Due to Prestressing Force

$$\Delta_3 = -\frac{Xel^2}{8EI} = -\frac{240,000 \times 35.01 \times (150 \times 12)^2}{8 \times 30 \times 10^6 \times 66,273}$$

$$= -\frac{113432.4}{66,273.16} = -1.71 \text{ in.}$$

4. Total Deflection

$$\Delta_t = \Delta_1 + \Delta_2 + \Delta_3 = +2.08 + 0.62 - 1.71 = 0.99 \text{ in.}$$

Permissible deflection

$$\Delta_p = \frac{1}{800} = \frac{150 \times 12}{800} = 2.25 \text{ in.} \qquad 0.99 \text{ in.} < 2.25 \text{ in.}$$

F. Design of Shear Connectors

1. Shear Values for Different Load Positions

Influence lines for shear are shown in Fig. 5.30.
Loadings:

Lane loading $q = 1/2 \times 0.64 \times 1.27 \times 1.18 = 0.48$ k/ft.
Concentrated load $P = 1/2 \times 28 \times 1.27 \times 1.18 = 20.98$ k.

Table 5.6 indicates values of shear from different positions of loads.

a. Section from 0 to 25'

$$Q = 60.9 \times 19.055 = 1{,}160.45 \text{ in.}^3$$

$$S_1 = \frac{VQ}{I_{cp}} = \frac{56.98 \times 1{,}160.45}{87{,}800} = 0.75 \text{ k/in.}$$

b. Section from 25' to 50'

$$S_2 = \frac{46.97 \times 1{,}160.45}{87{,}800} = 0.62 \text{ k/in.}$$

c. Section from 50' to 75'

$$S_3 = \frac{40.97 \times 1{,}160.45}{87{,}800} = 0.54 \text{ k/in.}$$

2. Selected Welded Studs

$$d = 3/4'' \qquad H = 6'' \qquad H/d = 6/0.75 = 8 > 4$$
$$\alpha = 10.6 \text{ for } 500{,}000 \text{ cycles}$$
$$Z_r = \alpha d^2 = 10.6 \times 0.562 = 5.96 \text{ k}$$

3. Pitch for Studs (assume two studs per row)

a. From 0 to 25'

$$p = \frac{2Z_r}{S_1} = \frac{2 \times 5.96}{0.75} = 15.89'' < 24'' \qquad \text{use } 15''$$

b. From 25' to 50'

$$p = \frac{2 \times 5.96}{0.62} = 19.22'' \qquad \text{use } 19''$$

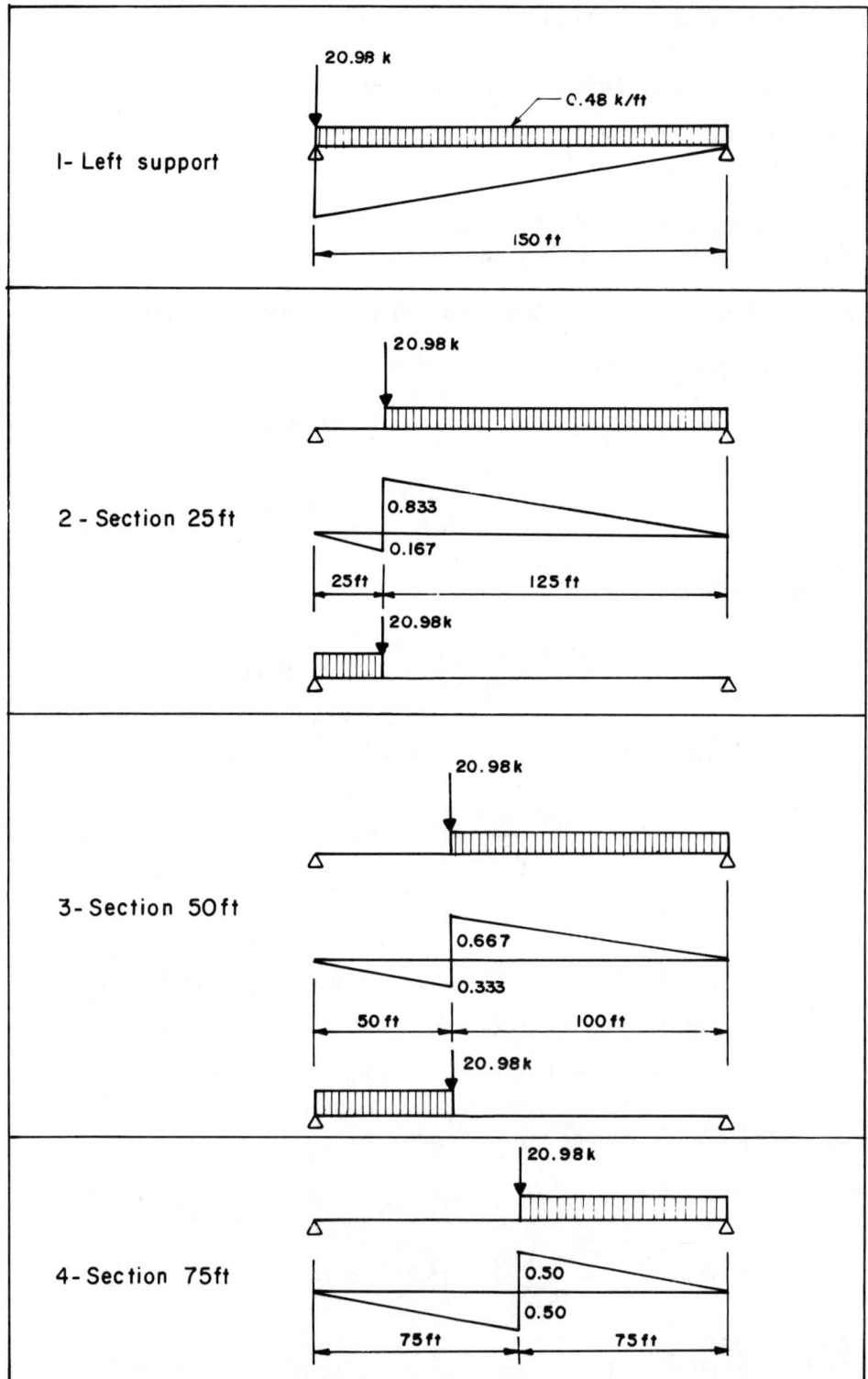

Fig. 5.30   Influence lines for shear.

Table 5.6

Left Support Range	V_{max} (kips)	V_{min} (kips)	Summary V (kips)
Left support	$20.98 + \dfrac{0.48 \times 1 \times 150}{2} = 56.98$	—	56.98
25'	$20.98 \times 0.833 + \dfrac{0.48 \times 0.833 \times 125}{2}$ $= 42.47$	$-20.98 \times 0.167 - \dfrac{0.48 \times 0.167 \times 25}{2}$ $= -4.50$	46.97
50'	$20.98 \times 0.667 + \dfrac{0.48 \times 0.667 \times 100}{2}$ $= 29.99$	$-20.98 \times 0.333 - \dfrac{0.48 \times 0.333 \times 50}{2}$ $= -10.98$	40.97
75'	$20.98 \times 0.50 + \dfrac{0.48 \times 0.50 \times 75}{2} = 1949$	$-20.98 \times 0.5 - \dfrac{0.48 \times 0.5 \times 75}{2}$ $= -19.49$	38.98

c. From 50' to 75'

$$p = \frac{2 \times 5.96}{0.54} = 22'' \quad \text{use } 22''$$

4. Studs Provided

$$2 \left(\frac{25 \times 12}{15} + \frac{25 \times 12}{19} + \frac{25 \times 12}{22} \right) = 98$$

For whole span: $2 \times 98 = 196$ studs.

5. Ultimate Strength

$$P_1 = A_s F_y = 70 \times 36 = 2520 \text{ k}$$

$$P_2 = 0.85 f_c bc = 0.85 \times 3 \times 84 \times 7.25 = 1553 \text{ k}$$

$$E_c = W^{3/2} 33\sqrt{f'_c} = \sqrt{145^3} \times 33\sqrt{3000} = 315 \times 10^4 \text{ psi}$$

$$= 3150 \text{ ksi}$$

$$S_u = 0.4 d^2 \sqrt{f'_c E_c} = 0.4 \times 0.75^2 \sqrt{3 \times 3150} = 21.87 \text{ k}$$

$$N = \frac{P_2}{\phi S_u} = \frac{1553}{0.85 \times 21.87} = 83.54$$

Actually, $98 > 83.54$, therefore the number of studs is satisfactory.

G. Design of Stiffeners

The following material is according to the AASHTO Specification, 1983.

1. Transverse Intermediate Stiffeners

$$\frac{D}{150} = \frac{60}{150} = 0.4'' \quad \text{but } 0.5'' > 0.4''$$

The average calculated unit shearing stress in the cross section of the web plate is

$$F_V = \frac{5.625 \times 10^7}{(D/t_w)^2} < \frac{F_y}{3} \tag{4.98}$$

or

$$F_V = \frac{5.625 \times 10^7}{(60/0.5)^2} = 3,906.3 < \frac{36,000}{3}$$

Therefore, intermediate stiffeners may be omitted.

2. *Bearing Stiffeners*

a. *The Thickness of the Bearing Stiffener Plates*

$$\delta = \frac{b'}{12} \sqrt{\frac{F_y}{33,000}} = \frac{9}{12} \sqrt{\frac{36,000}{33,000}} = 0.78'' \qquad (4.104)$$

Use $\delta = 3/4''$.

b. *Check Buckling of Bearing Stiffeners (Fig. 5.31)*

Width $= 18.05 \times 0.5 = 9''$

$$I_x = 2 \times \frac{1}{12} \times 0.75 \times 9^3 + 2 \times 0.75 \times 9 \times 9.25^2 + \frac{1}{12} \times 9 \times 0.5^3$$

$$= 1246.31 \text{ in.}^4$$

$$A = 2 \times 9 \times 0.75 + 9 \times 0.5 = 18 \text{ in.}^2 \qquad r_x = \sqrt{\frac{1246.31}{18}} = 8.32 \text{ in.}$$

$$I_y = 2 \times \frac{1}{12} \times 9 \times 0.75^3 + \frac{1}{12} \times 0.5 \times 9^3 = 31.01 \text{ in.}^4$$

$$r_y = \sqrt{\frac{31.01}{18}} = 1.31 \text{ in.}$$

c. *Reactions*

Dead load: $(0.909 + 0.338) \times 150/2 \qquad = 93.5 \text{ k}$

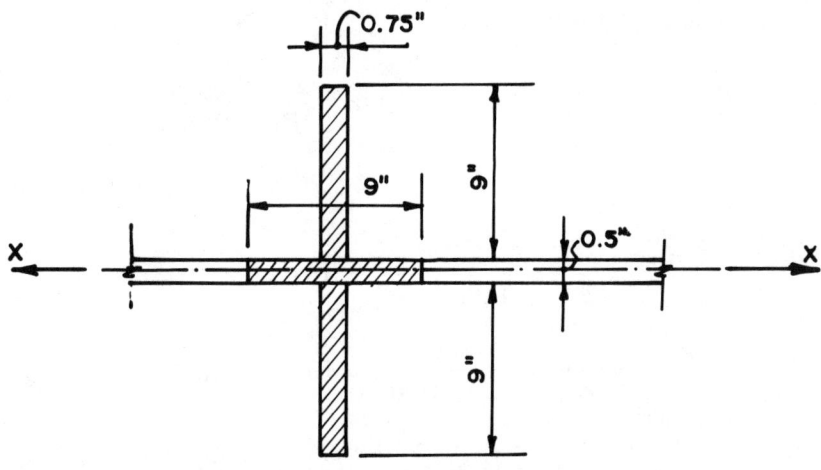

Fig. 5.31 Bearing stiffeners.

From Table 5.6, LL reaction equal to

56.98 k divided by impact factor is $56.98/1.18 = \underline{48.3\ k}$

Total $= 141.8\ k$

d. $$C_c = \sqrt{2\pi^2 \frac{E}{F_y}} = \sqrt{2 \times 3.14^2 \times \frac{30 \times 10^6}{36 \times 10^3}} = 128.6$$

$$\frac{KL}{r} = \frac{0.75 \times 60}{1.31} = 34.35 < 128.6$$

$$F_a = \frac{F_y}{F \cdot S}\left[1 - \frac{(KL/r)^2}{4\pi^2 E}F_y\right]$$

$$= \frac{36{,}000}{2}\left[1 - \frac{34.35^2 \times 36 \times 10^3}{4 \times 3.14^2 \times 30 \times 10^6}\right] = 17{,}354\ \text{psi}$$

$$\sigma = \frac{141{,}800}{18} = 7{,}877.8\ \text{psi} < 17{,}354\ \text{psi}$$

H. Prestress Loss due to Creep

1.

a. Stress in concrete slab due to dead load $0.660 - 0.006 = 0.654$ ksi $= 654$ psi
b. Modulus of elasticity for concrete $= 4.26 \times 10^6$ psi
c. Stress in concrete slab due to prestressing $0.04 + 0.005 = 0.045$ ksi $= 45$ psi
d. Final creep coefficient $\phi_n = 3.0$
e. Prestressing force applied $= 240$ k

2. Calculation of Prestress Loss

$$\Delta\epsilon = \epsilon_{co}^{DL}\,\phi_n - \epsilon_{co}^{PR}\,\phi_n + \epsilon_{co}^{PR}\frac{L}{2F_0} + \epsilon_{co}^{PR}\frac{L}{F_0} \tag{5.45}$$

where

$$\epsilon_{co}^{DL} = \frac{\sigma_0^{DL}}{E_{co}} = \frac{654}{4.26 \times 10^6} = 153.52 \times 10^{-6}$$

$$\epsilon_{co}^{PR} = \frac{\sigma_0^{PR}}{E_{co}} = \frac{45}{4.26 \times 10^6} = 10.56 \times 10^{-6}$$

$$\frac{L}{F_0} \approx \frac{e_0 - e_t}{e_0} = \frac{35.01 - 32}{35.01} = 0.086$$

$$\Delta\epsilon = 153.52 \times 10^{-6} \times 3 - 10.56 \times 10^{-6} \times 3 + 10.56$$
$$\times 10^{-6} \times 0.043 + 10.56 \times 10^{-6} \times 0.086$$
$$= 430.24 \times 10^{-6}$$

$$T^{CR} = C^{CR} = E_{co}A_c\Delta\epsilon \tag{5.50}$$

$$A_c = 7.25 \times 7 \times 12 = 609 \text{ in.}^2$$

$$T^{CR} = C^{CR} = 4.26 \times 10^6 \times 609 \times 430.24 \times 10^{-6} = 1.12 \times 10^6 \text{ lb}$$

Moment $M_{CP}^{CR} = C^{CR}(y_{cpo} - d/2)$ \hfill (5.51)
$$= 1.12 \times 10^6 (31.24 - 3.625) = 30.9 \times 10^6 \text{ lb/in.}$$

$$\Delta F_{CR} = \frac{2M_{DL+SD}^{CR} \, xe_0}{3\left[e_0^2 + \dfrac{I_{cp}}{A_{cp}} + \dfrac{EI_{cp}}{E_t A_t}\right]}$$

where $A_t = 1.2$ in.^{2}, $A_{cp} = 90.3$ in.^{2}, and

$$\Delta F_{CR} = \frac{2 \times 30.9 \times 10^6 \times 35.01}{3\left(35.01^2 + \dfrac{66,273.16}{90.3} + \dfrac{66,273,16}{1.2}\right)} = 1,2611 \text{ lb}$$

$$\text{Loss of prestress due to creep} = 12.6 \text{ k}$$
$$\text{Percentage loss} = \frac{12.6}{240} \times 100\% = 5.25\%$$

5.4 EXAMPLES OF PRESTRESSED COMPOSITE BRIDGES

5.4.1 Project by Coff, United States

Figure 5.32 shows a prestressed highway bridge proposed by Coff.[25] The main plate girders are prestressed by 24 tendons, each having a diameter of 60 mm. These tendons are relatively closely spaced and have polygonal configuration.

5.4.2 Lauffen Bridge, Germany

Figure 5.33 shows an elevation and cross section of the highway bridge at Lauffen over the Neckar Canal. This bridge has a span of 111.50 ft (34 m) and was built in 1955.[26]

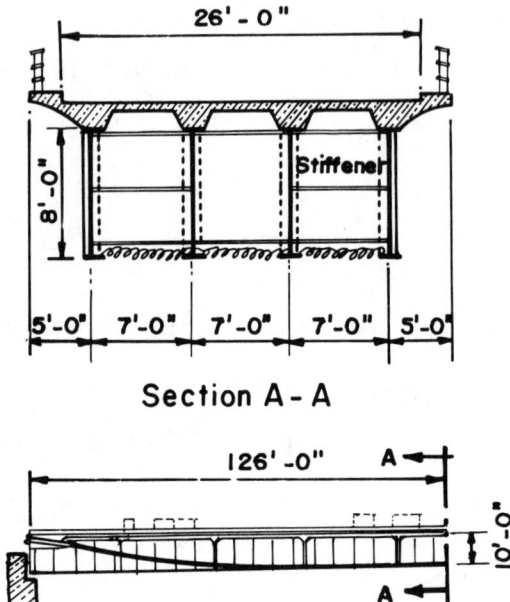

Section A - A

Fig. 5.32 Prestressed bridge, by L. Coff.

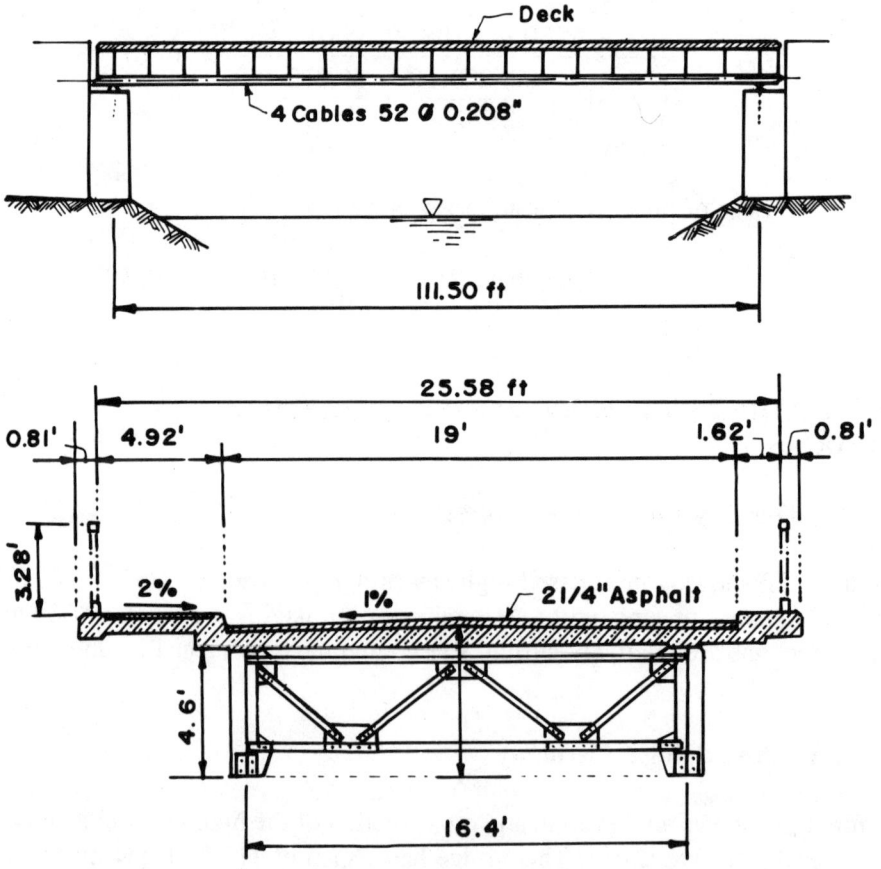

Fig. 5.33 Lauffen Bridge, elevation and cross section.

In cross section, the bridge consists of two plate girders supporting a reinforced concrete deck. Each plate girder was prestressed by four straight tendons. Each tendon consisted of 52 wires, each having a diameter of 0.208 in. (5.3 mm) and a box-shaped chord placed at the bottom of each girder.

To protect against corrosion, each box was filled with asphalt. After the construction of transverse bracing between the girders, they were then poststressed by hydraulic jacks, after which the deck was concreted. Poststressing reduced the stresses at the top chord by 28% and at the bottom chord by 61%.

5.4.3 Railway Bridge, Mexico

In 1961, five short-span railway bridges incorporating prestressed steel I beams, were completed in Mexico (Fig. 5.34).[27] The upper and lower flanges of a built-up steel beam comprise 6 × 6 in. angles welded onto a web plate. Spiral shear connectors were welded inside the upper flange so as to tie together the beams and cast-in-place concrete throughout. A welded steel plate closes the bottom flange of each beam. The lower flange flares at each end of the beam to accommodate anchors for the prestressing cable. The Swiss BBRV stressing system is used for the anchorages. The bridges vary in size, with spans ranging from 39 to 82 ft. The longest has prestressed girders 8.7 ft deep and spaced at 6.9 ft c/c with a prestressing force of 100 tons per girder.

The savings on this type of bridge, as compared with conventional I-beam spans, range up to 55% in weight of steel and up to 25% in cost.

5.4.4 Bellingham Delta Type Bridge, United States

In 1966, in Whatcom County, Washington, near Bellingham, a 150-ft (46 m) span bridge consisting of a concrete deck supported by two delta girders was built for logging loads and other heavy traffic (Fig. 5.35).[28]

The bottom flange of each girder is comprised of plates 24 in. wide. These plates are 1 in. thick for 20 ft each way from the center, then 3/4 in. thick for

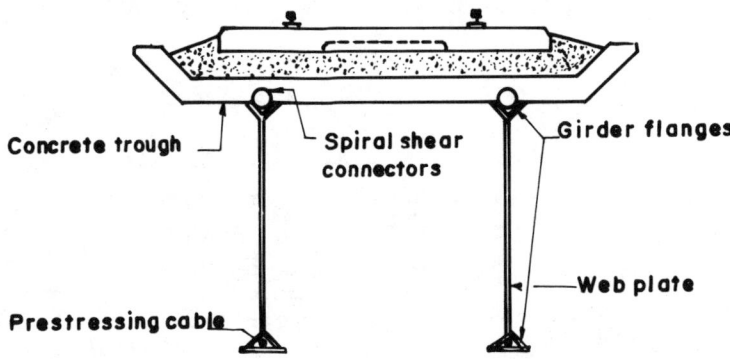

Fig. 5.34   Cross-sectional view of a typical prestressed railway bridge Mexico.

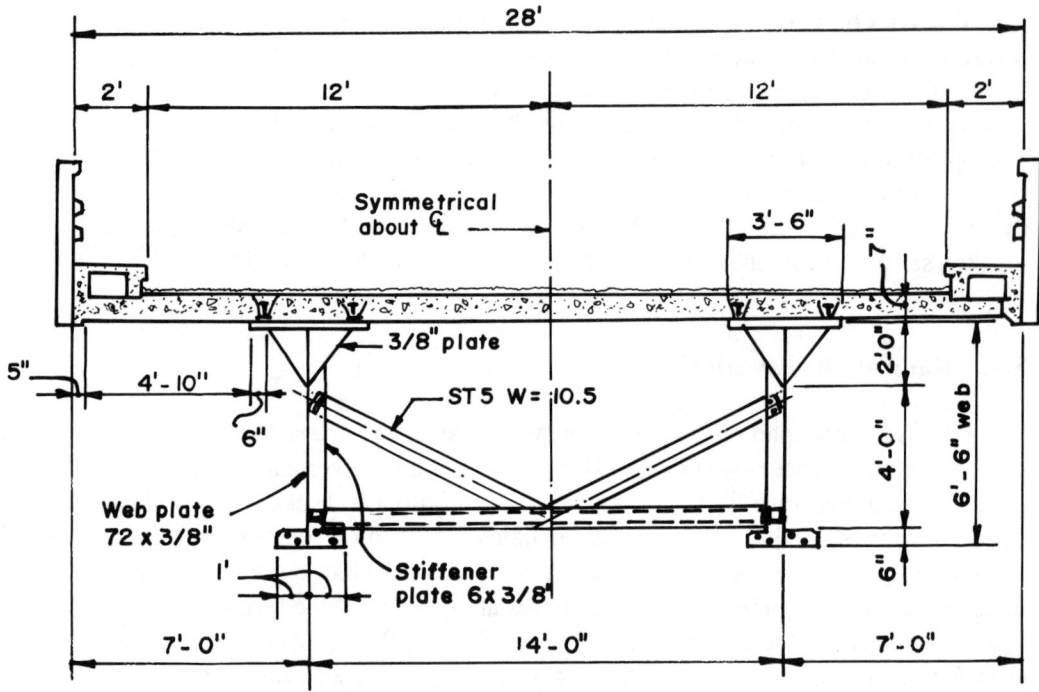

Fig. 5.35   Cross-sectional view of a bridge in Whatcom County, Washington.

20 ft more, and finally, 1/2 in. thick for 35 ft at the girder ends. These plates are centered under and welded onto the web plates. At their outer edges, 6 × 4 × 3/8 in., the angles are welded onto them, the 6-in. legs being downturned toward the compression flange and the 4-in. legs inturned toward the webs.

At each girder end, 1/4-in. "pattern plates," punched for the desired positioning of the prestressed strands, close off the ends of the two compartments.

The fabricated girders were set onto a prestressed bed. Then 36 1/2-in. round prestressing strands per girder were introduced through the "pattern plates" and on into the interior, being directed through the 6-in. top gaps at the girder ends and the long open mid-span space between them (Fig. 5.36).

A total of 72 cables, 1/2 in. in diameter, were pretensioned to 22,000 lb each, making the aggregate total compressive force just under 800,000 lb. Then, the

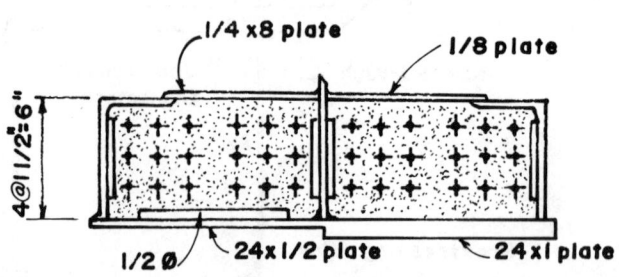

Fig. 5.36   Whatcom County Bridge. A detailed cross-sectional area of a bottom flange and strand.

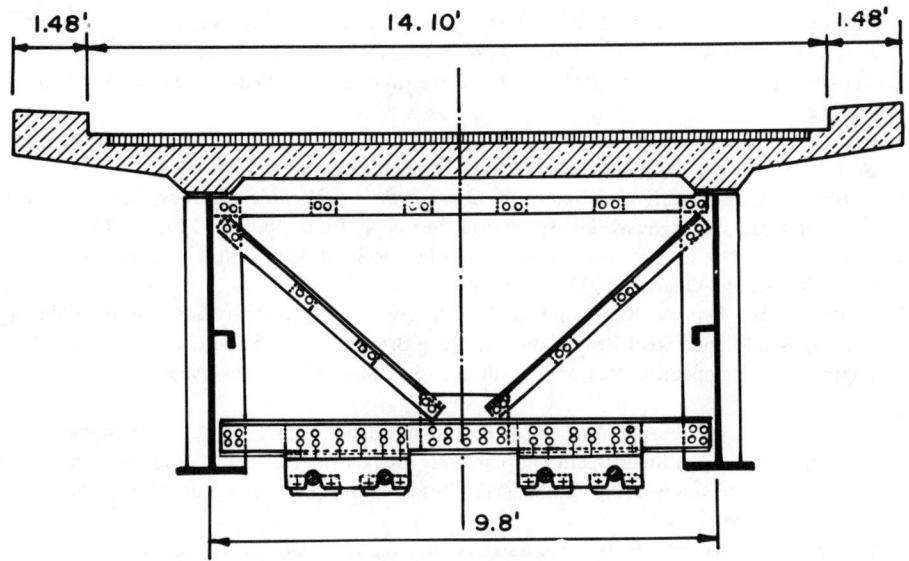

Fig. 5.37 Bridge at Ischl, cross section.

high early-strength concrete was fed in directly through the open tops and vibrated into place. Next the heat of steam curing was applied. When test cylinders showed a strength of 5,000 psi, the strands were burned off from the bed's abutments and the girders were then prepared for testing.

The concrete, confined within its steel jacket, is restrained from lateral dilation and consequently has a greater-than-normal axial strength. The strand within the concrete and within the steel shell is completely protected from air, moisture, and rust.

5.4.5 Ischl Bridge, Austria

The first prestressed composite plate girder bridge was built in 1961 in Austria at Ischl, having a span of 154 ft (47 m).[29] In cross section the bridge's reinforced concrete deck has a total width of 17 ft (5.2 m) and thickness of 8.7 in. Spacing of the two plate girders is 9.6 ft (3 m) (Fig. 5.37).

The bridge is prestressed by four high-strength cables connected by anchorages to the heavy transverse beams located at both ends of the bridge.

REFERENCES

1. Dischinger, F., "Composite Steel Bridges Prestressed by High Strength Cables," Der Bauingenieur, no. 11, pp. 321–322; no. 12, pp. 364–376, 1949, (in German).
2. Coff, L., Composite Floor of Metal and Concrete, U.S. Patent 2, 510 958, 1950.
3. Szilard, R., "Design of Prestressed Composite Steel Structures," *Journal of the Structural Division, Proceedings of the ASCE*, November 1959, pp. 97–123.
4. Hoadley, P. G., "Behaviour of Prestressed Composite Steel Beams," *Journal of the Structural Division, Proceedings of the ASCE*, June 1963, pp. 21–34.

5. Reagan, R. S., and Krahl, N. W., "Behavior of Prestressed Composite Beams," *Journal of the Structural Division, Proceedings of the ASCE*, December 1967, pp. 87–108.
6. Knowles, P. R., *Composite Steel and Concrete Construction*, Wiley, New York, 1973, pp. 68–91.
7. Roik, K., "Methods of Prestressing Continuous Composite Girders," *Proc. Conf. on Steel Bridges*, British Constructional Steelwork Association, London, June 1968.
8. Gibshman, E. E., *Design of Composite Construction in Highway Bridges*, Scientific-Technical Literature on Highway Transportation, Moscow, 1956, pp. 117–139, (in Russian).
9. Anand, S. C., "A Prestressed Composite Girder for Short Span Bridge," *Engineering Journal AISC*, Second Quarter, 1973, pp. 36–42
10. Klaiber, F. W., Dunker, K. F., and Sanders, Jr., W. W., "Feasibility Study of Strengthening Existing Single Span Steel Beam Concrete Deck Bridge," Final Report, Department of Civil Engineering, Engineering Research Institute, Iowa State University, Ames, June 1981, pp. 1–141.
11. Dunker, K. F., Klaiber, F. W., Beck, B. L., and Sanders, Jr., W. W., "Strengthening of Existing Single Span Steel Beam and Concrete Deck Bridges," Final Report—Part II, Department of Civil Engineering, Engineering Research Institute, Iowa State University, Ames, March 1985, pp. 1–146.
12. Dunker, K. F., Klaiber, F. W., and Sanders, Jr., W. W., "Design Manual for Strengthening Single Span Composite Bridges by Post-Tensioning," Final Report—Part III, Department of Civil Engineering, Engineering Research Institute, Ames, March 1985, pp. 1–101.
13. Dunker, K. F., "Strengthening of Simple Span Composite Bridges by Post-Tensioning," Ph.D. Dissertation, Iowa State University, Ames, 1985, pp. 1–247.
14. Troitsky, M. S., Zielinski, Z. A., and Nouraeyan, A., "Analytical and Experimental Study on Composite Prestressed and Poststressed Steel Girder Bridges," *Can. Soc. for Civ. Engr., Proceedings*, vol. I, May 19–22, 1987, Montreal, pp. 159–169.
15. *Standard Specifications for Highway Bridges, AASHTO*, 12th ed., sec. 1.7.48, 1977, p. 194.
16. Szilard, op. cit.
17. Belenya, E., and Streletzkii et al., *Metal Structures*, 2nd ed., vol. II, Stroiizdat, Moscow, 1982, pp. 364–366 (in Russian).
18. Magnel, G., "Creep of Steel and Concrete in Relation to Prestressed Concrete," *Proc. Am. Conc. Inst.*, February 1948, pp. 484–500.
19. Lin, T. Y., and Burns, N. H., *Design of Prestressed Concrete Structures*, 3d ed., Wiley, New York, 1981.
20. Szilard, op. cit.
21. Ibid.
22. Dischinger, op. cit.
23. Lin and Burns, op. cit.
24. Leonhardt, F., *Practical Application of Prestressed Concrete*, State Edition of Literature for Construction and Architecture, Moscow, 1957, pp. 406–409 (in Russian).
25. Anon. American Engineering Studies Prestressing of Structural Steel, Civil Engineering, ASCE, no. 11, 1950.
26. Burkhardt, E., "Highway Bridge Over the Shipping Canal at Lauffen," Die Bautechnik, vol. 32, no. 7, 1955, pp. 238–241, (in German).
27. Anon. "Mexico Opens New Rail Line to Pacific," *Engineering News-Record*, December 14, 1961, pp. 30, 34.
28. Hoadley, H. M., "Steel Bridge Girders with Prestressed Composite Tension Flanges," *Civil Engineering, ASCE*, May 1966, pp. 70–72.
29. Krapfenbauer, R., "Prestressed Steel Bridge at Ischl and the Project of Prestressed Steel Bridge Over the Mur at Graz." International Special Convention "Prestressed Steel Constructions" from 10 to 13 September 1963. *Wissenschaftliche Veroffentichungen aus der Fakultät Bauwesen der Technische Universität Dresden*, no. 30, 1964, pp. 848–851.

Chapter 6
Prestressed Continuous Composite Girders

6.1 INTRODUCTION

The prestressing of continuous steel bridges was proposed by Dischinger in 1949.[1] Dischinger designed and analyzed the prestressing of highway and railway steel bridges of different systems such as continuous plate and box girders, suspension, and arch-type bridges. Continuous prestressed steel bridges analyzed by Dischinger and having cross sections of multiple-cell box type are discussed in more detail in Chapter 7.

The depth of main girders of continuous composite structures is usually made somewhat smaller than in simple spans. At spans smaller than 300 ft it is useful, because of construction problems, to design the girders of constant depth in the range of 1/20 to 1/30 of the span. At spans greater than 300 ft the girders are designed to have increased depth above the intermediate supports. The depth above the intermediate supports is about 1/15 to 1/20 of the span, whereas the depth at mid-span is about 1/40 to 1/50. In continuous span beam bridges, it is recommended to use the length of the end spans equal to 20 or 35% of mid spans.

Based on Dischinger's ideas, the Montabaur Bridge in Germany was built in 1957, and since that time the idea of prestressing continuous steel bridges has been used in bridge engineering. Prestressing of continuous bridges is applied to obtain economy in steel and prevent the creation of cracks in the concrete deck.

6.1.1 Geometry and Structural Characteristics of Continuous Plate Girders

Prestressing and regulation are often applied to composite continuous girders for the economy in steel, which is obtained by full participation of materials and sometimes by substitution of medium steel by high-strength reinforcing steel, and also to protect the concrete slab against the formation of cracks. The meth-

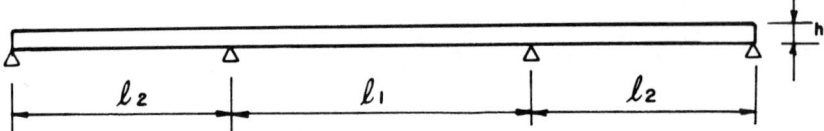

Fig. 6.1 Continuous girder having parallel chords.

ods of prestressing and regulation are different. Most often applied is bending of the steel girder before connection with the concrete slab, usually by using permanent or temporary supports. For larger spans prestressing is applied by longitudinal high-strength steel tendons at zones of negative moments after the concrete slab is connected to the steel girder.

Also economical are hybrid-type girders, in which the most stressed sections of the steel chords are built from high-strength steel, but the web and less stressed sections of the chords are constructed from mild steel.

The geometry of continuous girders having parallel chords is given in Fig. 6.1.

The geometry of continuous girders having increased height above the intermediate supports are shown in Figs. 6.2(a) and (b). These configurations of the girders provide some economy in steel.

The heights of main girders as a function of the span are shown in the Table 6.1. The limits are given for the depth from the bottom edge of the bottom chord to the top of concrete slab. Therefore, the depth of the steel girder is somewhat smaller.

For continuous composite prestressed steel plate girders, the participation of the concrete slab under compression together with the steel top chord permits a substantial reduction in their cross sections. To some degree it also helps to increase the rigidities of other steel elements of the main girders and to improve the dynamic characteristics of the bridge system.

In cross section a composite girder having the deck slab on the top chord generally consists of two main girders (Fig. 6.3). For large spans, a cross section may consist of single or multiple box girders, each having as its bottom flange a stiffened plate (Fig. 6.3(e)).

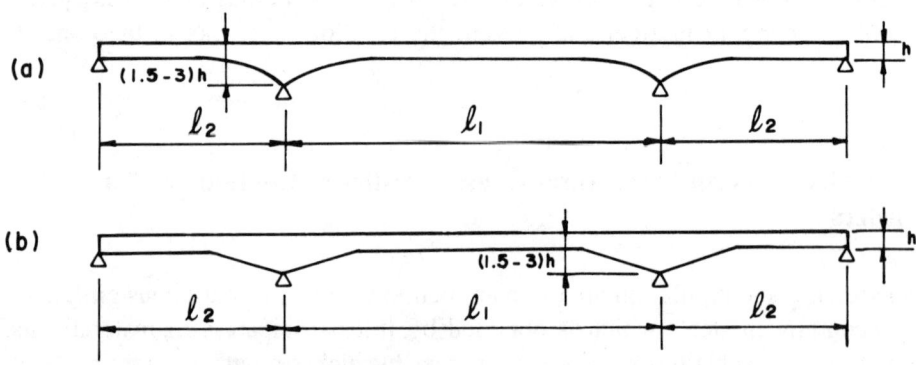

Fig. 6.2 Continuous girders having changeable heights.

Table 6.1 Geometry of Continuous Composite Girders

Scheme	Configurations and Heights of Highway Bridges	
	h_1/l_1	l_2/l_1
Fig. 6.1	1/15–1/20	0.3–1.0
Fig. 6.2	1/20–1/25	0.4–0.8

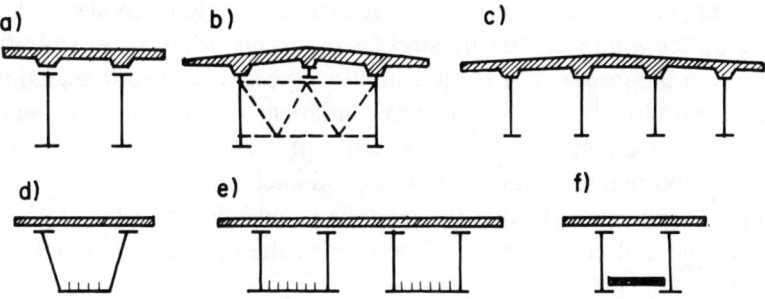

Fig. 6.3   Cross sections of composite plate girders.

Greater numbers of plate girders may be useful at the greater width of the bridge and also for bridges having small spans, less than 130 ft (40 m). The optimum depth of the main girders at spans having many girders is smaller than at the spans having only two girders, and at the box cross section is smaller than at the open cross section.

There are three schemes of deck-type plate girder bridges: when the deck slab is supported only by the main girders (Fig. 6.4(a)), by main and transverse beams (Fig. 6.4(c)), and by main and longitudinal beams (Fig. 6.4(b)).

Composite plate girders are most often used for highway bridges having spans of 390–430 ft (120–130 m) and for railway deck-type bridges having spans of 115–230 ft (35–70 m).

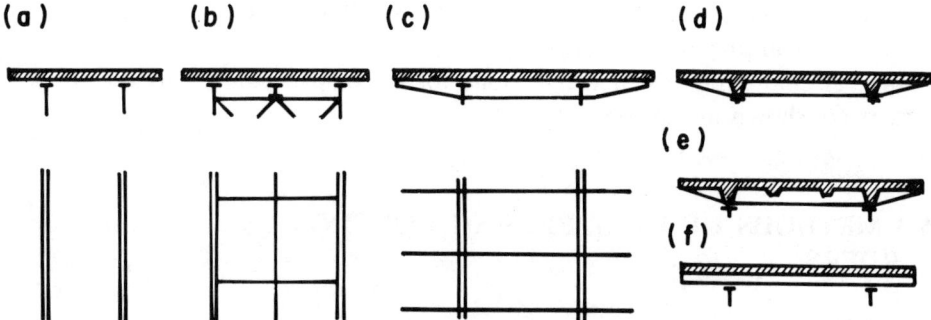

Fig. 6.4   Schemes of composite decks of plate girders.

6.2 TENDON CONFIGURATIONS OF PRESTRESSED CONTINUOUS PLATE GIRDERS

We now consider the various applicable configurations of the tendons for continuous plate girders.[2] Generally, the tendon geometry changes in accordance with the resulting tensile stresses along the girder. In the case of continuous span bridges, the position of the applied concentrated live load creates the maximum bending stresses at the extreme fibers of the section, due to negative moments above the intermediate supports and positive moments in the spans.

The configurations of tendons are related to the shape of moment diagrams. To get better performance from prestressed cables, the tendons should be placed along the girder where the tensile stresses exist and as far as possible from the neutral axis of the girder. The tendon should have a variable cross-sectional area in accordance with the variable bending moment that produces tension at the extreme fibers of the girder section. The use of such a system is expensive due to the use of too many additional saddles and anchorages. Other considerations in tendon configuration are that the tendons should be protected from environmental damage and corrosion by placement of the cables in covered spaces or by isolation.

The most common kinds of tendon configuration for a two-span continuous prestressed girder are shown in Fig. 6.5 and are as follows.[3]

1. A straight short tendon above middle support, which reduces the tension due to negative moment
2. Separate straight short tendons placed at the positive and negative moment areas
3. Continuous polygonal tendons having straight parts at the positive moments and a triangular shape at the negative moments
4. A continuous polygonal tendon, having straight parts at the positive and negative moment regions

The inclined parts of the continuous polygonal tendons at the two ends as well as at the intermediate supports (Fig. 6.5(c), (d), and (e)) induce vertical forces in the opposite direction to the support reactions, and as a result, the shear forces are reduced in accordance with the vertical components of prestressing forces. On the other hand, inclined tendons require saddles, which raise the cost and create losses in prestressing force.

Typical configurations of the tendons for three-span continuous prestressed girders are shown in Fig. 6.6.[4]

6.3 METHODS OF PRESTRESSING CONTINUOUS GIRDERS

The following sections describe different fabrication methods for prestressing continuous girders.

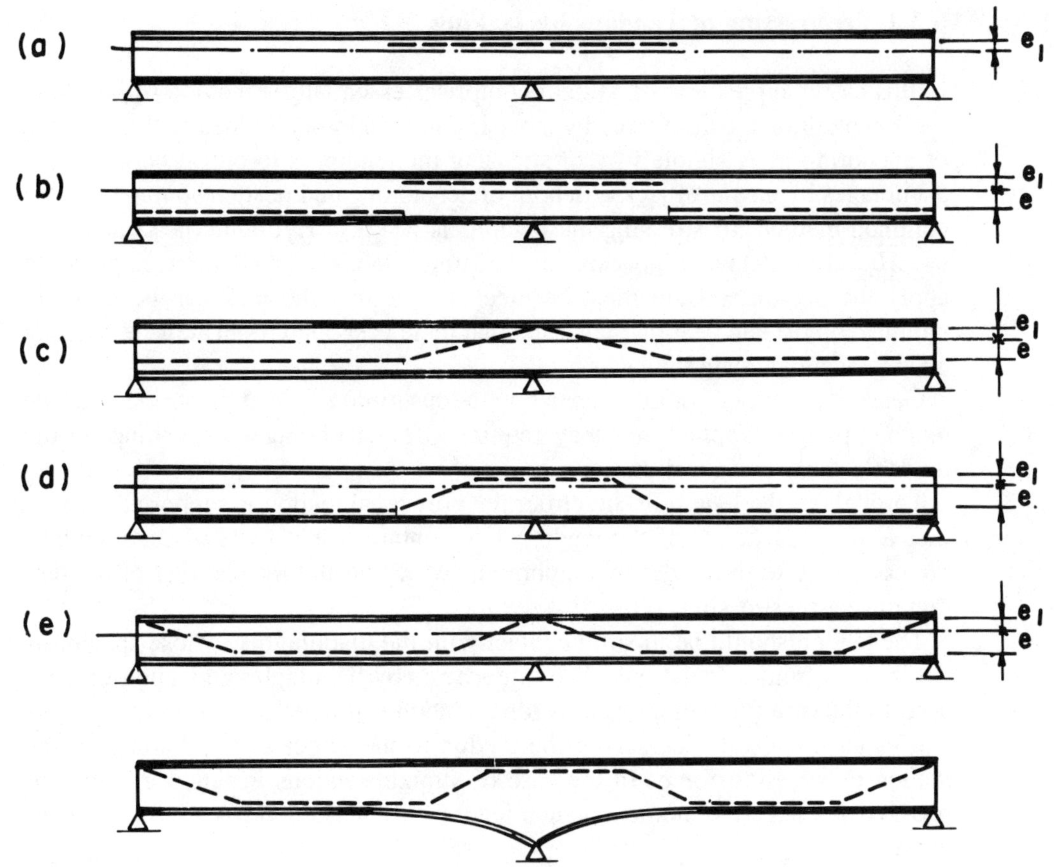

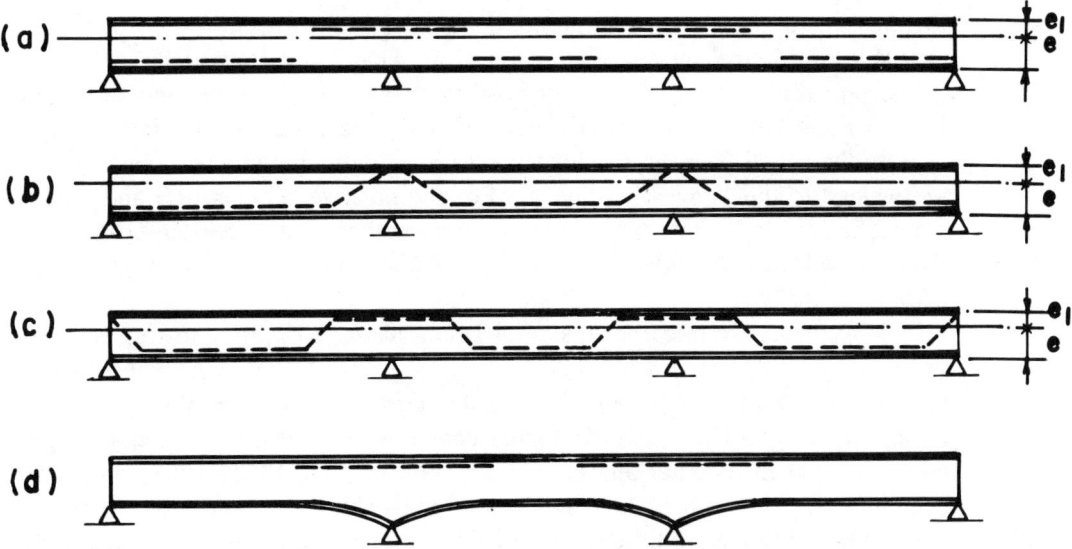

Fig. 6.5   Typical kinds of tendon configurations for two-span continuous prestressed girders.

Fig. 6.6   Typical kinds of tendon configurations for three-span continuous prestressed girders.

6.3.1 Prestressing of Tendons by Jacking

In this case the prestressing system comprises essentially a method of introducing longitudinal tensile forces by stressing the tendon combined with a method of anchoring it. A simple way of stressing the tendon is to pull it between two anchorages by means of jacks. In both pretensioning and posttensioning the most common method for stressing the tendons is jacking. Hydraulic jacks are often used because of their high capacity and the relatively small force required to apply the prestress. Care must be taken to see that the jack can be properly mounted on the end bearing plates and that there is enough room at the tensioning ends to accommodate the jacks.

Generally, for two- or three-span continuous girders, direct prestressing of the tendons may be applied and may require a few anchorages depending on the tendon's shape and location.

Throughout the length of the girder the polygonal tendon is supported by saddles at the bent points of the tendon. Horizontal straight parts of such tendons are connected to the girder by diaphragms which ensure the stability of the tendon during prestressing.

The tendon should be in solid contact with the diaphragms, whose spacing is determined from calculations. The girder rests by the diaphragms upon the tendon. In the case of continuous polygonal tendon end faces of a girder, the girder carries anchorages which secure the tendon to the girder and transmit to it the tendon prestressing force. In the case of straight tendons installed in zones of positive and negative moments, such tendons are tensioned having separate anchorages.

6.3.2 Prestressing by High-Strength Steel Cover Plates

Prestressing of two-span continuous composite highway girders utilizing the available rolled steel section was analyzed by Anand and Talesstichi.[5]

The prestressing mechanisms for positive and negative regions are shown in Figs. 6.7 and 6.8. The jacking forces, P, are applied at one-third points in stage I. With the rolled beam in the loaded position, high-strength steel cover plates are welded to the flanges in stage II. For the positive regions, AD and CE, a plate is welded to the top flange only, whereas in the negative region, DBE, each flange is welded with a plate. In this stage, initial stresses in the cover plates are zero.

Jacking forces are released in stage III, and in stage IV, the prestressed steel beam is turned upside down for the positive regions, AD and CE, whereas it is left as it is for the center negative region, DBE. The stresses do not change during this stage. The prestressed steel beams of the positive and negative regions are spliced together and cast in concrete in stage V, and this composite beam is ready to be subjected to superimposed live and dead loads. It is assumed that in the positive region the weight of the concrete is carried by the steel cross section and the superimposed loads by the composite beam. Since concrete re-

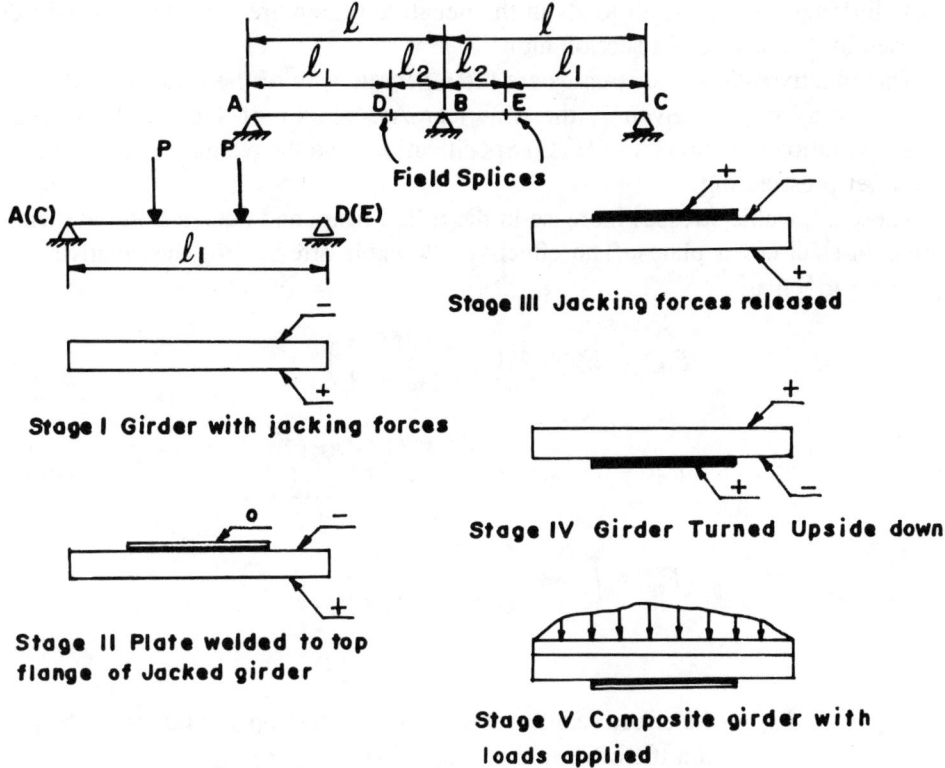

Fig. 6.7 Prestressing mechanism for positive moment. (Courtesy ASCE, Journal of the Structural Division)

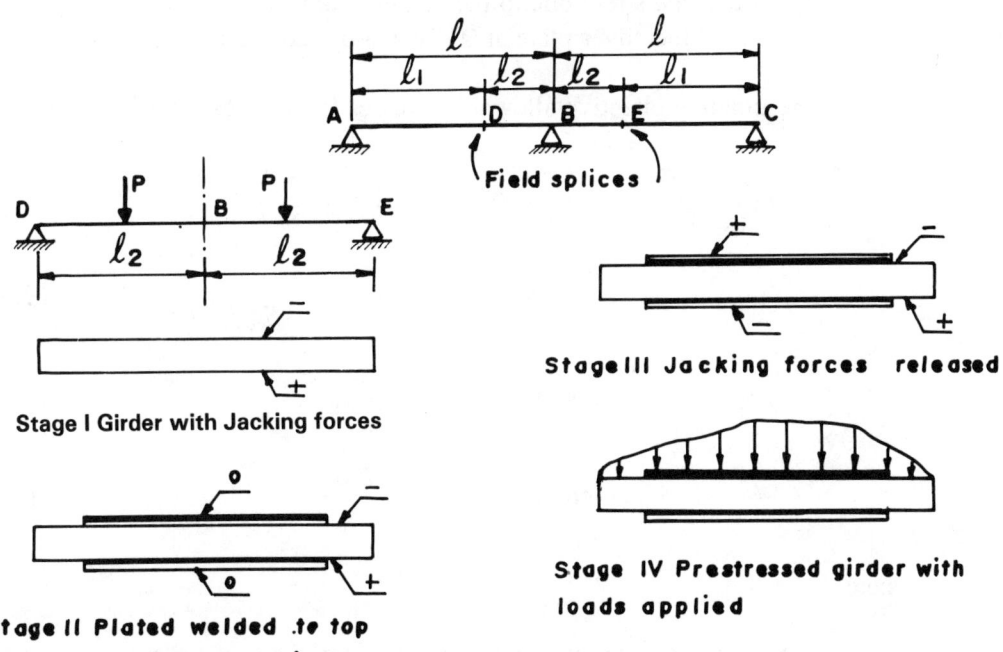

Fig. 6.8 Prestressing mechanism for negative moment. (Courtesy ASCE, Journal of the Structural Division)

sists little tensile stress, all loads in the negative region are considered as being carried by the steel cross section alone.

The effective allowable stresses are the algebraic sum of the maximum allowable bending stresses given by the American Association of State Highway and Transportation Officials (AASHTO) Specifications and the residual stresses present after prestressing.

These allowable stresses increase in the rolled beam and decrease in the high-strength steel cover plates. The effective allowable stresses for the positive region are given as[6]

$$F_{ebb} = F_a + F_b\left(\frac{2.625\xi}{2.625\xi + 1}\right) \tag{6.1}$$

$$F_{etb} = -F_a - F_b\left(\frac{2.175\xi - 0.17}{2.175\xi + 0.83}\right) \tag{6.2}$$

$$F_{ep} = F_{ap} - \left(\frac{F_b}{2.418\xi + 0.923}\right) \tag{6.3}$$

where

F_{etb}, F_{ebb}, F_{ep} = the effective allowable stresses at the top and bottom of beam and the cover plate, respectively

F_a, F_{ap} = the allowable stresses in the beam and the cover plate, respectively

F_b = the stress due to the jacking moment

ξ = the cover plate area divided by the beam area

The corresponding effective allowable stresses in the negative region are:

$$F_{etb} = F_a + F_b\left(1 - \frac{S_b}{S_{tb}}\right) \tag{6.4}$$

$$F_{ebb} = -F_a - F_b\left(1 - \frac{S_b}{S_{bb}}\right) \tag{6.5}$$

$$F_{ep}\,(\text{top}) = F_{ap} - \frac{F_b S_b}{S_p} \tag{6.6}$$

$$F_{ep}\,(\text{bottom}) = -F_{ap} + \frac{F_b S_b}{S_p} \tag{6.7}$$

where

S_b = the section modulus of the beam alone

S_{tb}, S_{bb}, S_p = the section moduli at the top and bottom fibers of the beam and the plate of the total beam plate cross section

6.3.3 Prestressing by Redistribution of Bending Moments

By changing the support levels at continuous girders, it is possible to redistribute the bending moments and consequently to redistribute the stresses in cross sections of the girder. This artificial method allowing the most useful or expedient combination of stresses in the carrying members of a girder is called *the regulation of stresses*.

The regulation method of changing support levels leads to the artificial creation of additional bending moments which are added algebraically to the moments in a girder section. This method, consisting in changing the support levels at continuous girders, is discussed in detail in Chapter 2, Section 2.5.

6.4 STATIC INDETERMINACY OF CONTINUOUS PRESTRESSED PLATE GIRDERS

Continuous prestressed plate girders are statically indeterminate systems. In such systems prestressing may be created by tensioning of the tendons or displacement of the supports. In some cases it is useful to apply prestressing simultaneously with the displacements. Statically indeterminate plate girders are analyzed by the general rules of structural mechanics. Each tendon increases the static indeterminacy of the system by one unit. In the general case of n times the statically indeterminate system having K tendons, the system is $(n + K)$ times statically indeterminate.

During analysis by the force displacement method it is convenient to use for the redundants of the basic system forces in the tendons X_i (incremental forces in the tendons). Force displacement equations are as follows:

$$\delta_{11}X_1 + \delta_{12}X_2 + \cdots + \Delta_{1P} = 0 \tag{6.8}$$

$$\delta_{21}X_1 + \delta_{22}X_2 + \cdots + \Delta_{2P} = 0 \tag{6.9}$$

The coefficients at unknowns are calculated by the conventional formulas where

δ_{11} = displacement of point 1, due to $X_1 = 1$
δ_{12} = displacement of point 1, due to $X_2 = 1$
δ_{21} = displacement of point 2, due to $X_1 = 1$
δ_{22} = displacement of point 2, due to $X_2 = 1$
Δ_{1P} = total displacement of point 1, due to external load

An important part of the design is the choice of tendons to be prestressed. The optimal prestressing force, considering the economy and cost of material, is determined by trial and error.

During the analysis of every statically indeterminate system, it is first necessary to assume cross sections of the girder and tendons. During the first approximation it is possible to design the structure without prestressing, to choose the cross-section and accept prestressing forces for the tendons, considering stability of the parts compressed under prestressing. It is important to establish the sequence of prestressing and application of dead loading to the structure. In many cases it is useful to apply part of the dead load initially and to apply prestressing after.

6.5 ANALYSIS OF CONTINUOUS PRESTRESSED PLATE GIRDERS

6.5.1 General Data

The technical literature on the structural analysis of continuous prestressed steel plate girder bridges reveals that very little research has been performed in this area.

The flexibility method and the complementary virtual work method are recommended for the solution of prestressed continuous steel girders, which provide the analysis of the incremental prestressing force.

The incremental prestressing forces are introduced by the applied live loads and superimposed dead loads to the steel girder bridge prestressed by tendons. The tendons elongate under the incremental loads, and correspondingly the prestressed force increases through the tendons.

The prestressed continuous steel girder is an indeterminate system, and the number of required equations depends on the degree of indeterminacy of the system.

At the continuous girders tendons are placed at zones of tensile stresses, namely in spans along the bottom chord and at the supports along the top chord (Fig. 6.9(a)). Tendons may be uninterrupted, passing from the bottom chord to the top, as in Fig. 6.9(b). However, such a tendon configuration is somewhat complicated, and for this reason separate rectilinear tendons are more often installed.

Let us consider continuous girder having two equal spans prestressed by three tendons: two in the spans and one above the middle support (Fig. 6.10(a)). The girder is under uniformly distributed load p. The analysis may consist of the following three parts:

1. The girder is analyzed by the conventional method without considering the action of the tendons, and the resultant moment diagram is shown in Fig. 6.10(b).

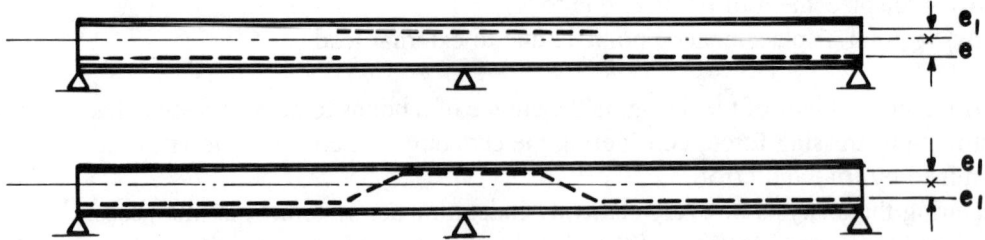

Fig. 6.9 Configuration of tendons at continuous girders.

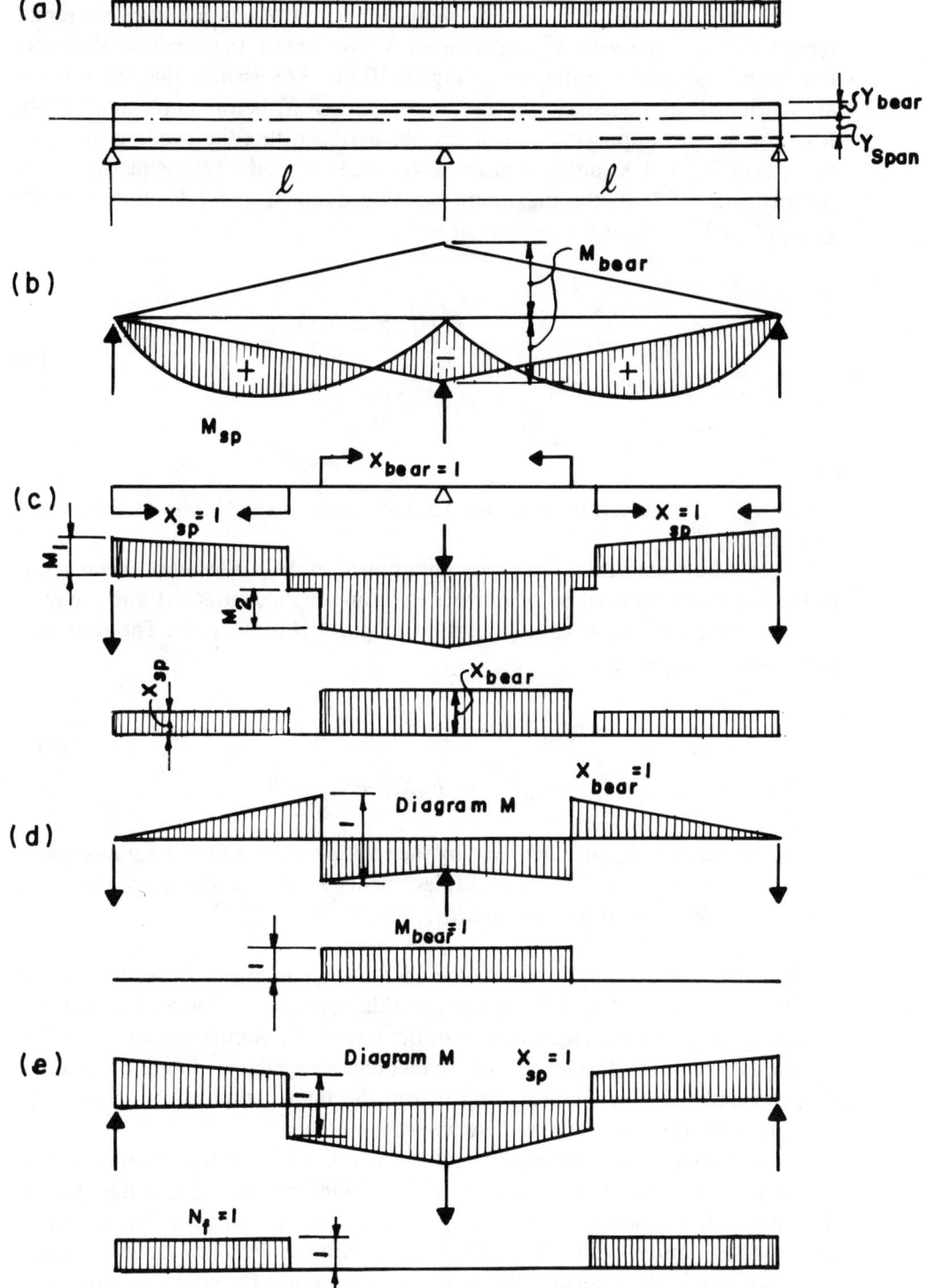

Fig. 6.10 Calculation of a continuous beam.

2. Further, the two-span girder is analyzed under the action of the prestressed tendon forces in the span X_{sp} and support X_s. We obtain the corresponding moments and normal force diagrams (Fig. 6.10(c)). We assume that the tendons are tensioned simultaneously by the forces X_{sp} and X_s before application of the load. At the first approximation, it may be assumed that the bending moments that forces X_{sp} and X_s produce at the anchorages have the following values: in the span, 25–30% of the maximum bending moment under loading over the girder, and 40–50% at the support or

$$X_{sp} = \frac{M_{sp}}{y_{sp}}(0.25 \text{ to } 0.3)$$

$$X_s = \frac{M_s}{y_s}(0.4 \text{ to } 0.5)$$

(6.10)

where y_{sp}, y_s = distances from the tendons to the neutral axis of the girder, respectively.

3. Also the incremental forces are determined in the tendons under the action of loading p. As the basic system, we are considering a continuous girder having two unknown tendon forces, X_{sp1} and X_{s1} (Figs. 6.10(c) and (d). The force displacement equations are

$$\delta_{spsp}X_{sp1} + \delta_{sps}X_{s1} + \delta_{sp1} = 0$$

$$\delta_{ssp}X_{sp1} + \delta_{ss}X_{s1} + \delta_{sp} = 0$$

(6.11)

The displacements δ_{ik} are calculated by the multiplication of the diagrams shown in Figs. 6.10(c) and 6.10(d), and displacements δ_{1p} by the multiplication of the diagrams in Figs. 6.10(b), (c), and (d).

The calculated bending moments are added to the moments under the loading M_p, the prestressing forces X_{sp} and X_s, and the incremental forces X_{sp1} and X_{si}.

During design of the cross section of the girder, the normal forces in the tendons are considered. The lengths of the tendons are established after the moment diagram under loading, considering that at the interruption of the tendon, the cross section of the girder may take an acting moment.

At the continuous two-span girder, it is possible to obtain the maximum effect under prestressing by installation of single rectilinear tendon above the support. The substantial economy in metal, up to 33%, may be obtained during design of a two-span girder with a single tendon above the support and having different cross sections in the span and above the support, or a symmetrical section in the span having moment of inertia I_{sp} and an asymmetric section above the support having moment of inertia I_s along length of the tendon (Fig. 6.11).

The stresses at the girder cross section are checked.

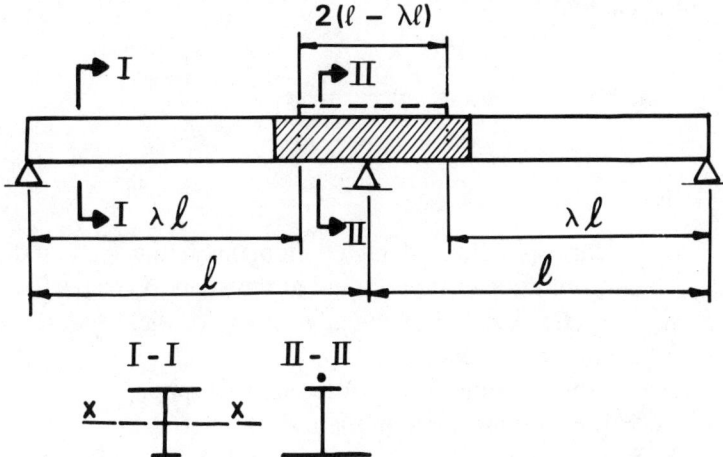

Fig. 6.11 Scheme of two-span continuous girder having changeable cross sections and tendon above support.

a. Stresses at section above mid-support:

$$\delta_t = \frac{M_s^b}{S_t} - \frac{n_2 X + X_1}{A_s} - \frac{M_s^{X+X_1}}{S_t} \leq f_{\text{all}} \qquad (6.12)$$

$$\delta_b = \frac{M_s^b}{S_b} - \frac{n_2 X + X_1}{A_s} + \frac{M_s^{X+X_1}}{S_b} \leq f_{\text{all}} \qquad (6.13)$$

b. Stresses in the span:

$$\delta_t = \delta_b = \frac{\pm M_{sp}^b}{S} \pm \frac{M_{sp}^{X+X_1}}{S} \leq f_{\text{all}} \qquad (6.14)$$

c. Stresses at the location of the anchorage at a distance of λl from the end support:

$$_t\delta_{\lambda l(\text{left})} = \frac{-M_{\lambda l}^b}{S_t} + \frac{M_{\lambda l}^{X+X_1}}{S_t} \leq f_{\text{all}} \qquad (6.15)$$

$$_b\delta_{\lambda l(\text{left})} = \frac{M_{\lambda l}^b}{S_b} - \frac{M_{\lambda l}^{X+X_1}}{S_b} \leq f_{\text{all}} \qquad (6.16)$$

$$_t\delta_{\lambda l(\text{right})} = \frac{M_{\lambda l}^b}{S_t} - \frac{n_2 X + X_1}{A_s} - \frac{M_{\lambda l}^{X+X_1}}{S_t} \leq f_{\text{all}} \qquad (6.17)$$

$$_b\delta_{\lambda l(\text{right})} = \frac{-M_{\lambda l}^b}{S_b} - \frac{n_2 X + X_1}{A_s} + \frac{M_{\lambda l}^{X+X_1}}{S_b} \leq f_{\text{all}} \qquad (6.18)$$

d. Stress in the tendon:

$$\delta_{td} = \frac{n_1 X + X_1}{A_t} \leq f_{\text{all}} \tag{6.19}$$

where

$M_{sp}^b, M_{\lambda l}^b, M_s^b$ = corresponding calculated moments in the span, at distance λl from the end support and at the support, respectively

S_t, S_b = section moduli of the edge fibers of the cross section of the girder above support

A_s = cross section of the girder above support

X = prestressing force in the tendon

X_1 = incremental force in the tendon for the nonbeneficial loading of the span and for the support

$n_1 = 1.1$ = coefficient of the overloading

$n_2 = 0.9$ = coefficient of the reduction of prestressing due to incorrect control.

Formulas for the determination of the incremental forces in the tendon under different loadings were evaluated from the analysis of the continuous two-span girder of changeable cross section, and optimum prestressing forces in the tendons were obtained from the condition of equality of maximal stresses in the span and at the support.

6.5.2 Continuous Two-Span Girders

To demonstrate the analysis of tendon configurations, the energy method gives the exact solution for calculating the moments and incremental prestressing force due to live loads. Brodka et al. presents the solution for the special case of loadings and position of tendons.[7] Generally, the tendon geometry changes in accordance with the resulting tensile stresses along the girder. In the case of continuous-span bridges, the position of the applied concentrated live load creating the maximum bending stresses at the extreme fibers of the section, due to the so-called negative and positive moments, is not applied at the mid-span.

In this section a general method of analysis is presented for various configurations. To make it more applicable to the analysis of prestressed girder bridges, a computer program has been developed to calculate the incremental prestress force due to live loads and the moments at critical sections.

The high-strength steel cables (tendons) are attached to the girder such that when prestressed, it counteracts with the existing tensile stresses due to the applied loading systems.

The configurations of tendons are related to the shape of moment diagrams. To obtain better performance from the prestressed cables, the tendons should be placed along the girder where the tensile stresses exist, at a distance as far as possible from the neutral axis of the girder. The tendon should have a variable cross-sectional area in accordance with the variable bending moment that produces tension at the extreme fibers of the girder section. To create such a system of tendon placement is not only expensive due to the use of too many additional saddles and anchorages, but it is also impractical. Other considerations in tendon configuration are that the tendons should be protected from environmental damage and corrosion, by placing the cables in covered spaces or by isolation.

The most common tendon configurations are shown in Fig. 6.12 for two-span continuous prestressed girder bridges. A general analysis of the incremental prestressed force is presented in the following set of equations for the various configurations illustrated in Fig. 6.12. The output results of the computer program indicate the incremental prestressing force and the maximum negative moment. The following three cases are considered.

Case 1:

$$BM = \frac{\Delta_{1LL}\Delta_{22} - \Delta_{2LL}\Delta_{12}}{\Delta_{12}^2 - \Delta_{11}\Delta_{22}} \qquad (6.20)$$

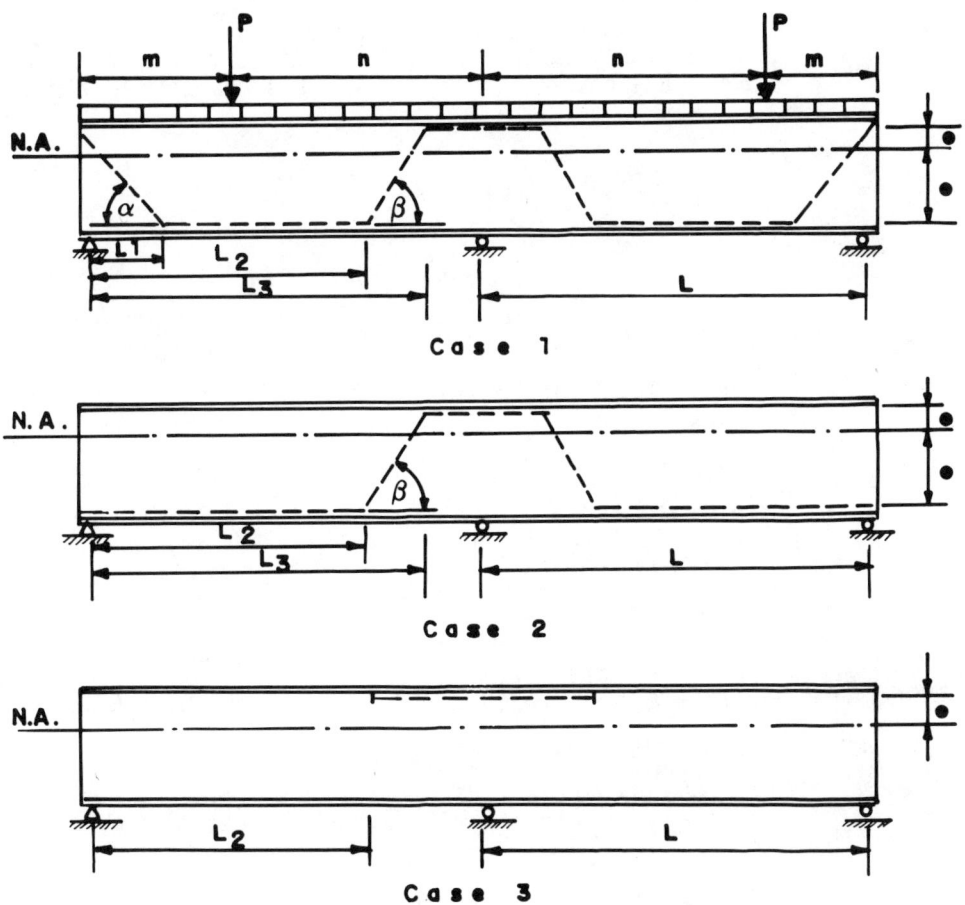

Fig. 6.12 Continuous two-span girders.

$$\Delta X = \frac{\Delta_{2LL}\Delta_{11} - \Delta_{1LL}\Delta_{12}}{\Delta_{12}^2 - \Delta_{12}\Delta_{22}} \qquad (6.21)$$

$$a = e_1 \qquad b = e + e_1\frac{L_1}{L} \qquad c = e = e_1\frac{L_2}{L} \qquad d = e_1\left(1 - \frac{L_3}{L}\right) \quad (6.22)$$

$$\Delta_{11} = \frac{2L}{3EI}$$

$$\Delta_{12} = \frac{1}{3EIL}\left[L_1^2(-a - c) + L^2{}_2(b + d) + L^2{}_3c + L^2(-d)\right.$$
$$\left. + L_1L_2(b - c) + L_2L_3(c + d) + LL_3(-d)\right] \qquad (6.23)$$

$$\Delta_{22} = \frac{2}{EI}\left[\frac{L_1}{6}(2a^2 - 2ab + 2b^2) + \frac{L_2 - L_1}{6}(2b^2 + 2bc + 2c^2)\right.$$
$$\left. + \frac{L_3 - L_2}{6}(2c^2 - 2cd + 2d^2) + \frac{L - L_3}{6}(2d^2)\right] \qquad (6.24)$$
$$+ \frac{2L}{EA} + \frac{2L_1}{E_tA_t\cos^3\alpha} + \frac{2(L_2 - L_1)}{E_tA_t} + \frac{2(L_3 - L_2)}{E_tA_t\cos^3\beta} + \frac{2(L - L_3)}{E_tA_t}$$

$$\Delta^q_{1LL} = -\frac{qL^3}{24EI} \qquad (6.25)$$

$$\Delta^P_{1LL} = -\frac{P}{EIL^2}\left(\frac{m^3n}{3} + \frac{m^2n^2}{2} + \frac{mn^3}{6}\right) \qquad (6.26)$$

$$\Delta_{1LL} = \Delta^q_{1LL} + \Delta^P_{1LL} \qquad (6.27)$$

$$\Delta^q_{2LL} = \frac{q}{2EI}\left\{aL_1^2\left(\frac{L}{6} - \frac{L_1}{12}\right)b\left[(L_1\right.\right.$$

$$+ L_2)\left(-\frac{LL_2}{6}\right) + \frac{L_2}{12}(L^2{}_2 + L_1L_2 + L_1^2)\right]$$

$$+ \left[(L_1 + L_2)\frac{LL_1}{6} - \frac{L_1}{12}(L_1^2 + L_1L_2 + L^2{}_2) + (L_3 + L_2)\right. \qquad (6.28)$$

$$\cdot \left(-\frac{LL_3}{6}\right) + \frac{L_3}{12}(L^2{}_3 + L_2L_3 + L^2{}_2) + d\left[(L_2 + L_3)\left(\frac{-LL_2}{6}\right)\right.$$

$$+ \frac{L_2}{12}(L^2{}_3 + L_2L_3 + L^2{}_2) + \frac{L}{12}(-L^2{}_3 + L^2 + LL_3)\right]\right\}$$

$$\Delta_{2LL}^{P} = \frac{P}{EIL}\left\{ n\left[(a - 2b)\frac{L_1^2}{6} + \left(\frac{b - c}{L_2 - L_1}\right)\left(\frac{m^3}{3} + \frac{L_1^3}{6} - \frac{L_1 m^2}{2}\right)\right.\right.$$

$$- \frac{b}{2}(m^2 - L_1^2)\right] + m\left[\frac{(b - c)(L_2 - m)}{L_2 - L_1}\right.$$

$$\cdot\left((L_2 + m)\left(\frac{L}{2} + \frac{L_1}{2} - \frac{L_2}{3}\right)\right.$$

$$\left.- LL_1 - \frac{m^2}{3}\right) + b(L_2 - m)\left(-L + \frac{L_2}{2} + \frac{m}{2}\right)$$

$$+ (c + d)\left((L_3 + L_2)\left(\frac{L}{2} + \frac{L_2}{6}\right) - LL_2 - \frac{L_3^2}{3}\right)$$

$$\left.\left.+ c(L_3 - L_2)\left(-L + \frac{L_3}{2} + \frac{L_2}{2}\right) + \frac{d}{3}(L - L_3)^2\right]\right\}$$

(6.29)

$$\Delta_{2LL} = \Delta_{2LL}^{q} + \Delta_{2LL}^{P}$$

(6.30)

Case 2: Similar equations are used, with $L_1 = 0$, $a = 0$, $\cos \alpha = 1$.

Case 3: The same equations as for case 1, and taking $L_1 = 0$, $L_3 = L_2$, $\cos\alpha = \cos\beta = 1$.

Stresses

1. Mid-Span

$$\text{Top fiber stresses} = -\frac{M^+}{S_{top}} + \frac{(X + \Delta X)e}{S_{top}} - \frac{(X + \Delta X)}{A} \quad (6.31)$$

$$\text{Bottom fiber stresses} = \frac{M^+}{S_{bot}} - \frac{(X + \Delta X)e}{S_{bot}} - \frac{(X + \Delta X)}{A} \quad (6.32)$$

2. Middle Support

$$\text{Top fiber stresses} = +\frac{M^-}{S_{top}} - \frac{(X + \Delta X)e_1}{S_{top}} - \frac{(X + \Delta X)}{A} \quad (6.33)$$

$$\text{Bottom fiber stresses} = -\frac{M^-}{S_{bot}} + \frac{(X + \Delta X)e_1}{S_{bot}} - \frac{(X + \Delta X)}{A} \quad (6.34)$$

where

M^+, M^- = the design positive and negative moments

X, ΔX = the prestressing and incremental prestressing force

The tendon configurations in Fig. 6.12 show a general idea of tendon place-ment, and when necessary, the various cases may be combined for the sake of economy. The inclined part of the cables at the two ends has the same signifi-cance as for the simply supported case; it counteracts the maximum shear force at the supports. In case 1, the tendons are placed to counteract the tension at the extreme fibers of the section; at the positive moment region and at the middle support (negative moment region). In order to avoid the use of too many saddles and anchorages, the bottom set of tendons is placed horizontally at the tension region of the span. This reduces the cost of fixing the saddles, diminishes the prestressing losses due to saddle friction and anchorages, and prevents the oc-currence of secondary stresses resulting from these fixtures.

The second case, having a horizontal part close to the two end supports, does not induce any vertical component of prestressed force. However, it slightly reduces the maximum bending moment at the middle support, and the cost of extra anchorages is saved.

Case 3, as illustrated in Fig. 6.12, has tendons extended only at the top tension region, which counteracts the applied negative moment. In this case, the nega-tive bending moment is increased by 2–3%, and the incremental prestressed force is 15–20% of the similar value for case 1.

A comparison of the various configurations as seen in Fig. 6.12, and consid-eration of the computer results, shows that the tendon placement in case 1 pre-sents more satisfactory results than the other two systems. Where there exists a significant negative moment at the middle support, it is advantageous to combine cases 1 and 3. Case 2 can be employed where the maximum shear force at the end supports is significant.

6.5.3 Continuous Three-Span Girders

The objective of this section is to introduce the detailed static analysis of con-tinuous steel girders consisting of three and more spans prestressed by tendons (high-strength steel cables). The section presents the static analysis procedure of continuous steel girders prestressed by high-strength steel cables (tendons). The analysis represents the most general approach for investigating prestressed con-tinuous steel girders of variable span lengths and under arbitrary loading inten-sity and position.

The flexibility method is the basis for the analysis of the incremental prestress-ing force and the bending moments at the middle supports, which constitute the unknowns of the indeterminate structure. The flexibility coefficients are com-puted by the virtual work method. The given equations may be computerized for the analysis of girders having uniform cross section along the girder.

The static analysis covers the elastic range, and the principles of linear distri-bution of stresses and superposition of forces are valid. The analysis is based on the flexibility and virtual work methods. The indeterminate structure is converted to the determinate by assuming the redundants as the incremental prestressing force and the negative bending moments at the middle supports. The simultaneous equations of unknown redundants are written (flexibility method), and the deformation coefficients are calculated by the virtual work method. Matrix analysis is employed for the solution of redundants.

For various tendon configurations and three-span girders the incremental prestressing force and the negative bending moments (the unknown redundants) are formulated. The shear force and the bending moments at any section along the girder can then be calculated by the application of static equilibrium equations.

1. Basic Assumptions

a. The analysis of prestressed steel girders is performed in the elastic range. The stresses at each section are linearly distributed across the section. The principle of superposition of forces and deformations and other general assumptions applicable to elastic analysis are valid.

b. The assumed tendon configuration is a general placement in which the tendons effectively contribute in reducing the conventional stresses.

c. The position of the neutral axis is assumed to be in the same level along the girder, with a uniform cross section throughout the girder.

d. The uniformly distributed load is applied over the entire length of each span. The concentrated loads are applied within the horizontal position of the tendons placed at the region of positive bending moment.

e. Various spans are loaded such that the maximum bending moments, either positive or negative, are reached. The maximum incremental prestressing force occurs when all the spans are loaded.

f. The negative bending moment induces tension at the top fibers of the section and compression at the bottom fibers, whereas the positive moment has a reverse effect on the girder.

g. The selected sign convention indicates a negative sign for compression and a positive sign for tension stresses.

In this section the main concepts of prestressing by high-strength steel cables (tendons) are considered. The pretensioned or posttensioned tendons induce bending and axial compressive stresses proportional to prestressing force and the tendon eccentricity from the neutral axis, counteracting the stresses applied by service loads.

The effect of combined stresses due to the applied loads and prestressing forces in the elastic stage for an asymmetrical section are illustrated in Fig. 6.13. At various stages of loading stresses should not exceed the limits of allowable stresses. The two identified stages for stress control correspond to (1) stresses due to loads before prestressing is carried out (dead load only), as in equations (6.35a) and (6.35b), and (2) final summation of stresses under the application of service loads including dead and live loads and prestressing, as in equations (6.36a) and (6.36b).

$$+ \frac{M_{DL}y_t}{I} \leq \text{allowable stress in tension} \qquad (6.35a)$$

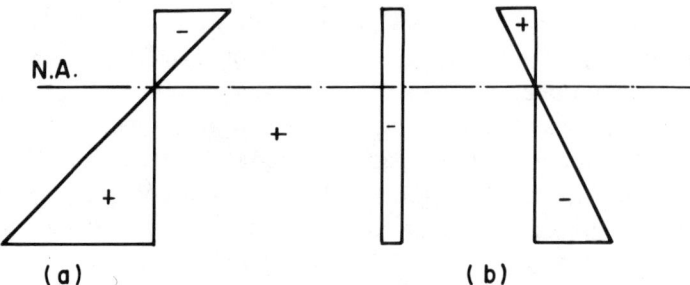

N.A.

(a)   (b)

Fig. 6.13 Stress distribution diagrams. (a) Stresses due to service loading. (b) Component stress diagrams corresponding to prestressing force.

$$-\frac{M_{DL}y_c}{I} \le \text{allowable stress in compression} \quad (6.35b)$$

$$-\frac{P}{A} - \frac{P \cdot e_t \cdot y_t}{I} + \frac{M \cdot y_t}{I} \le \text{allowable stress in tension} \quad (6.36a)$$

$$-\frac{P}{A} + \frac{P \cdot e_c \cdot y_c}{I} - \frac{M \cdot y_c}{I} \le \text{allowable stress in compression} \quad (6.36b)$$

where

P = effective prestressing force, including the incremental prestressing force

e_t, e_c = tendon eccentricity in the tension and compression zones, respectively

y_t = distance of the neutral axis to the extreme flange fibers in tension

y_c = distance of the neutral axis to the extreme flange fibers in compression

I = moment of inertia of the section

A = sectional area of the member

M = maximum bending moment due to the applied service loads

M_{DL} = maximum bending moment due to dead load only

2. Tendon Configuration and Section Geometry

To achieve the full advantage of prestressing in saving materials or increasing the capacity of the girder, careful consideration should be given to the selection of tendon configurations and the shape of the girder cross section.

For the effective use of prestressed cables, the tendons are placed as far as possible from the neutral axis at the tension zone of the section along the girder span. Although the most advantageous placement is a curved tendon corresponding to the bending moment diagram, practically it is difficult (if at all possible) to create such a placement system.

Equation (6.36a) indicates that any increase in tendon eccentricity (e), accordingly increases the countereffect of the induced stresses by prestressed tendons. In the case of final compression stresses in equation (6.36b) the tendon eccentricity (e) is limited to the following condition: Substitute $I/A = r^2$ in equation (6.36b) and rearrange it as

$$\frac{P}{A}\left(-1 + \frac{e \cdot y_c}{r^2}\right) - \frac{M \cdot y_c}{I} \leq f_{all} \text{ in compression} \qquad (6.36c)$$

The first expression, which is the stress contribution of prestressed tendons, indicates that if

$$\left(-1 + \frac{e \cdot y_c}{r^2}\right) \geq 0 \qquad (6.37)$$

the effect of conventional stresses corresponding to the applied service loadings ($M \cdot Y_c/I$) can be reduced. Therefore the limit of prestressing eccentricity is expressed as:

$$e \geq \frac{r^2}{y_c} \qquad (6.38)$$

Asymmetrical sections are among the most suitable sections for prestressing, having their neutral axis closer to the extreme compression fibers ($Y_c \geq Y_t$), as shown in Fig. 6.14.

Symmetrical sections may be prestressed, but economic gain is limited (significantly less than prestressing asymmetrical sections), depending on the position of the tendons. Figure 6.15 shows the stress distribution for symmetrical sections.

In a comparison of two sections having the same cross-sectional area, the prestressing capacity of the T section is greater than the I section, due to the higher tensile stress of the T section which is to be compensated by the prestressing force (Figs. 6.13 and 6.15).

3. Static Analysis

The application of variable prestressing forces including a bending moment and an axial force along the girder increases the degree of indeterminacy of the structure by one. The flexibility method simplifies the analysis of such indeterminate structures. The redundants are taken as (1) the incremental prestressing force and (2) the bending moment at the middle supports (negative bending moments).

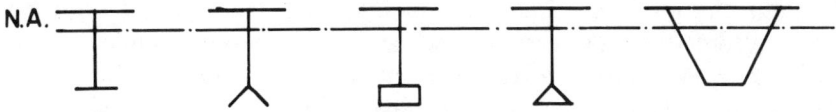

Fig. 6.14   The typical asymmetrical sections used for prestressed girders.

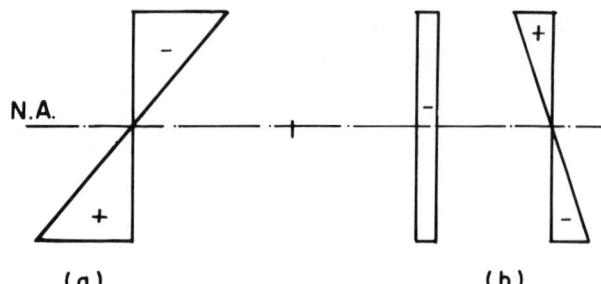

N.A.

Fig. 6.15 The stress distribution of a symmetrical section. (a) Stress corresponding to service load. (b) Stress diagrams due to prestressing.

(a)   **(b)**

The flexibility equations are written by using the principle of superposition of deformations at the points under consideration. For example, the sum of deformations at the point of redundant number 1 (at the first middle support from the left) are given as:

$$d_{11} \cdot M_1 + d_{12} \cdot M_2 + \cdots + d_{1,ns} \cdot X = D_{1L} \qquad ns = \text{number of spans}$$

where d_{1i} is the deformation at point 1 caused by a unit load corresponding to M_i at i and D_{1L} is the deformation at the same point corresponding to the applied service loads. Similarly, additional equations are written as the simultaneous equations (6.39).

$$d_{11} \cdot M_1 + d_{12} \cdot M_2 + \cdots + d_{1j} \cdot X = -D_{1L}$$
$$d_{21} \cdot M_1 + d_{22} \cdot M_2 + \cdots + d_{2j} \cdot X = -D_{2L} \qquad (6.39)$$
$$\cdots$$
$$d_{i1} \cdot M_1 + d_{i2} \cdot M_2 + \cdots + d_{ij} \cdot X = -D_{iL}$$

where $i = 1, ns$, and $j = 1, ns$.

The flexibility coefficients d_{ij} and D_{iL} are calculated by the virtual work method, by equating the internal work performed by the internal induced stresses along the girder and the external work done by a unit load and the corresponding deformations. The expression for the flexibility coefficients is derived as:

$$d_{ij} = \int \frac{M_i M_j}{E \cdot I} \, dx + \int \frac{N_i N_j}{E \cdot A} \, dx \qquad (6.40)$$

where M is the bending moment and N is the axial force. Appendix I illustrates the flexibility coefficients for the continuous span girder having two equal side spans of l_1 and typical intermediate spans of l_m.

The corresponding bending moment and the axial force diagrams for a three-span continuous girder are illustrated in Figs. 6.16 and 6.17. Figure 6.16(a) shows the geometry of the assumed girder and the tendon configurations. The released structure is shown in Fig. 6.16(b). Figures 6.16(c) and 6.16(d) demonstrate the bending-moment diagrams of the released structure due to a concentrated load at distance m and a uniformly distributed load over a typical span.

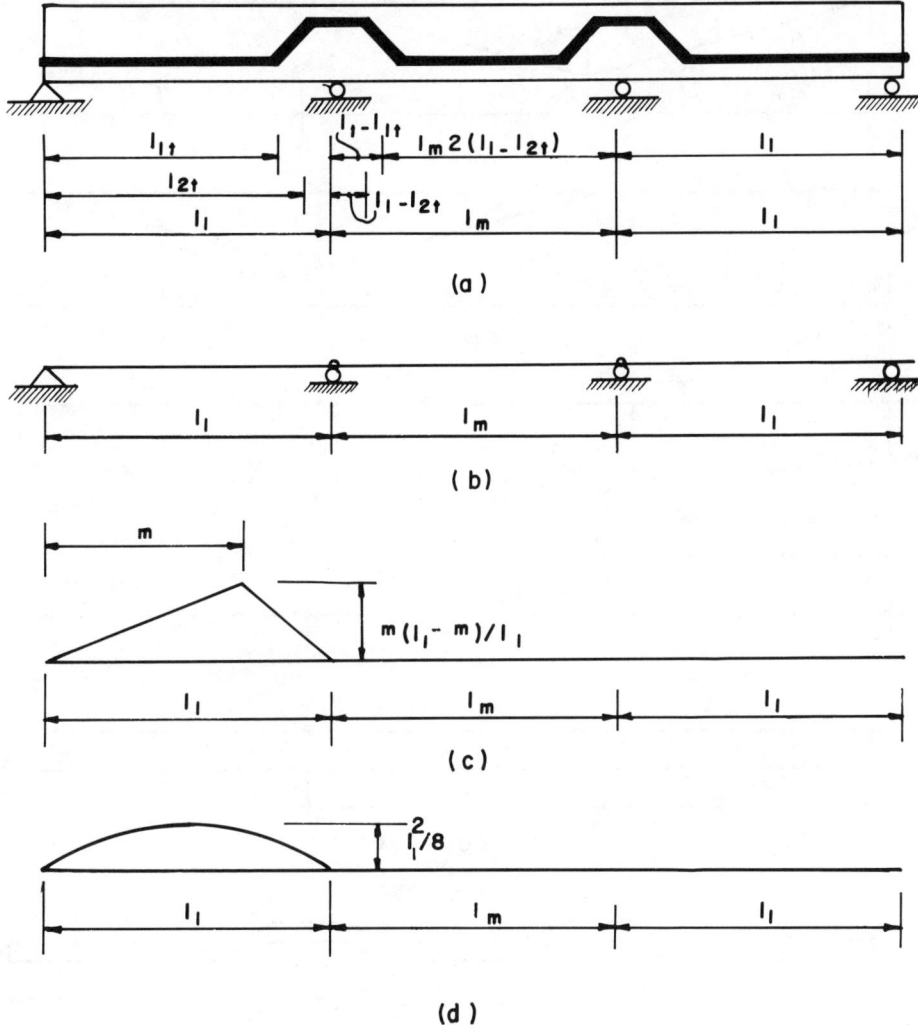

Fig. 6.16 Tendon placement and bending moment diagrams. (a) The three-span girder pre-stressed by tendon. (b) The released structure. (c) A typical bending moment diagram due to a concentrated load at m. (d) A typical bending-moment diagram due to a uniformly distributed load.

Figures 6.17(a) through 6.17(c) illustrate the bending-moment diagrams due to a unit arbitrary load at the corresponding redundant. The effects of a unit pre-stressing axial force along the entire girder are shown in Figs. 6.17(d) and 6.17(e).

For the purpose of the analysis of deformation coefficients (d_{ij} and D_{iL}), the general equations for bending moments and axial forces at any section, x, over the entire length of the girder can be written as the set of equations (6.41) through (6.43).

Equations (6.41a) and (6.41b) indicate the moments at any section along the girder caused by a unit negative moment at the ith internal support, that is, $i = 1, ns - 1$.

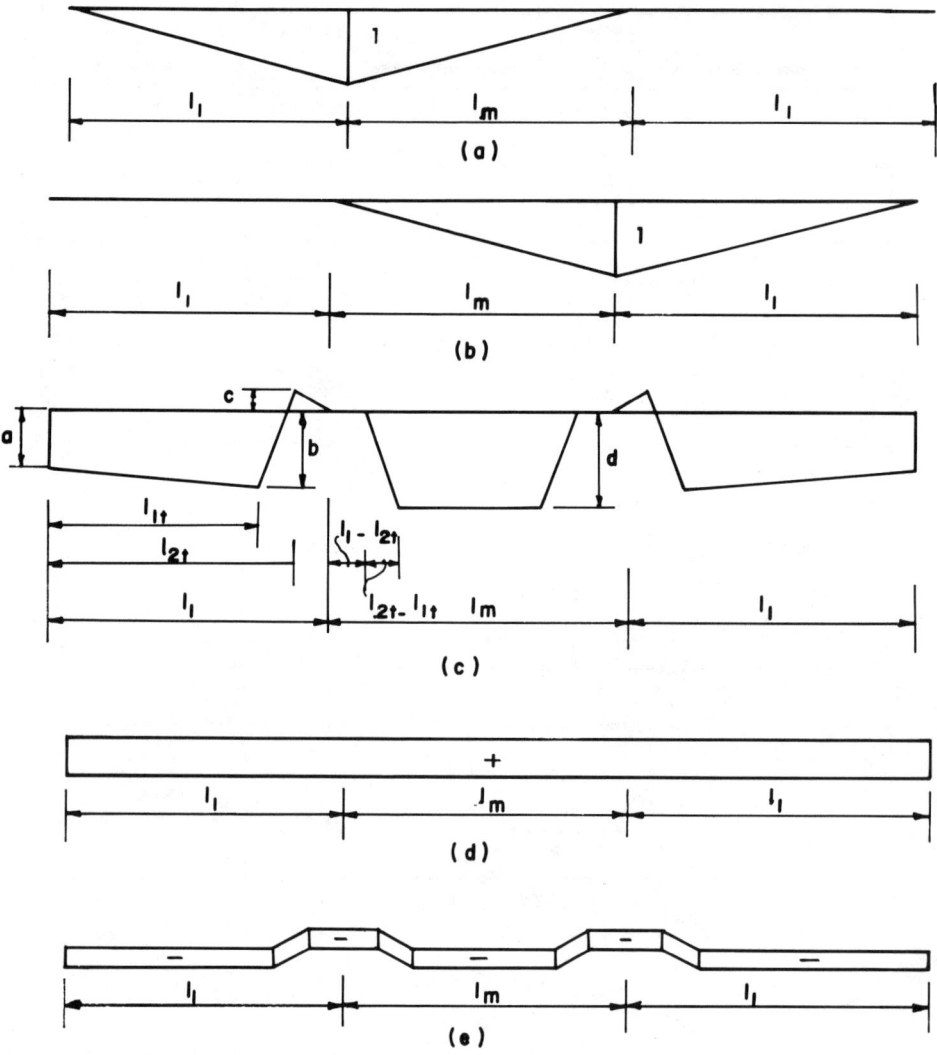

Fig. 6.17 The bending moment diagrams due to (a) a unit moment at the intermediate support 1; (b) a unit moment at support 2; (c) a unit prestressing force along the tendon and the axial force diagram due to: (d) a unit force along the girder; (e) a unit axial force along the tendon.

$$M_i = -\frac{x}{l_i} \qquad \text{for the interval} \quad 0 \geq x \geq l_i \qquad (6.41a)$$

$$M_i = -\left(1 - \frac{x}{l_{i+1}}\right) \quad \text{for the interval} \quad 0 \geq x \geq l_{i+1} \qquad (6.41b)$$

Equations (6.42a) through (6.42h) illustrate moments for various intervals along the girder corresponding to a unit prestressing axial force applied by the tendons. M_1 is the equation for the moment distribution along the first span and the last one ($M_{ns-1P=M^1}$), whereas M_k indicates the moment equations for any span k. The length of the straight tendons over the internal supports ($l_1 - l_{2t}$) are equal for all spans (Fig. 6.16a)

$$M_1 = -\left(a + \frac{b - a}{l_{1t}} \cdot x\right) \qquad 0 \geq x \geq l_{1t} \qquad (6.42\text{a})$$

$$M_1 = -\left[b - \frac{b + c}{l_{2t} - l_{1t}}(x - l_{1t})\right] \qquad l_{1t} \geq x \geq l_1 \qquad (6.42\text{b})$$

$$M_1 = -\frac{c}{l_1 - l_{2t}}(x - l_1) \qquad l_{2t} \geq x \geq l_1 \qquad (6.42\text{c})$$

$$M_k = 0 \qquad 0 \geq x \geq l_1 - l_{2t} \qquad (6.42\text{d})$$

$$M_k = -\frac{d}{l_{2t} - l_t}[x - (l_1 - l_{2t})] \qquad l_1 - l_{2t} \geq x \geq l_1 - l_{2t} \qquad (6.42\text{e})$$

$$M_k = -d \qquad l_1 - l_{1t} \geq x \geq l_k - (l_1 - l_{1t}) \qquad (6.42\text{f})$$

$$M_k = -\frac{d}{l_{1t} - l_{2t}}(x - l_k + l_1 - l_{2t}) \qquad l_k - (l_1 - L_{1t}) \leq x \leq l_k \\ - (l_1 - l_{2t}) \qquad (6.42\text{g})$$

$$M_k = 0 \qquad l_k - (l_1 - l_{2t}) \leq x \leq l_k \qquad (6.42\text{h})$$

in which $k = 2, ns - 1$; $a = {}_{1t}$; $b = e_{1t} + e_{1c}(l_{1t}/l_1)$; and $d = e_{kt} + e_{kc}$.

The moment equation (6.43a) demonstrates the change in moment over span j due to a uniformly distributed load of unit intensity. Similarly, equations (6.43b) and (6.43c) specify the moments for span j caused by a concentrated load at distance m from the support $n - 1$. The spans are loaded such that the maximum negative moment is created at the intermediate supports.

$$M_{kL} = \frac{(l_k \cdot x - x^2)}{2} \quad \text{for the interval} \quad 0 \geq x \geq l_k \qquad (6.43\text{a})$$

$$M_{kL} = \frac{(l_k - m) \cdot x}{l} \quad \text{for the interval} \quad 0 \geq x \geq m \qquad (6.43\text{b})$$

$$M_{kL} = \frac{m \cdot (l_k - x)}{l_k} \quad \text{for the interval} \quad m \geq x \geq l_k \qquad (6.43\text{c})$$

for all values of $k = 1, ns$.

The simultaneous equations (6.39) in a matrix form are shown as equation (6.44).

$$
\begin{bmatrix}
d_{11} & d_{12} & \cdot & \cdot & d_{1j} \\
d_{12} & d_{22} & \cdot & \cdot & d_{2j} \\
\cdot & \cdot & \cdot & & \cdot \\
d_{il} & d_{i2} & \cdot & \cdot & d_{ij}
\end{bmatrix}
*
\begin{Bmatrix}
M_1 \\
M_2 \\
\cdot \\
X
\end{Bmatrix}
=
\begin{Bmatrix}
D_{1L} \\
D_{2L} \\
- \quad \cdot \\
D_{iL}
\end{Bmatrix}
\qquad (6.44)
$$

Equation (6.44) can be written in simplified form as follows:

$$
[d_{ij}] \cdot \{M_{ns-1}, X\} = \{D_{iL}\} \qquad i = 1, ns \quad j = 1, ns \qquad (6.45)
$$

$$
\{M_{ns-1}, X\} = [d_{ij}]^{-1} \cdot \{D_{iL}\} \qquad (6.46)
$$

where $i = 1, ns - 1$; $j = 1, ns - 1$; $k = ns$; and $[d_{ij}]^{-1}$ is the inverse flexibility matrix.

The variable ns represents the number of spans, $ns - 1$ indicates the number of intermediate supports or negative bending moment at supports, and ns corresponds to the incremental prestressing force. The subscript L relates to the deformations caused by the applied service loads.

When the unknown redundants, the bending moments at the intermediate supports, and the incremental prestressing force are calculated, the bending moment, the shear, and the axial force along the girder can be analyzed by using the equations of static equilibrium.

4. Summary

A criterion for the analysis of continuous-span steel girders prestressed by high-strength steel tendons has been presented. The proposed analysis procedure is generalized for multiple-span girders having various span lengths. The selected tendon configuration may be changed along the girder while the derived equations remain unchanged except for the parameters related to the tendons' placement. The analysis is based on the application of unit load systems with uniformly distributed or concentrated. Therefore the maximum values for negative bending moments and the incremental prestressed force corresponding to the various system of loadings which produce these maxima can be evaluated.

5. Appendix I

The flexibility coefficients for continuous-span girders prestressed by tendons of high-strength steel are presented here. The equations are generalized for multiple-span girders. The term ns indicates the number of spans and the subscript m defines the respective properties of the mth intermediate span.

$$
d_{11} = \frac{l_1}{3 \cdot E \cdot I_1} + \frac{l_m}{3 \cdot E \cdot I_m}
$$

$$d_{22} = \frac{2 \cdot l_m}{3 \cdot E \cdot I_m}$$

$$d_{kk} = d_{22} \qquad \text{where } k = 2, \, ns - 2 \text{ and for } ns \geq 4$$

$$d_{jj} = d_{11} \qquad \text{where } j = ns - 1$$

$$d_{12} = \frac{l_m}{6 \cdot E \cdot I_m}$$

$$d_{13} = 0$$

$$d_{1j} = 0 \qquad \text{where } j = 3, \, ns - 1$$

$$d_{ij} = d_{ji} \qquad \text{for all values of } i = 1, \, ns \text{ and } j = 1, \, ns$$

$$d_{1,ns} = \frac{1}{6 \cdot E \cdot I_1 \cdot l_1} \left[al_{1t}^2 + bl_{2t}^2 + bl_{1t}l_{2t} \right.$$

$$+ c(l_{1t}^2 - l_1^2 + l_{1t}l_{2t} - l_1 l_{2t}) \big]$$

$$+ \frac{d}{2 \cdot E \cdot I_m} (l_m - 2l_1 + l_{1t} + L_{2t})$$

$$d_{j,ns} = d_{1,ns} \qquad \text{for } j = ns - 1$$

$$d_{k,ns} = \frac{d}{E \cdot I_m} (l_m - 2l_1 + l_{1t} + l_{2t})$$

$$\text{where } k = 2, \, ns - 2 \text{ for all values of } ns \geq 4$$

$$d_{ns,ns} = \frac{2}{3 \cdot E \cdot I_1} \left[l_{1t}(a^2 + b^2 + ab) + (l_{2t} - l_{1t})(b^2 + c^2 - bc) \right.$$

$$+ c^2(l_1 - l_{2t}) \big] + \frac{d^2}{E \cdot I_m} \left[\frac{2}{3} (l_{2t} - l_{1t}) + l_m - 2(l_1 - l_{1t}) \right]$$

$$+ 2l_1 \left(\frac{1}{EA_1} + \frac{1}{E_t A_t} \right) + l_m(ns - 2) \left(\frac{1}{EA_m} + \frac{1}{E_t A_t} \right)$$

$$+ \frac{2(ns - 1)}{E_t A_t} \left(\frac{1}{\cos a^3} - 1 \right)$$

The flexibility coefficients d_{ij} are the deformations at section j due to a unit load at section i. The subscript ns is equal to the number of spans and at the same time refers to the incremental prestressing for X as the last redundant. The integers i through k are chosen as variable subscripts increasing by step one.

The deformations at the corresponding redundants i due to the actual loadings including the service loads and the prestressing force are defined as D_{iL}. The subscript U stands for uniformly distributed loads, and C refers to a concentrated load.

$$D_{1LU} = -\left(\frac{l_1^3}{24 \cdot E \cdot I_1} + \frac{l_m^3}{24 \cdot E \cdot I_m}\right)$$

$$D_{jLU} = -\left(\frac{2 \cdot l_m^3}{24 \cdot E \cdot I_m}\right) \quad \text{for } j = 2, \, ns - 2 \text{ and } ns \geq 4$$

$$D_{1LC} = -\frac{m(l_1^2 - m^2)}{6 \cdot E \cdot I_1 \cdot l_1} - \frac{n(l_m - n)(2 \cdot l_m - n)}{6 \cdot E \cdot I_m \cdot l_m}$$

$$D_{jLU} = -\frac{1}{6 \cdot E \cdot I_m \cdot l_m} \cdot \left[n(l_m^2 - n^2) + n(l_m - n)(2 \cdot l_m - n)\right]$$

$$\text{where } j = 2, \, ns - 2 \text{ for } ns \geq 4$$

$$D_{jLU} = D_{1LU} \quad \text{for } j = ns - 1$$

$$D_{jLC} = -\frac{n(l_m^2 - n^2)}{6 \cdot E \cdot I_m \cdot l_m} - \frac{m(l_1 - m)(2 \cdot l_1 - m)}{6 \cdot E \cdot I \cdot l_1} \quad \text{for } j = ns - 1$$

$$D_{iL} = D_{iLU} + D_{iLC} \quad \text{for } i = 1, \, ns$$

D_{iL} is the sum of deformations due to concentrated and distributed loadings.

$$D_{ns,LU1} = \frac{1}{2 \cdot E \cdot I}\left\{\left[\frac{l_1 \cdot l_{1t}^2}{6}(2 \cdot b + a) - \frac{l_{1t}^3}{12}(3 \cdot b + a)\right]\frac{l_{1t} + l_{2t}}{2}\right.$$

$$\cdot \left[b\left(l_1 \cdot l_{2t} + \frac{l_{1t}^2 + l_{2t}^2}{2}\right) + c\left(l_1 \cdot l_{1t} + \frac{l_{1t}^2 + l_{2t}^2}{2}\right)\right]$$

$$+ \left(\frac{l_{1t} + l_{2t}^2 + l_1 \cdot l_2}{3}\right)\left[-b(l + l_2) - c(l + l_1)\right]$$

$$\left. \cdot \, c \cdot \left[\frac{2 \cdot l_1}{3}(l_1^2 + l_{2t}^2 + l_1 \cdot l_{2t}) - \left(\frac{l_1 + l_{2t}}{4}\right)(3 \cdot l_1^2 + l_{2t}^2)\right]\right\}$$

where D_{nsLU1} corresponds to deformations in tendons due to a uniformly distributed load applied on the side spans.

$$D_{ns,LC} = -\frac{1}{E \cdot I}\left[m \cdot a\left(-\frac{m}{2} + \frac{m^2}{6 \cdot l_{1t}} + \frac{l_{1t}}{2} - \frac{l_{1t}^2}{6 \cdot l_1}\right)\right.$$

$$+ \, m \cdot b\left(-\frac{m^2}{6 \cdot l_{1t}} + \frac{l_{2t}}{2} - \frac{l_{2t}^2 + l_{1t} \cdot l_{2t}}{6 \cdot l_1}\right)$$

$$\left. + \, m \cdot c\left(\frac{l_{2t}}{6} + \frac{l_{1t}}{2} - \frac{l_{1t}^2 + l_{1t} \cdot l_{2t}}{6 \cdot l_1} - \frac{l_1}{3}\right)\right]$$

Similarly, D_{nsLC1} defines the deformations in tendons due to a concentrated load at distance m from the extreme left and right supports.

$$D_{ns,LUm} = -\frac{d}{12 \cdot EI_m} \Big\{ 4(l_m - l_1 - l_{2t}) \big[(l_1 - l_{1t})^2 + (l_1 - l_{2t})^2$$
$$+ (l_1 - l_{1t})(l_1 - l_{2t}) \big] - 3(2l_1 - l_{1t} - l_{2t}) \big[(l_1 - l_{1t})^2$$
$$+ (l_1 - l_{2t})^2 + 2l_m(l_1 - l_{2t}) \big] + (l_m - 2l_1 + 2l_{2t})$$
$$\cdot \big[l_m^2 + 2l_m(l_1 - l_{1t}) - 2(l_1 - l_{1t})^2 \big] \Big\}$$

The coefficient $D_{ns,LUm}$ represents the deformation of the mth intermediate span under the application of a unit uniformly distributed load.

$$D_{ns,LCm} = -\frac{d}{6EI_m} \big[(l_{2t} - l_{1t})(3l_1 - 2l_{1t} - l_{2t}) + 3n(l_m - n) - 3(l_1 - l_{1t})^2 \big]$$

$D_{ns,LCm}$ is the corresponding deformation of the prestressing tendons due to a unit concentrated load.

6.5.4 Continuous Girders Under Movable Loads. Prestressing by Redistribution of Bending Moments

1. Introduction

The regulation of forces in continuous girders permits the redistribution of bending moments. Depending on the lengths of the spans of continuous girders and loading, the maximum bending moments may originate at the end as well as at the intermediate spans of a continuous girder. Consider the end span $l_1 = al_n$ and intermediate span l_n which are loaded by the dead and live loads (Fig. 6.18).[8]

For the section ul_1 at the end span with maximum moment of bending moment at prestressed girder, we have

$$M_K + uM_K = M_u + M_a$$

But from the ratio

$$\frac{M_a}{ul^1} = \frac{M_1 - M_1^1}{1}$$

we have

$$M_a = u(M_1 - M_1^1)$$

Therefore,

$$M_K + uM_K = M_u + u(M_1 - M_1^1)$$

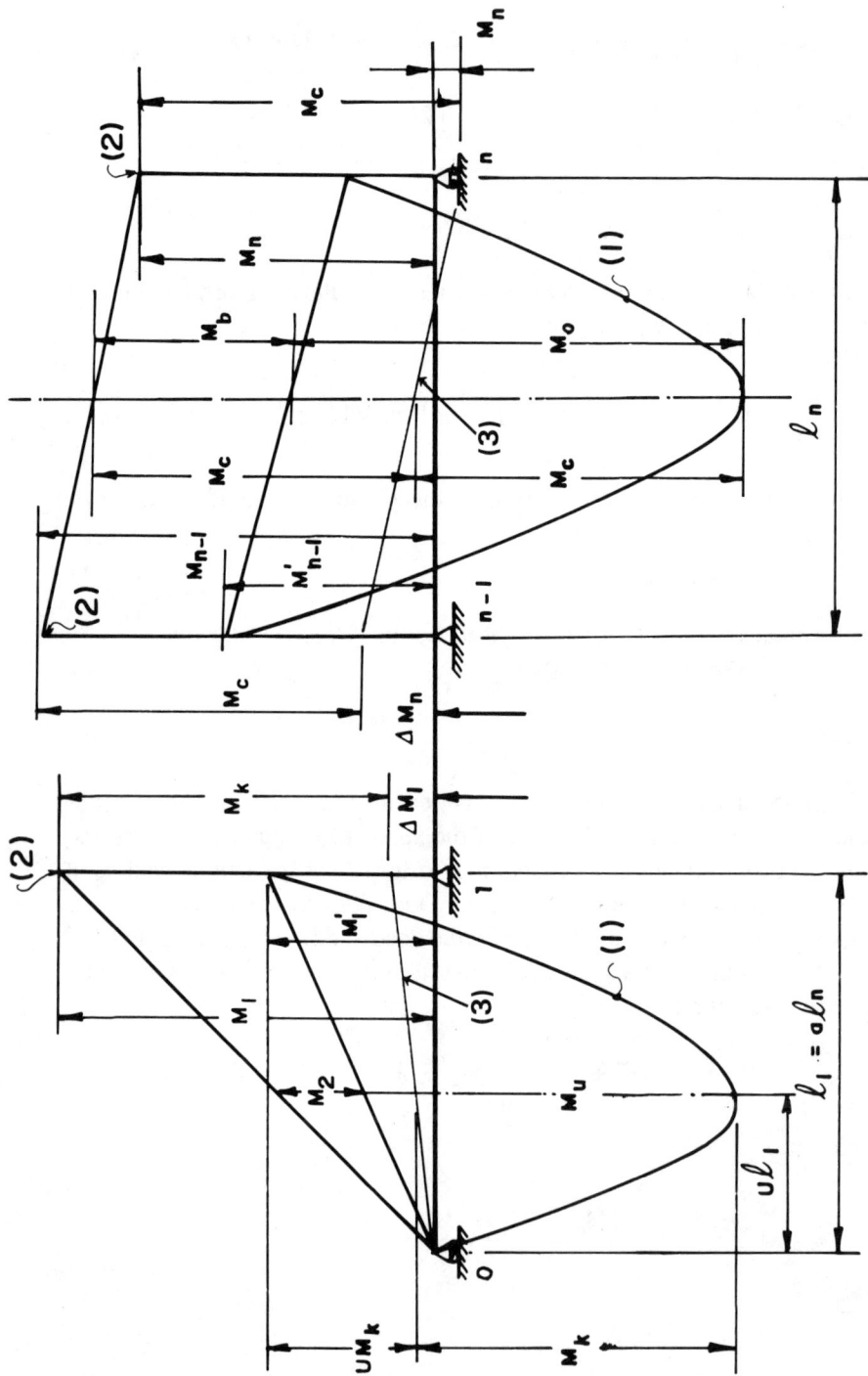

Fig. 6.18   Determination of bending moment at the end and intermediate spans of a continuous, prestressed girder. 1, Diagram of moments due to loading of influence lines; 2, maximum values of support methods; 3, diagram of required additional moments ΔM under prestressing.

from which the calculated value of the bending moment M_K at the end span for continuous prestressed girder is

$$M_K = \frac{1}{1 + u} \left[M_u + u(M_1 - M'_1) \right] \qquad (6.47)$$

For a section having maximum span moment under external load at intermediate span l_n, we have

$$2 M_c = M_0 + M_b = M_0 + 0.5 \left[(M_{n-1} - M'_{n-1}) + (M_n - M'_n) \right]$$

from which calculated value of bending moment M_c of intermediate span of continuous girder after prestressing is

$$M_c = 0.5 \left[M_0 + 0.5(M_{n-1} + M_n - M'_{n-1} - M'_n) \right] \qquad (6.48)$$

M_u, M_0 = bending moments under external loadings at section ul_1 of end span and maximum moment under external loading at the n span, considered as separated from the continuous girder, respectively

M_n = maximum value of bending moment at r support of continuous girder without prestressing

M'_n = bending moment at n support of continuous girder without prestressing at maximum moment at n span

Additional moment ΔM_n, which will be produced at n support of the girder during prestressing, will be determined by the formula

$$\Delta M_n = M - M_n = \theta M \qquad (6.49)$$

where

M = calculated bending moment at support n of prestressed girder, determined from the greater value of (6.47) or (6.48)

θ = coefficient which is defined by the ratio; $\theta = \Delta M / M$

2. Distributed Live Load

When the girder is loaded along its whole length by dead load g and along separate spans by the distributed live load entering into formulas (6.47) and (6.48), the values of the moments will be

$$M_u = (p' + g) \frac{ul_1^2}{2} - (p' + g) \frac{(ul_1)^2}{2}$$

$$= (p' + g) \frac{u}{2} (1 - u) a^2 l_n^2 \qquad (6.50)$$

$$M_n = (\alpha_{ng} g + \alpha_{np} p) \, l_n^2$$

$$M'_n = (\alpha_{ng} g + \alpha'_{np} p') l_n^2$$

where

p, p' = intensities of the live loads for the influence lines of support and span moments, respectively

α_{ng}, α_{np} = numerical coefficients in the formulas for bending moments at n support

α'_{np} = under dead load g and live loads p and p', respectively

By introducing (6.50) into (6.47) and (6.48), we obtain values of bending moments at the end and mid-spans under live loads, namely:

$$M_k = \frac{ua^2}{1 + u}\left[\frac{1 - u}{2} + x(\alpha_{1p} - \alpha'_{1p})\right]q' l_n^2 \tag{6.51}$$

$$M_c = \left\{0.0625 + \frac{x}{4}\left[(\alpha_{n-1,p} + \alpha_{np})1 - \alpha'_{n-1,p} - \alpha'_{np}\right]\right\}q' l_n^2 \tag{6.52}$$

where

$$x = \frac{p'}{q'} \qquad e = \frac{p}{p'} \qquad q = p' + g \tag{6.53}$$

The distance ul_1 from the end support to the section, at which the span moment of the prestressed girder will be maximum, may be determined from the condition of equivalence to zero the transverse force at this section, namely

$$Q = \frac{q'al_n}{2} - q'ual_n - \frac{M_k}{al_n} = 0$$

or   (6.54)

$$q'(1 - 2u)a^2 l_n^2 - 2M_k = 0$$

By introducing (6.51) into (6.54), after transformation, we obtain $u^2 + 2ud - 1 = 0$ and

$$u = \sqrt{d^2 + 1} - d \tag{6.55}$$

where

$$d = 1 + x(\alpha_{1_p} - \alpha'_{1_p}) \tag{6.56}$$

It follows from (6.51) and (6.52) that the minimal values of bending moments at prestressed girders, irrespective of the number of spans, is determined at $x = 0$, after the greater of the values

$$M_K = 0.0858\,ga^2l^2 \qquad M_c = 0.0625\,gl^2 \tag{6.57}$$

3. Concentrated Loads

When the girder is loaded by the system of movable concentrated loads, the maximum span moments at the first n spans (Fig. 6.18) will be

$$M_{\text{span}1} = M_u - uM_1'$$
$$M_{\text{span}n} = M_0 - 0.5(M_{n-1}' + M_n') \tag{6.58}$$

By introducing (6.58) into (6.47) and (6.48), we obtain the moments

$$M_K = \frac{1}{1+u}(M_{sp.1} + uM_1)$$

$$M_c = \frac{1}{1+u}[M_{sp.n} + 0.5(M_{n-1} + M_n)] \tag{6.59}$$

where any n moment is determined after formula

$$M_n = \alpha_{ng} g l_n^2 + \Sigma P_y \tag{6.60}$$

and the value u, determined from the condition $Q = 0$, is possible to use as equal to 0.4.

4. Vertical Displacements of Rigid Supports of the Girders During Prestressing

The values of the required vertical displacement Δ of separate supports during prestressing of the girders are determined from the conditions arising at the supports, namely, considering increase of moments

$$\Delta M_{n-1} \qquad \Delta M_n \ldots$$

$$\Delta M_{n-1} l_n + 2\Delta M_n(l_n + l_{n+1}) + \Delta M_{n+1} l_n \tag{6.61}$$

$$= -6E_a I\left[\frac{\Delta_{n-1}}{l_n} - \Delta_n\left(\frac{1}{l_n} + \frac{1}{l_{n+1}}\right) + \frac{\Delta_{n+1}}{l_{n+1}}\right]$$

where E_a and I are the module of elasticity of metal and moment of inertia of the girder cross section, respectively. At the girders that are symmetrical with respect to their mid-span, and under live load, the displacements will be also symmetrical (Fig. 6.19). At the two- and three-span symmetrical girders there will be one unknown displacement, and at the four- and five- span girders there will be two unknown displacements.

By introducing into formula (6.61) all given increases of moments ΔM and

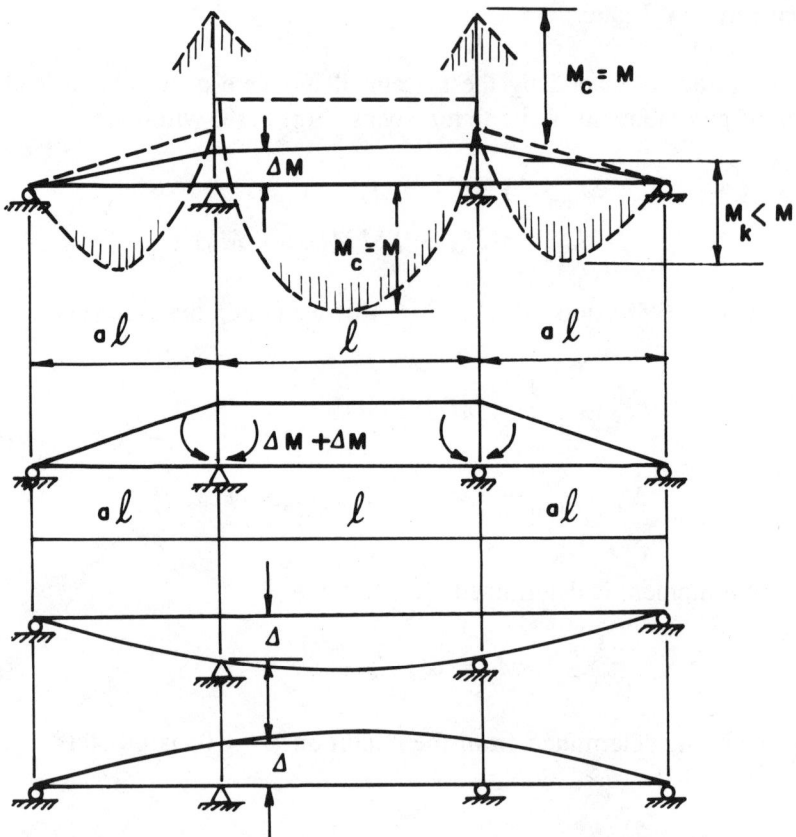

Fig. 6.19 Maximum values of the envelope of moments for a three-span continuous girder, diagram of additional moments ΔM and displacements Δ at the prestressed girder.

spans, and using for nondisplaced supports $\Delta = 0$, the required displacement may be expressed as follows:

$$\Delta = \frac{\epsilon \Delta M l^2}{E_a I} \tag{6.62}$$

5. Numerical Example

Design an interior prestressed girder for a two-lane highway bridge having two continuous spans, each one 100 ft, under H20-44 truck loading. Girders are of A36 steel. Cross section of this bridge is similar as used for Numerical Example 5.3, as are dead load and superimposed dead load.

   a. Dead Load + Superimposed Dead Load (Fig. 6.20)

$$g_{\text{tot}} = g_{DL} + g_{SDL} = 0.909 + 0.338 = 1.247 \text{ k/ft}$$

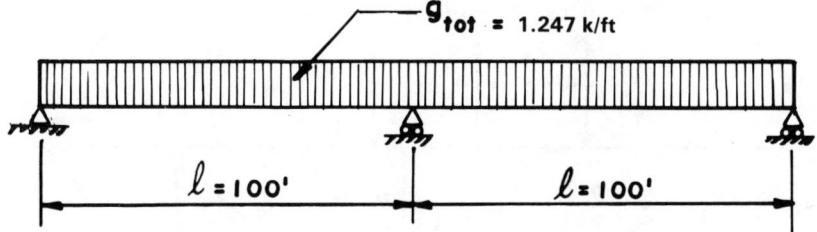

Fig. 6.20 Continuous two-span girder under total dead load g_{tot}.

b. Maximum Span Bending Moment

$$M_{DL} = 0.07gl^2 = 0.07 \times 1.247 \times 100^2 = 872.9 \text{ kft}$$

By loading the influence line for the span moment by truck (Fig. 6.21), we have

$$M_{LL} = \left[4 \times 0.135 + 16 \left(0.206 + 0.145 \right) \right] \times 100 = 615.6 \text{ kft}$$

c. Maximum Support Bending Moment

$$M_{DL} = -\frac{gl^2}{8} = -\frac{1.247 \times 100^2}{8} = -1558.75 \text{ kft}$$

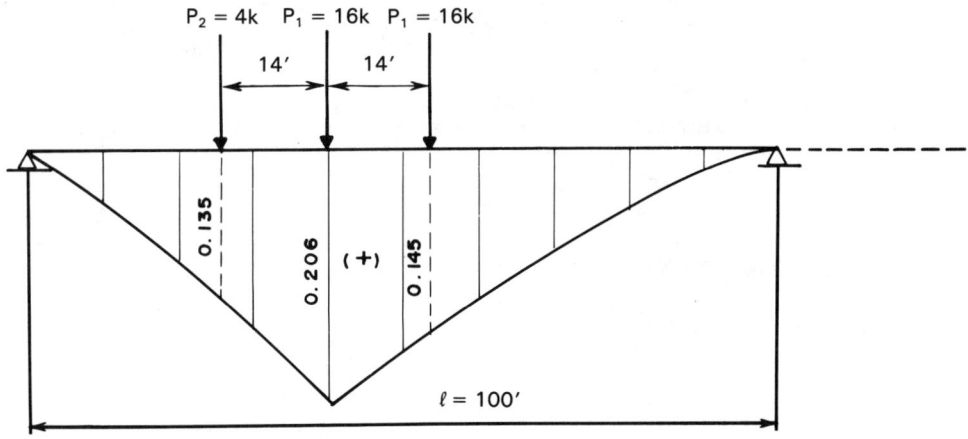

M = 4 x 0.135 + 16 (0.206 + 0.145) = 0.54 + 5.616 = 6.156 kft

M_{Span} = 0.351 $P_1\ell$ + 0.135 $P_2\ell$

Fig. 6.21 Influence line for maximum span moment.

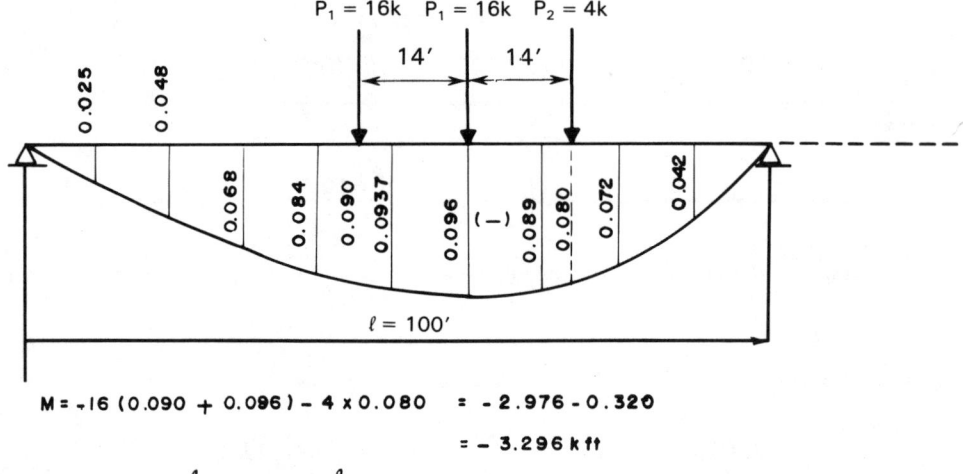

$$M = -16(0.090 + 0.096) - 4 \times 0.080 \quad = -2.976 - 0.320$$
$$= -3.296 \text{ kft}$$
$$M_1 = 0.186 \; P_1 \ell + 0.080 \; P_2 \ell$$

Fig. 6.22 Influence line for maximum support moment.

By loading the influence line for the support moment by truck (Fig. 6.22), we have

$$M_{\text{sup}} = \left[-4 \times 0.080 - 16(0.090 + 0.096) \right] \times 100 = -329.6 \text{ kft}$$

d. Coefficient for Live Load

$$\text{Fraction: } \frac{S}{5.5} = \frac{7.0}{5.5} = 1.27$$

$$\text{Impact: } I = \frac{50}{L + 125} = \frac{50}{100 + 125} = 0.22$$

$$I = 1.22$$

e. Total Span and Support Moments

$$M_{sp} = M_{DL} + M_{LL} \times 1.27 \times 1.22$$

$$= 872.9 + 615.6 \times 1.27 \times 1.22 = 872.9 + 953.8$$

$$= 1826.7 \text{ kft}$$

$$M_1 = -M_{DL} - M_{LL} \times 1.27 \times 1.22$$

$$= -1558.75 - 329.6 \times 1.27 \times 1.22 = -1558.75 - 510.7$$

$$= -2069.45 \text{ kft}$$

f. Using the First Formula (6.59)

We have

$$M_K = \frac{M_{sp} + uM_1}{1 + u} = \frac{1826.9 + 0.4 \times 2069.45}{1 + 0.4}$$

$$= \frac{2654.68}{1.4} = 1896.2 \text{ kft}$$

and $\Delta M = M_1 - M_K$, considering the absolute value of M_1

$$\Delta M = 2069.45 - 1896.2 = 173.25 \text{ kft}$$

g. Required Moment of Inertia of the Girder

$$S = \frac{M}{\sigma} = \frac{1826.7 \times 10^3 \times 12}{30 \times 10^3} = 730.76 \text{ in.}^3$$

Assuming required height of the girder

$$h = \frac{L}{30} = \frac{100 \times 12}{30} = 40 \text{ in.}$$

and required moment of inertia

$$I = S \times \frac{h}{2} = 730.76 \times \frac{40}{2} = 14615.2 \text{ in.}^4$$

Accepted cross section of the girder (Fig. 6.23)

$$I = 2 \times \frac{19 \times 1^3}{12} + 2 \times 19 \times 1 \times 19.5^2 + \frac{0.5 \times 28^3}{12}$$

$$= 3.16 + 14,449.5 + 914.67 = 15,367.33 \text{ in.}^4$$

$$= 15,367.33 > 14,615 \text{ in.}^4$$

h. Required Vertical Displacement of the Middle Support, after Formula (6.62)

$$\Delta = \frac{\Delta M l^2}{3EI} = \frac{173.25 \times 10^3 \times 12 \times 10^4 \times 12^2}{3 \times 30 \times 10^6 \times 15,367.33}$$

$$= \frac{29,9376}{138,305.97} = 2.16 \text{ in.}$$

Use $\Delta = 2 \ 1/4$ in.

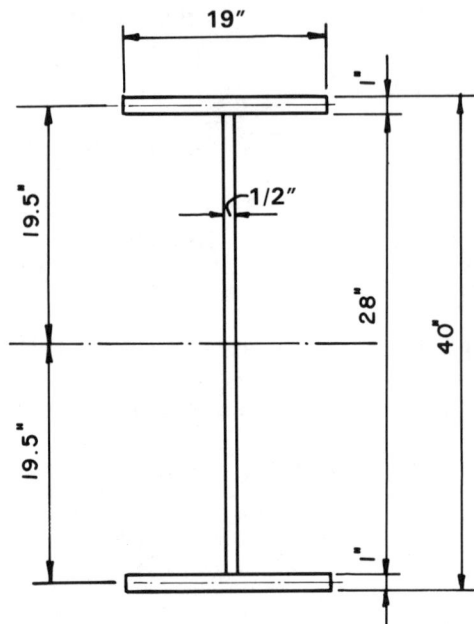

Fig. 6.23   Cross section of the girder.

6.6 METHODS OF DECK PRESTRESSING

The prestressing of continuous-girder decks is used to control cracking of the concrete by reducing the tensile stresses to an acceptable level or to avoid them completely. Methods of prestressing may be divided into the following systems:

1. Prestressing by the introduction of longitudinal forces by means of tendons (Fig. 6.24).[9,10] This may be achieved by prestressing tendons placed in the concrete slab or by truss action. The stressing of the tendons is done in most cases from one end, where recesses, cutouts, or anchor brackets are provided for the positioning of the stressing equipment.

2. Introduction of longitudinal forces, as shown in Fig. 6.25. The concrete slab is located in regions of the girder which are subject to tension from live loads. The concrete flange plate is connected at both ends with the steel beams and prestressed by means of jacks. The center gap is then packed, and after removal of the jacks, the cutouts are filled with concrete. Finally, the concrete slab is connected continuously with the steel flange.

3. Prestressing by the introduction of temporary supports, or internal redistribution of loads (Fig. 6.26). In this case two separate means of achieving the load distribution have been combined. Temporary supports engage the load-carrying capacity of the full composite cross section.

4. Prestressing by cambering of continuous composite girders, discussed by Roik.[11,12] By this method the steel girder is raised (Fig. 6.27), in which condition the slab is concreted. After the concrete has hardened, the composite girders are lowered to the position shown in Fig. 6.27 (b), which results in the compression of the slab. The magnitude of the precompression is limited by the allowable compressive stress in the concrete.

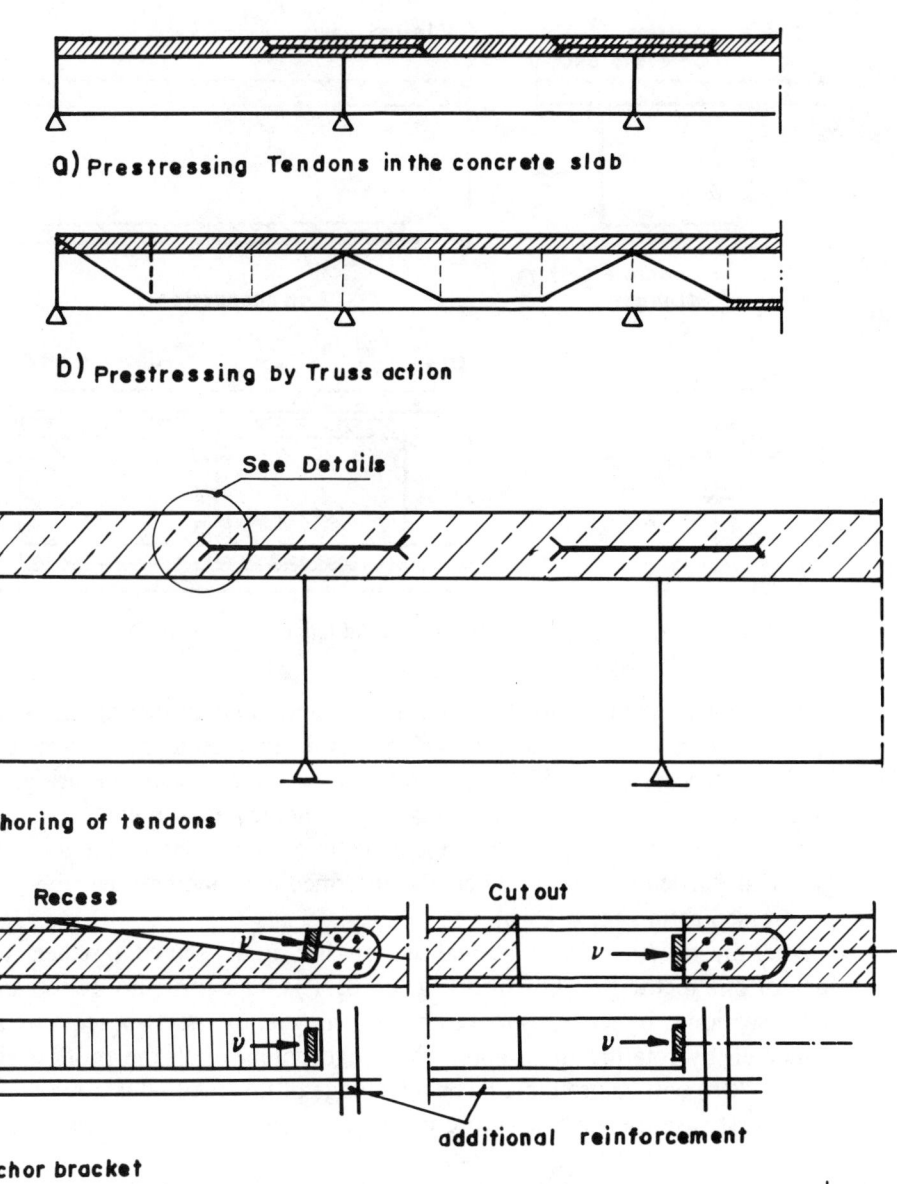

a) Prestressing Tendons in the concrete slab

b) Prestressing by Truss action

See Details

Achoring of tendons

Recess   Cut out

additional reinforcement

Anchor bracket

c) Stressing equipment is provided by :

(1) Recesses,   (2) Cut out , (3) Anchorages.

Fig. 6.24   Prestressing by the introduction of longitudinal forces.

The method of prestressing by cambering is, however, only economical when the bridge is of a certain total length. With bridges of greater length, the distances through which the superstructure must be lowered to achieve effective precompression are so great that the operation becomes too impractical.

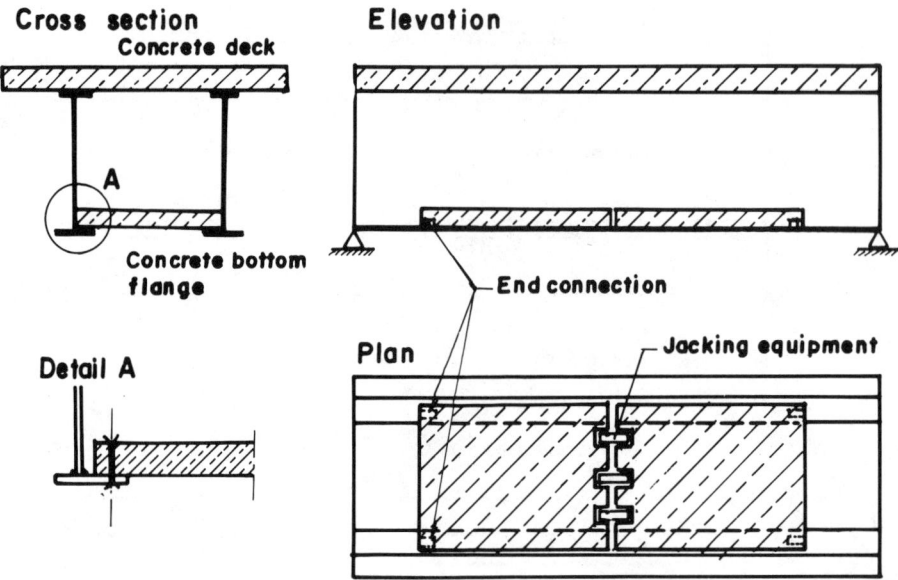

Fig. 6.25   Introduction of longitudinal forces.

As a typical example Roik considered a continuous bridge having a length of
8 × 164 ft = 1312 ft (400 m), which was designated to be prestressed by the
lowering process (Fig. 6.28(a)).[13] In this case the center of the bridge must be
lowered about 19.7 ft (6 m), and this may be achieved with extensive temporary
supports. However, this can be done economically by placing hinges in the steel
girder at particular points. Then the intermediate members can be separately
prestressed as in Fig. 6.28(b).

After prestressing, the individual sections are rigidly joined together. The basic
idea of this method is shown in Fig. 6.28. The superstructure is separated into
three sections, linked to one another by hinges. If the two sections are now
cambered by rotating them about the hinge points, then the profiles shown in
Fig. 6.28(b) will be produced. The advantage of this method is obvious, because
in the first case the superstructure must be lowered over all seven piers by 19.7

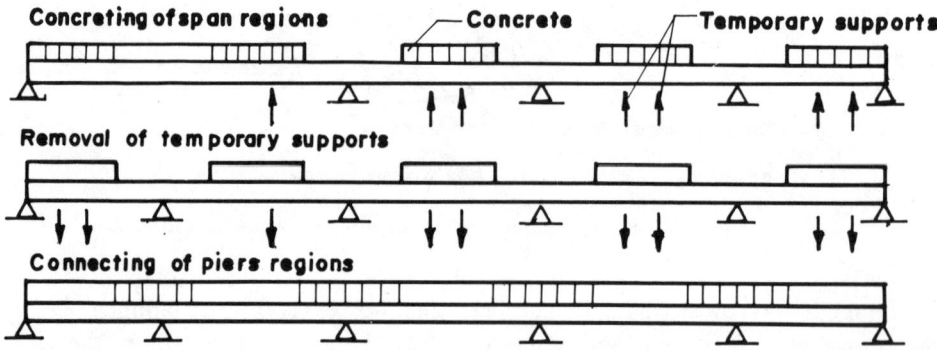

Fig. 6.26   Internal redistribution of loads.

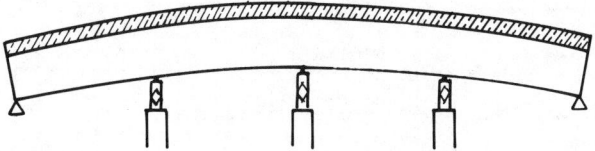

a) Concreting in raised condition

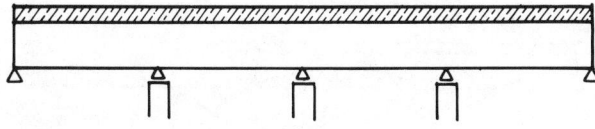

b) Lowering after concrete has matured

Fig. 6.27 Prestressing by cambering continuous spans. (Courtesy of the Prestressed Concrete Institute)

ft (6 m), whereas in the second case it is lowered over only three piers, by about 1.31 ft (0.40 m). The sequence of construction is shown in Fig. 6.29, together with the appropriate loading and deflection.

Similarly, at the hinge points, bending moments must be introduced during prestressing. This can be achieved in the following manner: In the gap in the concrete slab, situated over the hinge in the steel girder, hydraulic jacks are inserted which act upon both edges of the slab (Fig. 6.30). At the same time, the lower flanges of the steel girder are found together with splice plates to transmit the tensile forces that correspond to the compressive forces applied to the concrete slab by the jacks. This couple produces the bending moments on the cantilever ends of the composite girder. The hydraulic jacks in the concrete slab must be so arranged that they can satisfactorily transmit the desired compressive force. When the desired pressure has been attained, the jacks are halted and the gaps in the slab between the jacks are filled with concrete. When this concrete has matured, the jacks can be released and the pockets also filled with concrete as in Fig. 6.31.

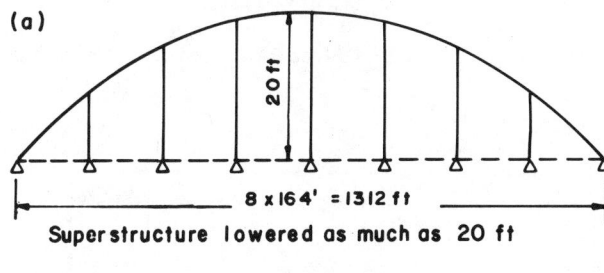

(a)

20 ft

8 x 164' = 1312 ft

Superstructure lowered as much as 20 ft

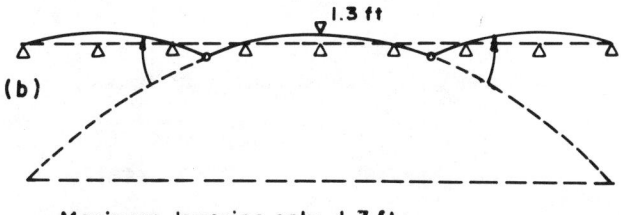

1.3 ft

(b)

Maximum lowering only 1.3 ft

Fig. 6.28 Prestressing by induced cambering. (Courtesy of the Prestressed Concrete Institute)

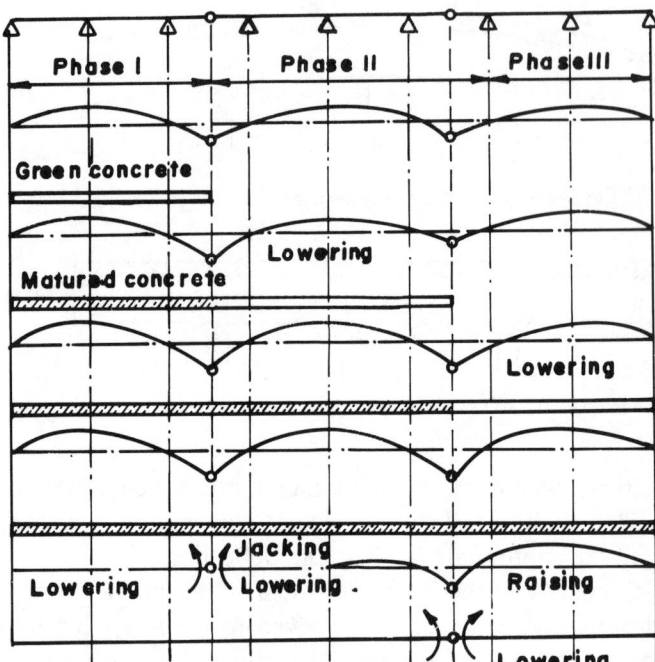

Fig. 6.29 Stages in prestressing with a temporary link system. (Courtesy of the Prestressed Concrete Institute)

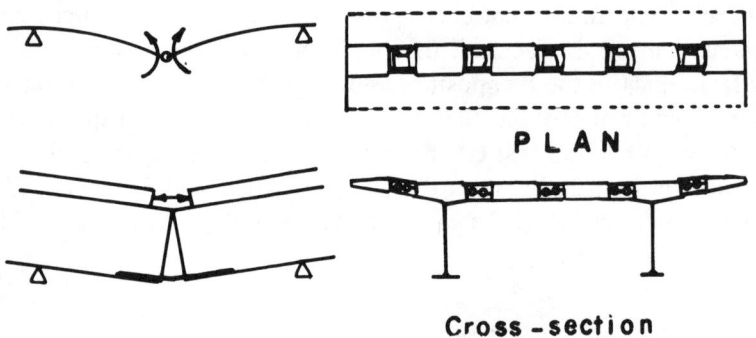

Fig. 6.30   Application of moment at hinge point. (Courtesy of the Prestressed Concrete Institute)

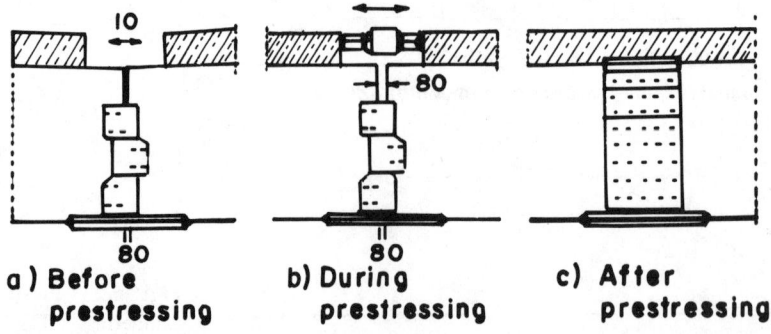

Fig. 6.31   Sequence for completion of hinge point. (Courtesy of the Prestressed Concrete Institute)

6.6.1 Precast Prestressed Concrete Bridge Deck Panels

According to the Special Report by the PCI Committee on Bridges, the use of precast panels has proven to be both economical and convenient.[14] The panels are used as a composite part of the completed deck. They replace the main bottom (positive moment) transverse deck reinforcement and also serve as a form surface for the cast-in-place concrete upper layer that contains the top of deck (negative moment) reinforcement.

A typical deck panel detail is shown in Fig. 6.32. Panel dimensions have varied, but practical ranges have been found to be: thickness, $2\frac{1}{2}$–$4\frac{1}{2}$in.; width, 4 ft or 8 ft; length, to provide a minimum of 3 in. of bearing on the beam flanges. Welded wired fabric is used for the longitudinal distribution of steel and can also serve as lifting loops. The reinforcement in the panels, placed perpendicular to the prestressing strands, is required for load transfer and strand development. Typical details are shown in Fig. 6.32.

Welded steel plate girder bridge details are shown in Fig. 6.33. Panels are cast on standard long line beds. The panel's top surface is roughened to provide mechanical bonding with the cast-in-place tapping. Shipping and erection in the field have proven to be straightforward. Proper stacking, handling, and erection procedures are extremely important to avoid cracking panels.

6.6.2 Prestressed Precast Concrete Slab, PPCS Method

A new composite girder using prestressed concrete slab, referred to as the PPCS method, was developed in 1986.[15,16] The details of the PPCS method are illustrated in Fig. 6.34, for which procedures are outlined as follows.

1. The precast concrete panels are set on the steel girders and their joints are connected with epoxy adhesive or cement mortar.

2. The precast concrete slab is prestressed in the direction of the steel girders by prestressing wires, strands, or bars.

3. The precast concrete slabs are combined with steel girders. Concrete or cement mortar is filled into the holes located at the shear connectors of steel girders.

4. A part of the prestress in prestressing wires, strands, or bars is released after the precast concrete slab is combined with the steel girders.

Thus, the resident compressive stresses can remain in the precast concrete slab and the negative bending moment can be induced in the composite girders. The former will improve the durability of the concrete slab, particularly by reducing girder depth or the cross-sectional area of the flange plates.

For the application of this method to highway bridges, cracking of the precast concrete slabs and corresponding fatigue behaviors of the composite girders under heavy vehicle traffic were investigated. Accordingly, experimental studies on five model girders were conducted systematically under repeated loads by

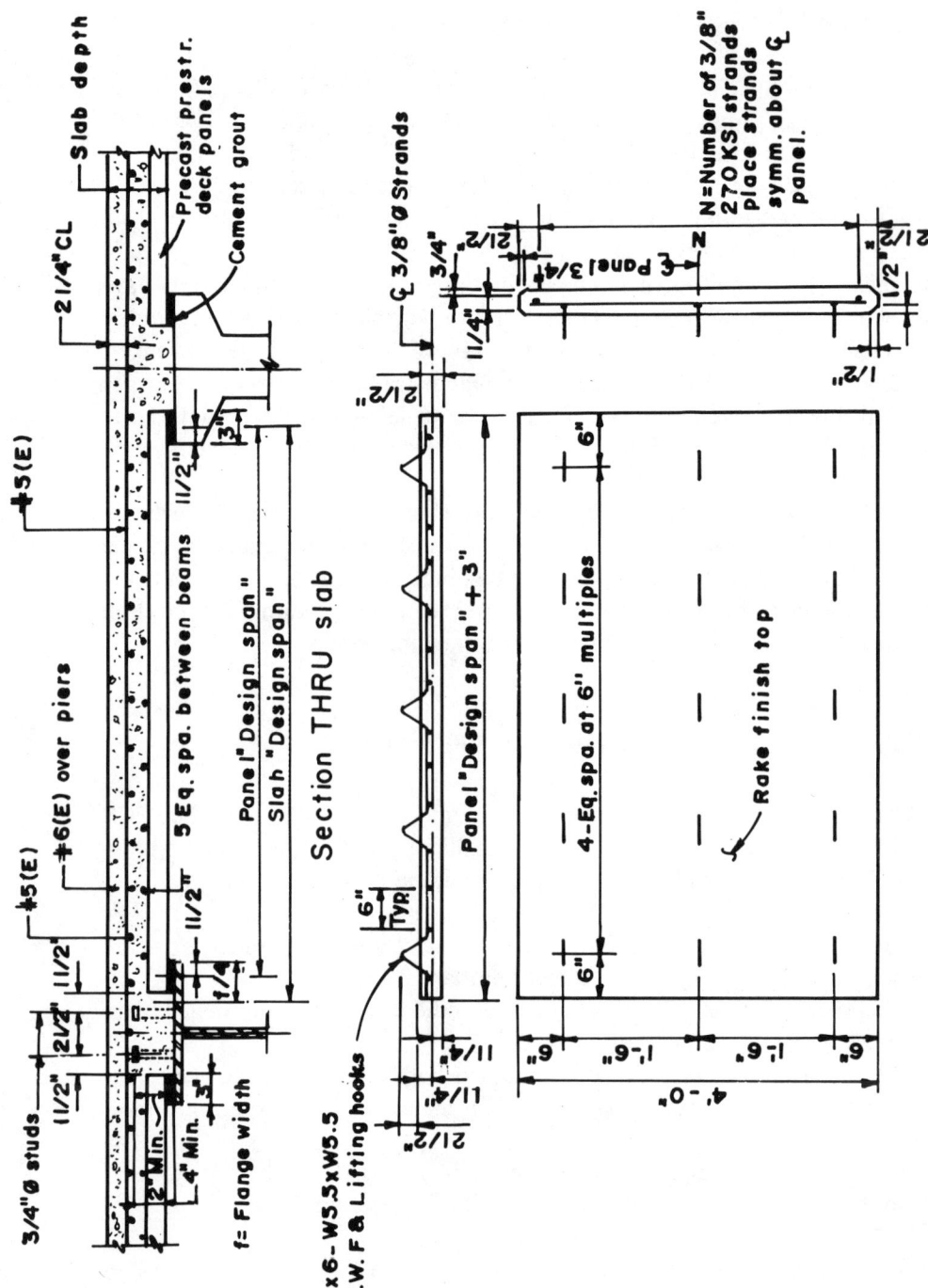

Fig. 6.32   Precast prestressed deck panel details. (Courtesy of the Prestressed Concrete Institute)

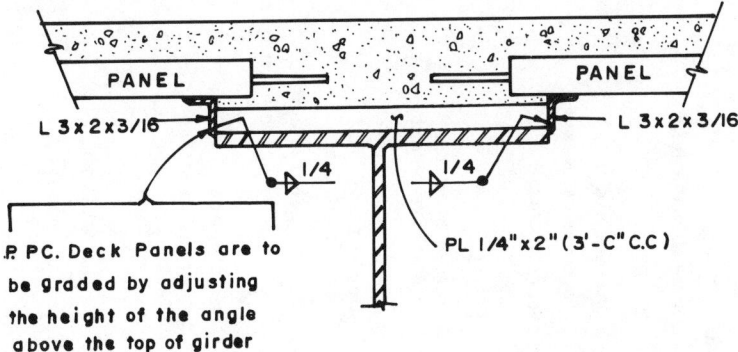

Fig. 6.32 (*Continued*)

making use of moving vehicles on the composite girder bridges built by the PPCS method. The main conclusions obtained from the test results can be summarized as follows:

1. The composite girder built by the PPCS method can behave as a complete composite beam with sufficient rigidity even in the case of moving vehicles.

2. The continuity conditions of axial displacement of the joints of precast concrete slabs can be molded due to prestress force, although the concrete slab built by the PPCS method is composed of a few discontinuous precast concrete slab panels.

3. The cracking of concrete slabs can be significantly controlled through the introduction of appropriate prestress forces rather than the improvement of strength and quality of the concrete materials.

4. The adjustments of longitudinal prestress forces in the PPCS method can control not only transverse cracking but also longitudinal cracking.

5. It is preferable to utilize two-directionally prestressed concrete slabs in the case where the concrete slab undergoes extremely heavy traffic flows.

6. Various test results for the durability and serviceability of concrete slab were reported in this paper.

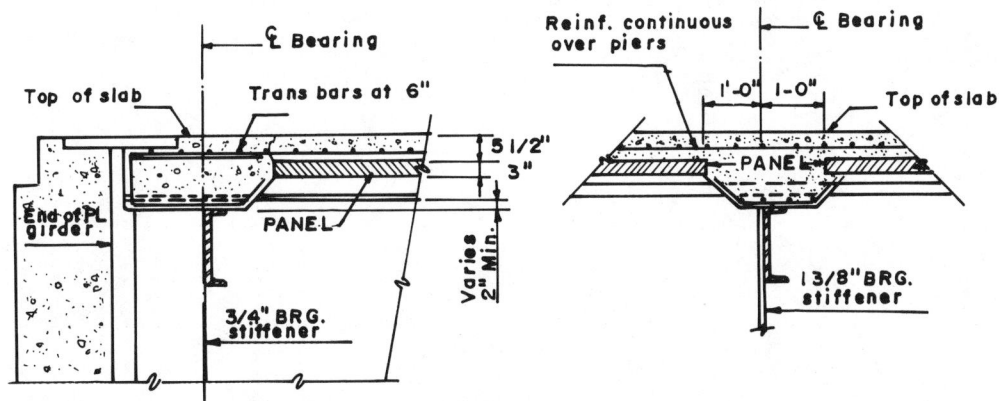

Fig. 6.33 Welded steel plate girder bridge details. (Courtesy of the Prestressed Concrete Institute)

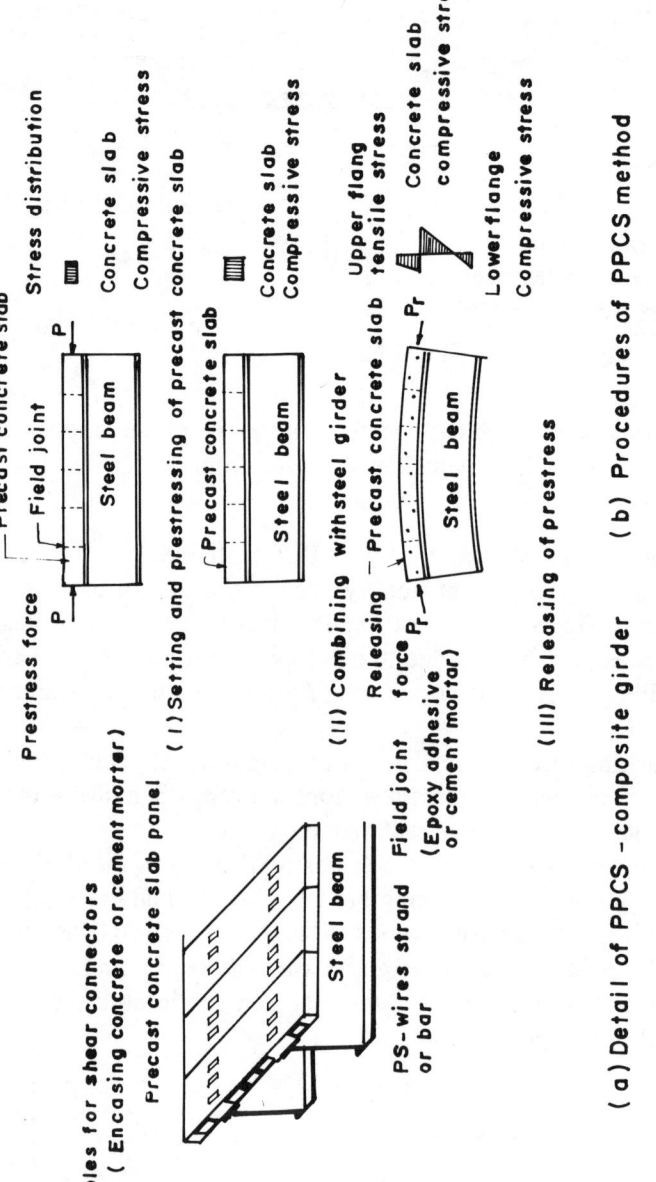

Fig. 6.34   Detail and procedures of the PPCS method.

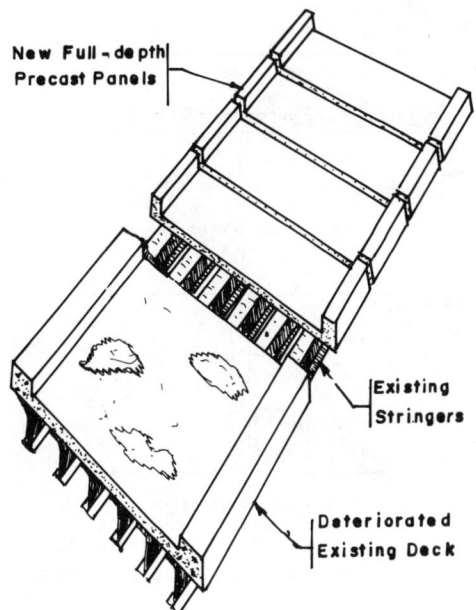

New Full-depth Precast Panels

Existing Stringers

Deteriorated Existing Deck

Fig. 6.35 Full-depth precast panel application. (Courtesy of the Prestressed Concrete Institute)

6.6.3 Precast Concrete Bridge Deck

The use of precast concrete for new bridge construction and for the rehabilitation of deteriorated bridges is economical and structurally applicable to modern structures.[17,18] Figure 6.35 schematically illustrates a method where old cast-in-place deck is segmentally removed and replaced by full-depth deck elements.

Precast elements can provide an additional advantage of greater durability over cast-in-place concrete. One important point to remember is that precast concrete becomes increasingly economical with repetitive elements. Figure 6.36 shows the basic load transfer requirements for a composite structure using precast elements. Figures 6.36(a) and 6.36(b) show the vertical load transfer and horizontal shear transfer, respectively, at the interface of the deck slab and stringer. Figures 6.36(c) and 6.36(d) show the horizontal in-plane forces and vertical shear transfer at the interface of adjacent deck slab panel.

A proper design must adequately address the following three criteria:

1. Tactical requirements related to schedule and traffic interference
2. Geometric fit-up problems
3. Load transfer, strength, and serviceability requirements

Full-depth precast concrete decks have been used for the rehabilitation and construction of numerous bridges since the late sixties. This construction method has been used successfully for both highway and railway bridges. All types of bridge superstructure, such as common deck over girder bridges, truss bridges, steel box girder bridges, suspension bridges, and cable-stayed bridges, have been built using precast decks.

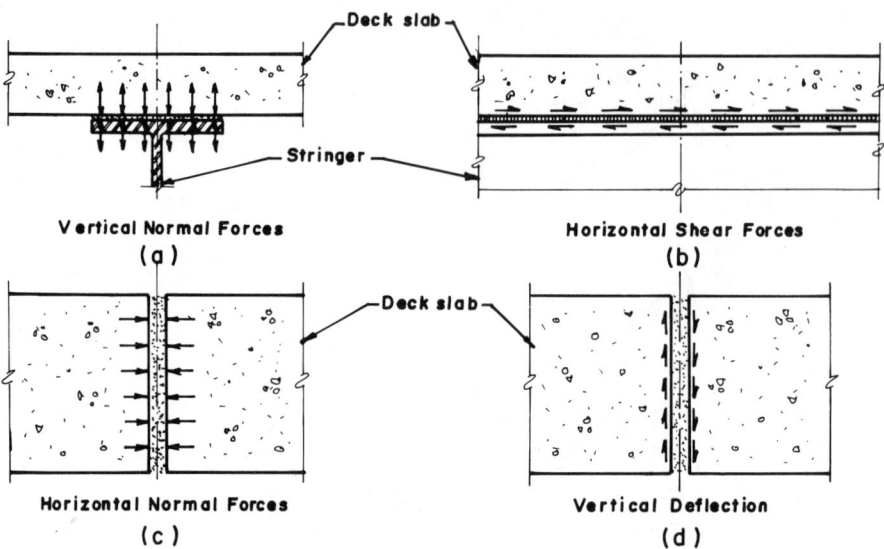

Fig. 6.36   Load transfer mechanisms of a composite deck. (Courtesy of the Prestressed Concrete Institute)

6.7 TYPICAL CONTINUOUS COMPOSITE PRESTRESSED BRIDGES

6.7.1 Montabaur Bridge, Germany

In 1957, a prestressed composite bridge, was built over the Auback Valley near Montabaur, Germany, in accordance with the ideas expressed by Dischinger.[19] The bridge is a continuous structure having three spans, and its main structural system consists of two main girders spaced 21 ft 4 in. apart and a continuous longitudinal internal girder, placed centrally, which cooperate fully with the main system after its structural connections have been established (Fig. 6.37).[20,21,22,23] A cross-sectional view of the bridge is shown in Fig. 6.38.

For esthetic reasons, the curved tendons were located at the internal sides of the girder webs in their guided grooves. The anchoring occurred at the main girders at their neutral axes.

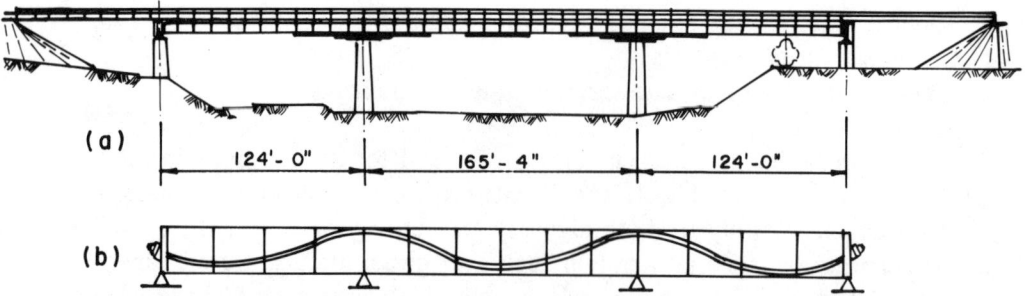

Fig. 6.37   Montabaur Bridge, Germany. (a) Longitudinal elevation. (b) An arrangement of the prestressing cable for the steelwork.

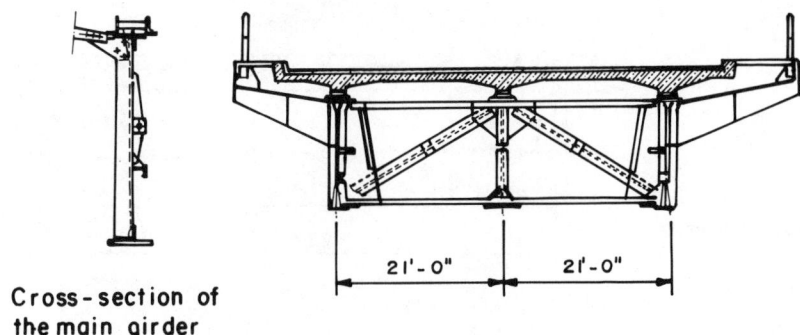

**Cross-section of
the main girder**

Fig. 6.38   Montabaur Bridge. A cross-sectional view.

A reinforced concrete slab was placed along its whole length and was supported by rollers fixed into the middle of the central span of the structural steelwork. For processing, two cables of 2.85-in. diameter each were used. These were patent-locked suspension bridge cables. To reduce any friction losses occurring during prestressing, both the grooves and the cables were thickly greased.

Tests were then carried out in order to determine whether the behavior of the structure agreed with that of the calculations. For the purpose of these tests, a whole steel structure was assembled on the ground. By measuring the flange deflections of the main girders during tensioning of the cables, it was thus possible to compare any theoretical as well as actual friction losses. Finally, the deformation of the webs of the main girders was measured and the actual displacement of the ends of the cables ascertained.

In addition to the deflections, the displacements of the cable ends were measured. The calculated cable elongations under an applied force of 221 tons was 12.2 in. at each cable end. To this had to be added the amount of the end slip, which was estimated to be about 0.3 in. The measured cable elongations were in close agreement with those of calculated values. The slip, however, was only about 0.15–0.20 in.

The results of the measurements showed that the actual deflections can be calculated quite satisfactorily. It was confirmed that the friction losses are reduced by repeatedly tensioning the cable and that the cable force becomes more and more equalized over its whole length.

On completion of the tests, the bridge was launched from its temporary position on the motorway over auxiliary trestle supports and was then lowered into its final position. A bridge deck was then constructed in three sections. About a week after the completion of each section, that portion of the deck slab concerned was transversely prestressed. Ten days after the last section had been completed, the longitudinal prestress was applied by round bars which were concreted into the slab in sheet-metal ducts and were given the required tensile stress by means of jacks.

As the Montabaur Bridge consists of two bridges of equal size situated side by side, an old one of conventional design and a new prestressed steel structure, the weights of these two bridges was compared. The new bridge shows a 33% saving in weight in comparison to that of the older structure. The percentage of

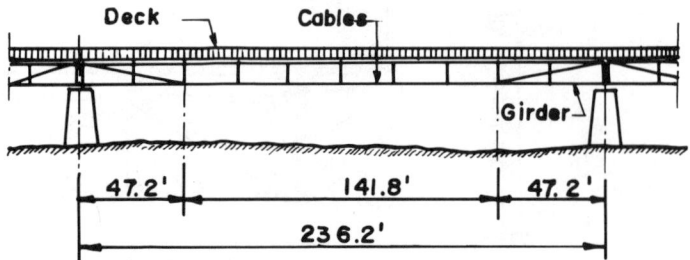

Fig. 6.39 North Bridge approaches. A typical prestressed span.

the high strength steel used in the old design amounts to 45% while only 17.25% of this grade of steel is used in the new bridge.

6.7.2 North Bridge Approaches at Düsseldorf, Germany

For the 6 × 236 ft continuous spans of the approaches to North Bridge over the River Rein at Düsseldorf, tensioned steel cables are of polygonal configuration (Fig. 6.39), supported by strengthened transverse bracing designed to resist rotation moments due to friction during tensioning of the cables.[24]

Each main girder is prestressed by four continuous strands, having a total length of 1541.6 ft and each having a diameter of 23/8 in. The cable are anchored to the steel girder and induce compression in the girder, similar to that in a cable-stayed bridge (Fig. 6.40). Details of the cable tiedown at the bottom chord of the bridge are shown in Fig. 6.41.

6.7.3 Bridge over the Tom River, USSR

A prestressed steel bridge over the Tom River in western Siberia was built in 1960.[25,26] The bridge crosses the river in five spans 240 + 3 × 358 + 240 ft (Fig. 6.42). In cross section, the roadway in 56 ft wide and there are two footpaths, each 7 ft 6 in. wide. In addition to highway traffic, the bridge is designed to carry a temporary railway siding (Fig. 6.43). The main girders vary in height from 18 ft at their mid-span, and they were assembled in place by riveting together 33–42 ft long beam elements which weight up to 11 tons each at their maximum. High-tensile structural steel used in construction has a yield point of about 22 tons/in.[2].

The reinforced concrete deck was monolithically constructed with steel girders of on-site concrete and shear plates which were also used as bearing jack supports during the stressing operations. The effect of interaction between the girders and the reinforced concrete deck was allowed for in the structural design.

The bridge was assembled simultaneously from both abutments, and the two halves met in the fourth span where a gap of 1 ft 8 in. (0.5 m) was left at mid-span. A similar gap was then produced in the middle of the second span, thus making the system statically determinate. The prestress was next applied at the top of the girders. When the stressing operations were completed, precast con-

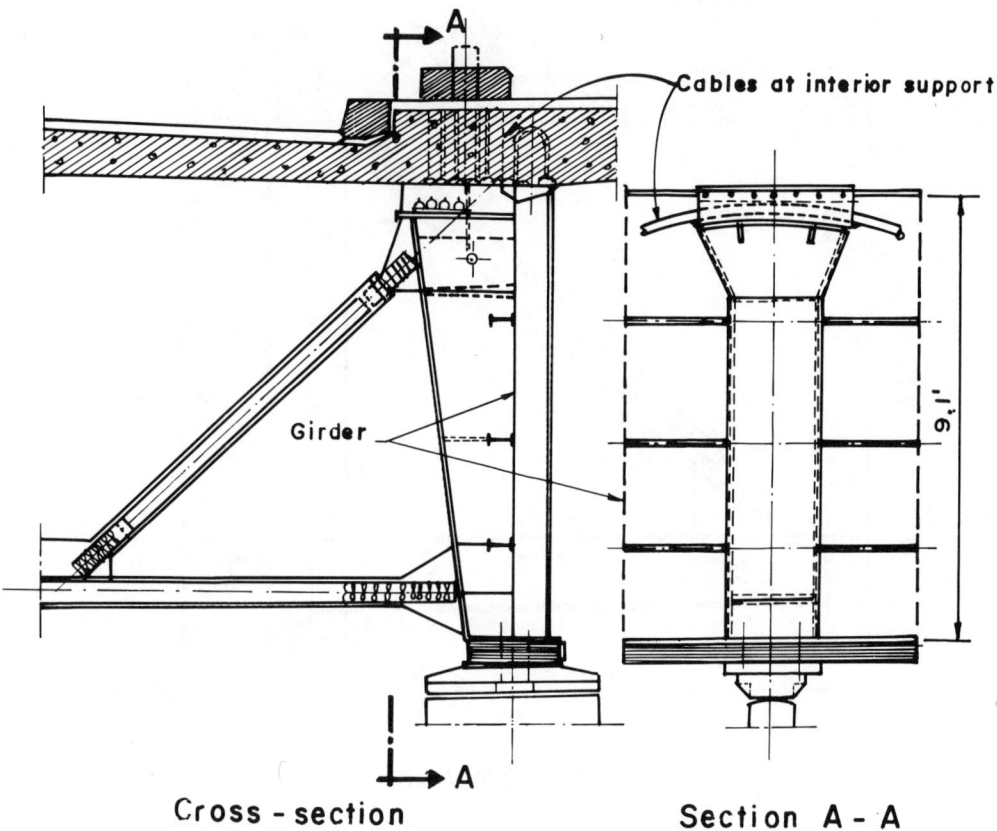

Cables at interior support

Girder

Cross - section   Section A - A

Fig. 6.40   North Bridge approaches. A cross-sectional view.

crete deck panels were positioned on the girders and a total temporary load of 57 tons was placed in eight points on each cantilever. On-site concrete was then poured in place to make the deck monolithic with three girders. When the temporary loading was removed, the girders straightened up, inducing compression in the deck and so counteracting the possible formation of cracks and adding a greater rigidity to the superstructure. The next operation was the closure of the two mid-span gaps, which converted the superstructure to a statically indeterminate system. It is suggested that for practical convenience, the temporary loading can best be applied by using containers of water.

The cables for prestressing the beams were assembled on the site from 0.118-in. (3-mm) diameter high-tensile wires, the diameter of the cable being 11/2 in. The cables were arranged in the form of a figure eight before they were put in place between the pair of anchors. There were 16 anchors on the top flange of each girder, of which 8 were permanent anchors fixed in position in the factory, and the remaining 8 were adjustable anchors welded in place on the site as the tensioning of the cables proceeded. Details of the anchors welded in place on the site are shown in Fig. 6.44.

In all, 320 cables were used in the bridge, and the arrangement of the cables over each intermediate support is shown in Fig. 6.45. The stress diagram in Fig. 6.45 shows the prestress in the top and bottom flange of a steel girder. As can

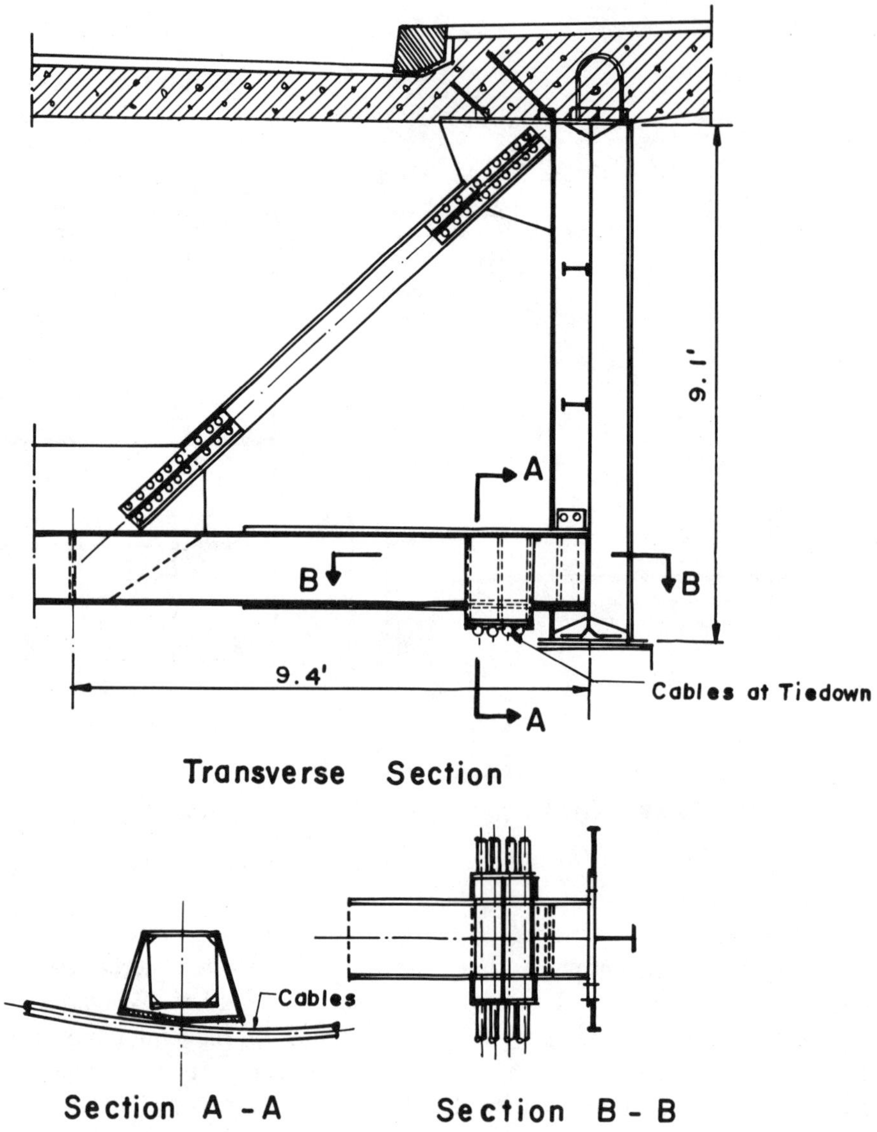

9.1'

9.4'

Cables at Tiedown

Transverse Section

Cables

Section A - A   **Section B - B**

Fig. 6.41   Detail of cable tiedown at the bottom chord of a bridge.

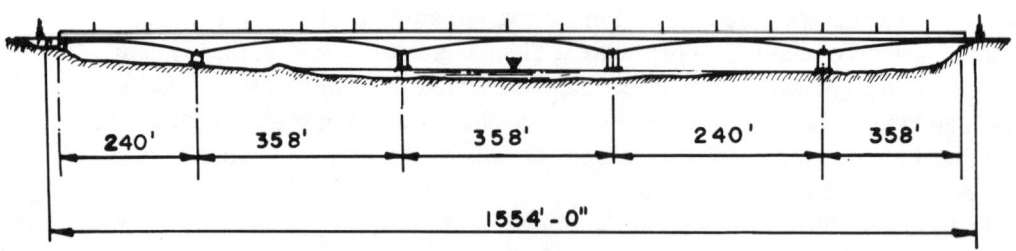

240'   358'   358'   240'   358'

1554'-0"

Fig. 6.42   Bridge over the Tom River. Elevation, USSR.

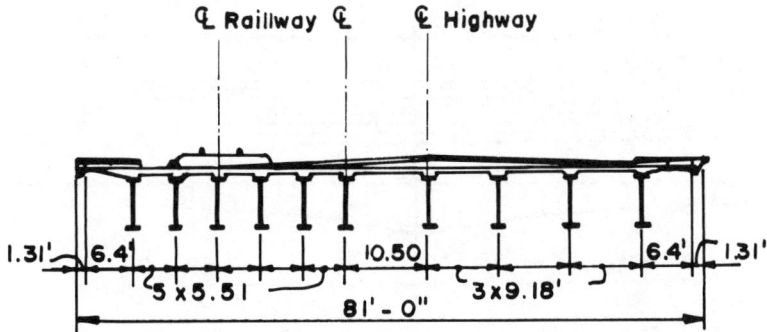

Fig. 6.43 Bridge over the Tom River. A cross-section.

be seen, the compressive stress in the top flange reaches about 38,400 psi (2,720 kg/cm²). In the bottom flange tensile stresses do not exceed 8,500 psi. There are eight staggered loops over the support, the distance between the relevant pair of anchors being about 186 ft.

Two prestressing methods were used, one for a test beam and another for the actual structure. The first method, shown in Fig. 6.46(a), allowed for the stressing of one cable only. The cable (1) was looped over a permanent anchor (2) at one end, over an adjustable anchor (3) at the other end. The 200-ton-capacity prestressing jack was placed between a static thrust bearer (4) which rested against a bearing support (5) and the sliding thrust bearer (6), transmitting the load to the adjustable anchor (3) through a prop (7). After the required stress in the cable was attained, the adjustable anchor was welded to the top flange of the steel girder.

The second method shown in Fig. 6.46(b), allowed for a simultaneous stressing of two cables. The prestressing force of 95 tons was applied to each cable,

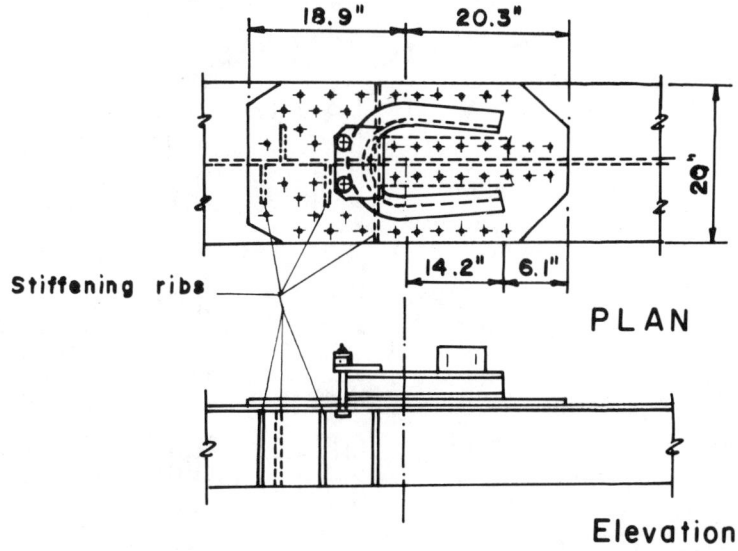

Fig. 6.44 Tom River Bridge. A plan and elevation of an anchor.

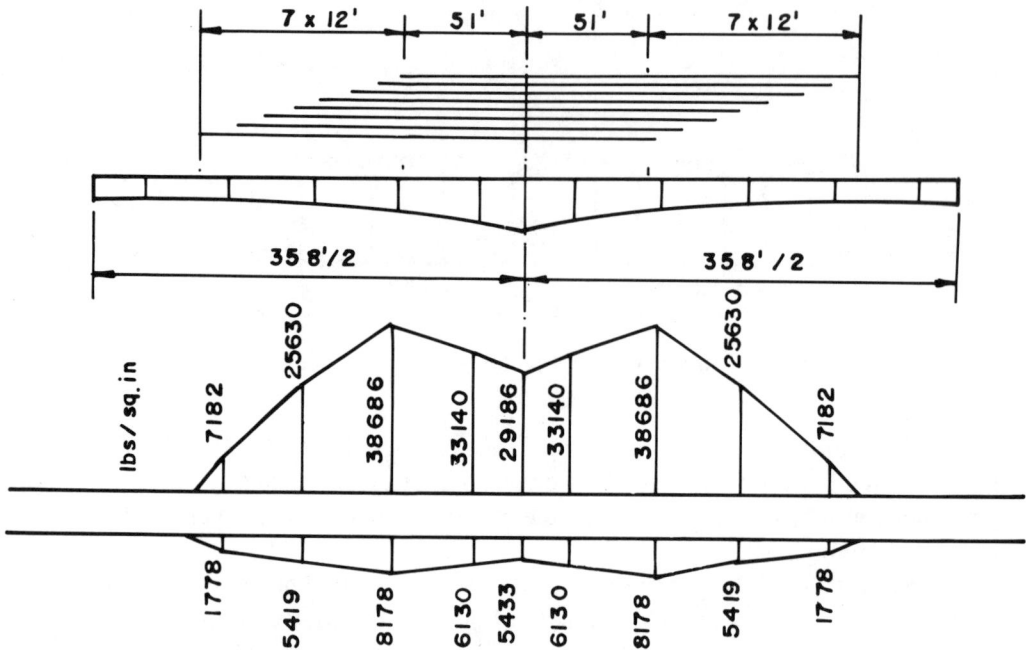

Fig. 6.45 Tom River Bridge. An arrangement of prestressing cables and a stress diagram for the top and bottom flanges of a girder.

and this produced a tensile stress of about 77,000 lb/in.[2] (54 kg/mm^2). During working conditions, this stress increases to 128,000 lb/in.[2] (90 kg/mm^2), which gives a safety factor of 2 against the breaking load.

It is estimated that about 500 tons of steel were saved by applying the prestress to the steel girders. This constitutes 10% of the total weight of steel used in the superstructure.

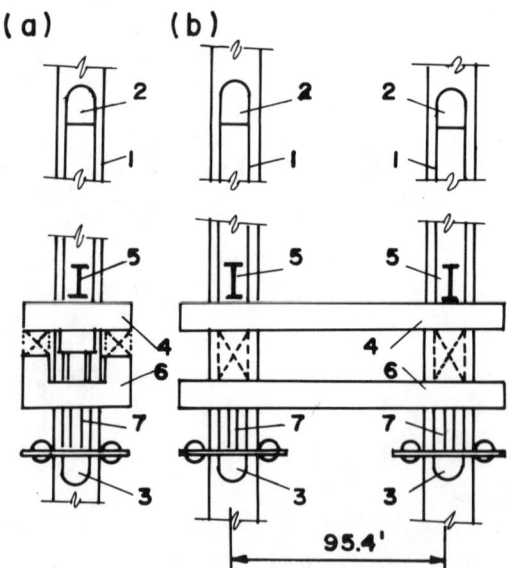

Fig. 6.46 Tom River Bridge. An arrangement for prestressing operations. (a) Prestressing of one cable. (b) Prestressing of two cables. 1, Cable; 2, permanent anchor; 3, adjustable anchor; 4, static thrust bearer; 5, bearing support; 6, sliding thrust bearer; 7, support.

Extensive experimental work has been carried out on the site, and the results show that the strength of the cable used in construction corresponds closely to an average strength of a single wire. Owing to the considerable length of cable, it is necessary to have 9 or 10 joints between the single wires. These are lap joints which are either soldered or wired together. The soundness of these joints has been proven in more than a hundred test specimens. An overall saving of up to 15 percent in the weight of steel can be achieved by the use of prestressing.

Tendons Compressing Top Flanges

In the USSR prestressing was applied to continuous composite bridge structures by tensioning bundles of high-strength wires connected to the top flanges of steel girders at zones under the action of negative moments. This tensioning was transferred only on steel girders by using special stop blocks and bundles concreted in longitudinal canals of precast reinforced concrete slabs. The compression of the reinforced concrete slabs was achieved by the application of another method of prestressing. In 1959–1960 this method was applied at the Tom River Bridge. The bridge, having continuous spans of 240 + 3 × 358 + 240 ft (73 + 3 × 1109 + 73 m), has two highway lanes 21 ft wide and two tracks for streetcars. The width of the bridge in cross section is 7.4 + 64.3 + 7.4 ft, and the deck is supported by 10 plate girders of changeable depth, having 7.5-ft abutments at the middle of large spans and 17.9 ft above intermediate supports. The erection of steel girders was performed by the cantilevering method. The sequence of prestressing and regulation of structure is shown in Fig. 6.47.

Tendons, each consisting of 125 wires having a diameter of 0.112 in. (3 mm) and a strength limit of 275 ksi (1900 MPa) and having loop shapes at the ends (Fig. 6.48) were tensioned in the statically determined bridge system, obtained after providing temporary gaps at the second and fourth spans. The bundles were prepared with the help of a winding machine having a diameter in the plan of about 131 ft and connected to the immovable and movable blocks. The spacing of these blocks was 186 ft.

The bundles were tensioned in pairs simultaneously on two neighboring girders by pulling the movable blocks by jacks. After finishing tensioning of

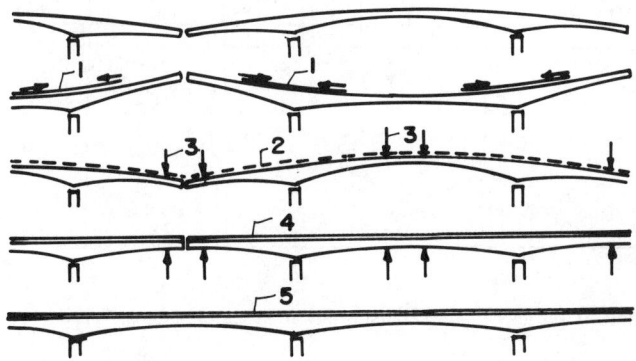

Fig. 6.47 Prestressing and regulation of a continuous bridge over the Tom River. 1, Prestressed cable bundles; 2, slabs not participating in the work; 3, temporary ballast loading; 4, slab participating in the work; 5, slab as part of the deck.

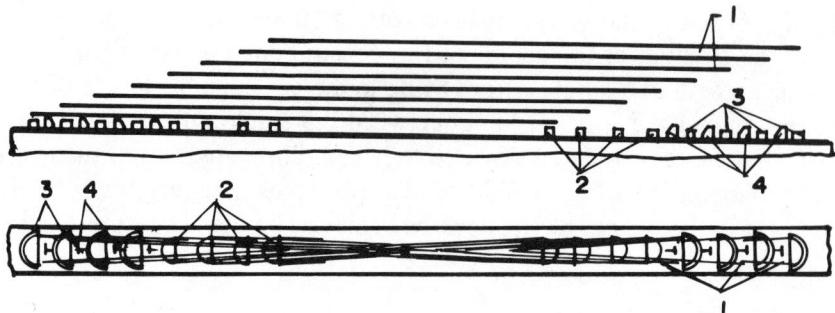

Fig. 6.48 Position of the high-strength reinforcing bars and anchorages on the top chord of the main girder of the bridge shown in Fig. 6-47: 1, Bundles of reinforcing bars; 2, unmovable anchors; 3, movable anchors; 4, anchors for jacks.

each pair of bundles, the holes were drilled and the movable blocks were riveted to the flange. The tensioning of the loop-type bundles was a complicated operation, and some loss in prestressing was observed.

The precast reinforced concrete slabs were placed after tensioning. The bottom part of the longitudinal joints in which the bundles were placed was filled with a plastic solution for protection of the bundles against corrosion, and upper parts of these joints, in which were placed plates of the blocks, were filled with the concrete. The compression of the reinforced concrete slab was performed after removal of the temporary ballast loading.

The prestressing of this structure gave a 10% savings in steel.

6.7.4 Aboshi Highway Bridge, Japan

The Aboshi Bridge, crossing Ibo River in Japan, was built in 1960 and consists of three parts: a continuous prestressed composite girder having three spans 105 + 105 + 105 ft (32 + 32 + 32 m), a Langer-type arch of 187 ft (57 m), and a simple composite girder span of 92 ft (28 m) (Fig. 6.49).[27] A cross-section of

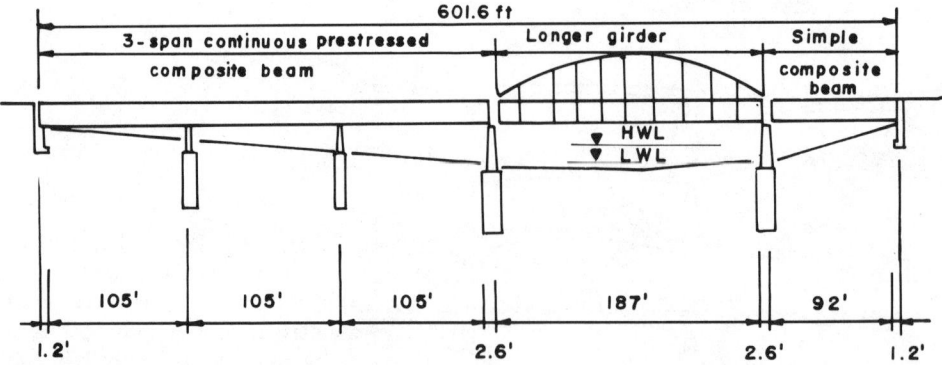

Fig. 6.49 Elevation of Aboshi Bridge. (Courtesy of the Transportation Research Board, Washington, DC)

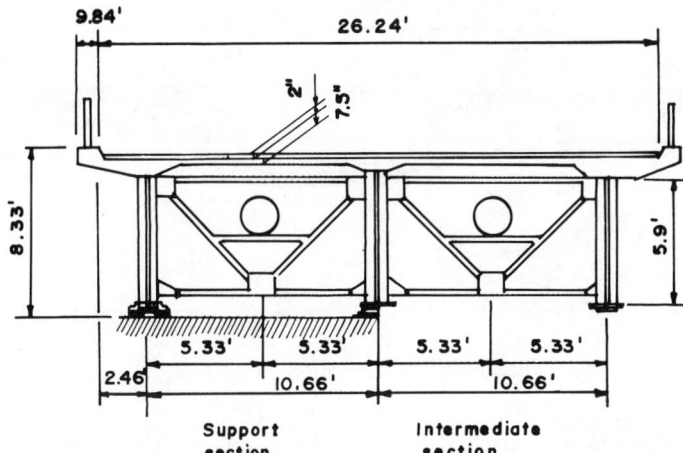

Fig. 6.50 Aboshi Bridge. Cross section of the continuous pre-stressed girder. (Courtesy of the Transportation Research Board, Washington, DC)

the continuous prestressed approach girder is shown in Fig. 6.50. This bridge has angle of skew equal to 69°27' (see Fig. 6.51). Actually this three-span continuous prestressed composite girder is pioneer work because it is the first bridge of this system built in Japan.

After erection of the continuous steel girder, the end supports were lowered 2 in. (5 cm) and the intermediate supports were raised 16 in. (40 cm). The end supports were prevented from springing up by the weights placed on them.

The concrete work was scheduled so that the slabs above the intermediate supports were done last. Three weeks after the concrete slab was finished, the compressive strength had reached 5000 psi (350 kg/cm^2). The supports of the composite girder were then lowered, using interlocking jacks. Adjustments were made every 3.4 in. (2 cm) so as to avoid unequal settlement and stressing. The last operations were pavement surfacing and the setting of the handrailings. Measurements of stress variation were made during erection to confirm the amount of prestress introduced and determine the amount of prestress decrease due to concrete shrinkage and creep. Measurements of the steel girder stress during the lowering of supports A and D and the raising of supports B and C (Fig. 6.51) show the measured values to be generally about 100–120% of the calculated ones.

A comparison of steel weight used for a continuous prestressed girder with

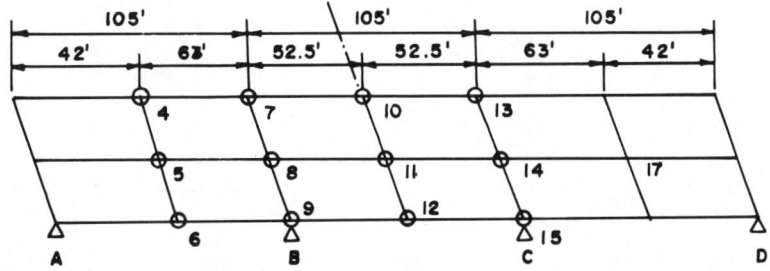

Fig. 6.51 Plan of the Aboshi Bridge. (Courtesy of the Transportation Research Board, Washington, DC)

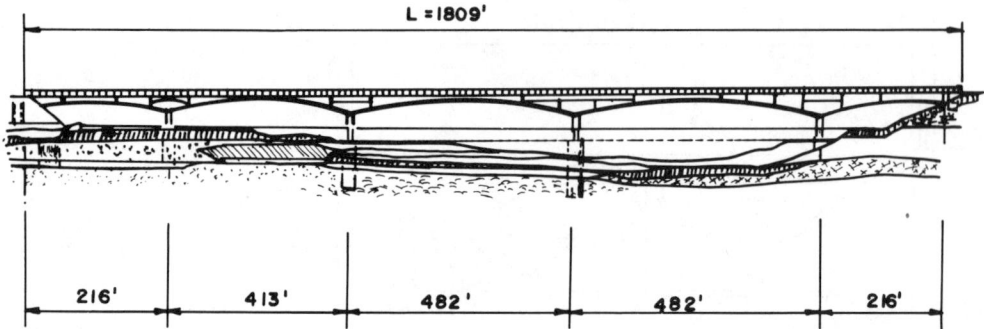

Fig. 6.52   Don River Bridge, USSR. Elevation.

that for a continuous conventional girder having the same spans indicates a 13.54% decrease in weight and an 8.4% decrease in construction cost for pre-stressed girder.

By the same method of prestressing, namely raising and settlement, the Kema Bridge was completed in 1960 as a continuous prestressed plate girder bridge having three spans, 130 + 115 + 179 ft. (39.5 + 35 + 54.4 m). It was the success of these two pioneer projects that induced application of continuous prestressed plate girder bridges in Japan.

6.7.5 Don River Bridge, USSR

The bridge crossing the Don River at Rostov, erected in 1963, is a prestressed continuous steel plate girder structure having five spans of 126 + 413 + 482 + 413 + 216 ft (Fig. 6.52).[28,29] The deck consists of precast reinforced concrete slabs acting as a composite with the plate girders. In cross section, the bridge consists of four welded main girders having changeable depth above the piers, the roadway width is 23 ft, and there are two sidewalks each 5.75 ft wide (Fig. 6.53).

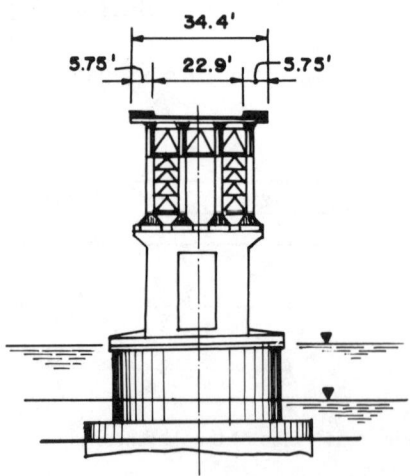

Fig. 6.53   Don River Bridge. Cross section.

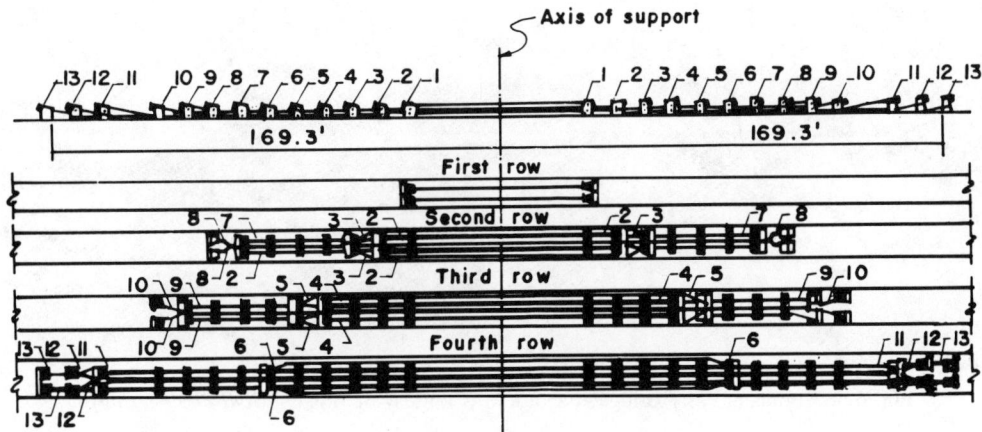

Fig. 6.54 Don River Bridge. Arrangement of strands above piers 3 and 4.

The top chords of the plate girders above the piers at sections under the action of large negative bending moments were prestressed by strand tendons of high-strength wire having a diameter of 0.196 in. The strands were installed one above the other and parallel. Above piers 2 and 5 the strands were placed in three rows, and above piers 3 and 4 the strands were placed in four rows (Fig. 6.54).

The erection of the superstructure was performed by the cantilever method, using temporary false work. To reduce static indeterminacy, and for convenience during prestressing installation and regulation, the erection of all five span structures was carried on two halves of the bridge as a statically determined system. For this purpose, two temporary hinges in the end spans were installed close to the locations of the zero bending moments under the uniform loading and transverse joint at the middle of the mid-span. The sequence of the above operations is shown in Fig. 6.55.

The reinforced concrete deck was prestressed after placing the temporary ballast of the mid-span and before concreting the joints between the precast concrete slabs. This ballast was removed after hardening the concrete in the joints.

The prestressing of the concrete deck prevents the formation of cracks in the

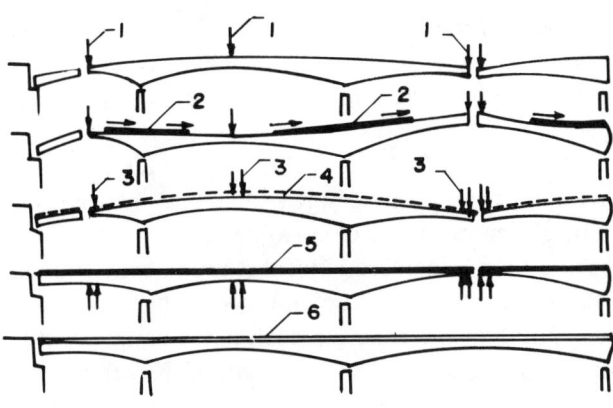

Fig. 6.55 Don River Bridge. Sequence of prestressing operations: 1, Placing of internal ballast; 2, prestressing of strands; 3, placing of second ballast; 4, nonstressed slab; 5, stressed slab; 6, deck slab.

Fig. 6.56   Bonners Ferry Bridge. Elevation. (Courtesy of ASCE—*Civil Engineering*)

deck slab and substantially increases the stiffness of the superstructure under live load.

The amount of structural steel for this bridge is only 75 lb/ft^2.

6.7.6 Bonners Ferry Bridge, United States

In 1984, in a competition for a new four-lane Bonners Ferry Bridge across the Kootenay River in Idaho, T. Y. Lin International Company designed a continuous cable-stressed steel plate girder structure, which was chosen from among eight bidders.

In elevation, the bridge is 1378 ft long, having ten spans varying from 100 ft to 155 ft (Fig. 6.56).[30,31,32,33]

The steel plate girders required two stages of cable stressing at their negative moment regions. The first state involved stressing the steel girder before placing the concrete deck to control the dead-load stress induced in the light top flange by the weight of one concrete deck.

In the second stage, the concrete was prestressed longitudinally to oppose the tensile stresses from the live-load composite action. The layout with just four girders and deck overhangs by 8 ft rather than the conventional 21/2 to 3 ft also permits a cost-saving slimming of the piers and narrowing of their hammer head caps.

In cross section, the 70-ft-wide bridge deck is attached to the steel girder by shear connectors and is always in compression, with the achievement of fully composite action over the entire length of the bridge, including its positive and negative areas (Fig. 6.57).

The four cable-stressed girders are set 18 ft c-c. To span the wide space between them, the deck slab is transversely posttensioned. At one end, where the roadway flares for an off ramp, girder spacing widens to 23 ft c-c.

In effect, the capacity of each girder is increased by substituting high-strength steel cable for a portion of the lower-strength structural steel. The design also called for longitudinal posttensioning of the concrete deck in the some negative moment areas to achieve full composite action for the length of the bridge.

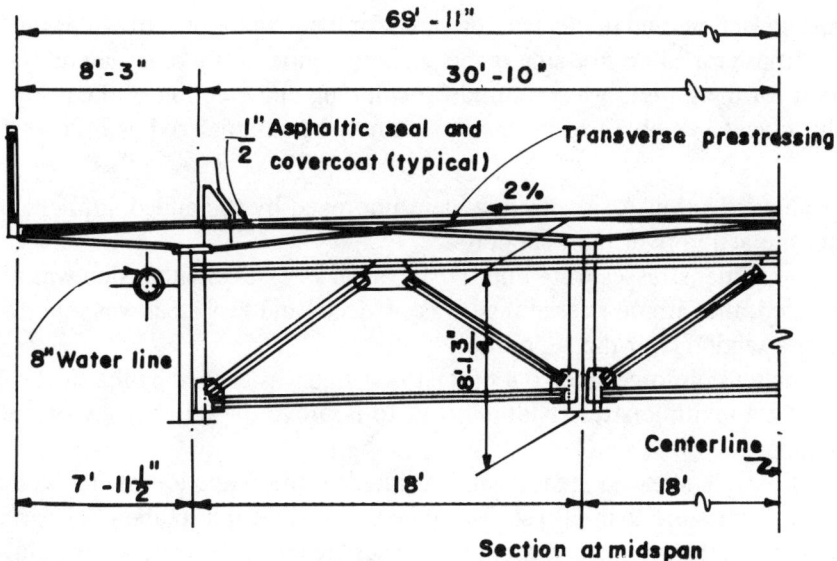

Fig. 6.57   Bonners Ferry Bridge. A cross section.

The full design potential of cable-stressed steel is shown in Figure 6.58. Figure 6.58(a) shows the replacement of the top flange, which is actually not needed, with that of a steel pipe containing the stressing cables. Shear connectors provide

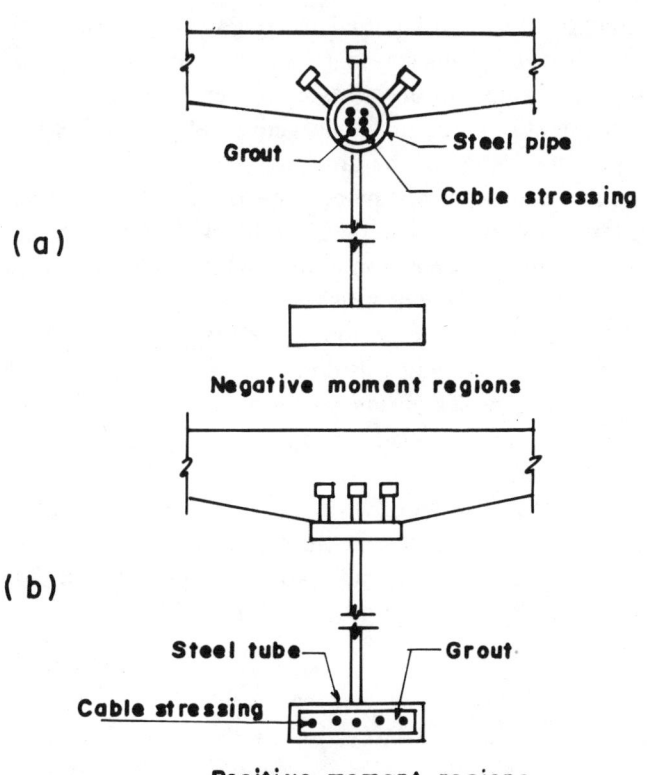

Fig. 6.58 Bonners Ferry Bridge. Design potential of cable-stressed steel.

composite action, and the rectangular tube for the lower flange provides the same dual purpose for cable stressing in the positive region. Temporary lateral stability support for the girders was required for shipping and erection of the bridge.

The advantages obtained in this project may be summarized as follows:

1. The deflections of the bridge were improved by the added stiffness of the composite action over their supports.

2. The girder stresses were improved to such an extent that tension was almost eliminated, the fatigue stress range was reduced, and less steel was required for the cross section over the supports.

3. The tension force taken out of the top flange was carried by the steel cables, providing a multiple stress path network to improve the redundancy of the tension flanges.

4. The cable-stressed steel design resulted in the reduction of the weight of the steel from more than 30 psf for a conventional design to about 24 psf.

5. Varying the prestressed section thickness reduced the concrete weight, thus cutting costs.

6. Design efficiencies also provided some significant cost savings for the contractor. Costs for fabrication normally run between 25 and 50% of the total, while erection accounts for 10 to 25%. Because of the fewer girders for Bonners Ferry and the attention to detailing, fabrication costs made up 24% of the total and erection costs came in at only 8%.

For a test run on the behavior of the girders under cable stressing, the fabricator erected in the shop a pair of 64-ft girders that span over one of the piers, complete with cross bracing and fittings for the stressing cables. The cables were strung through 4-in.-diameter steel pipe sections that were welded to split wide-flange beams and bolted to the top flange of the girder.

Erection required placing two cable-stressed pairs of girders over an end pier, then setting four fill-in girders between them and the abutment. Diagonal cables rigged to piers held the stressed girder in position until connections were made. That sequence was repeated until erection was finished.

Specifications called for longitudinal posttensioning of the girders over the piers to be carried out in two stages. The first was done before pouring the deck and was necessary only to control stresses during erection. The tendons for the second longitudinal posttensioning were embedded in the concrete and stressed after the deck was poured and cured.

In general, the cable-stressed steel concept offers an alternative to traditionally designed steel bridges. It promises savings in construction costs, better control of deflections, and improved stress distribution, and provides multiple load path redundancy and offers solutions to some of the problems of steel bridges.

REFERENCES

1. Dischinger, F., "Composite Steel Bridges Prestressed by High Strength Cables," Der Bauingenieur, nos. 11 and 12, 1949, (in German).

2. Troitsky, M. S., and Rabbani, N. F., "Tendon Configurations of Prestressed Steel Girder Bridges," CSCE Centennial Conference, May 19-22, 1987, Montreal, vol. 1, pp. 171-182.
3. Brodka, J., Jerka-Kulawinska, K., and Kwasniewski, M., "Prestressed Steel Girder, Statical Calculation," Verlags Gesellschaft Rudolf Muller, Cologne-Braunsfeld, 1968, pp. 13-14, (in German).
4. Brodka, J., and Klobukowski, J., "Prestressed Steel Constructions," Edition Wilhelm Ernst und Sohn, Berlin, 1969, p. 99, (in German).
5. Anand, S. C., and Talesstchi, A., "Prestressed Composite Steel Beam Bridge Girder," *Journal of the Structural Division, Proceedings of the ASCE*, March 1973, pp. 301-319.
6. Anand, S. C., and Fennell, C. J., "Prestressed Composite Continuous Bridge Girder," *Journal of the Structural Division, Proceedings of the ASCE*, January 1976, pp. 311-315.
7. Brodka, Jerka-Kulawinska, and Kwasniewski, op. cit., pp. 42-56.
8. Kunitskii, L. P., "Design of Prestressed Continuous Metal Beams Under Movable Load," *Construction and Architecture*, No. 9, 1968, pp. 3-11 (in Russian).
9. Roik, K., and Haensel, J., "Design and Construction Methods for Composite Girders in Europe," in *The Design of Steel Bridges*, ed. Rocky, K. C., and Evans, H. R., Granada, London, 1981, pp. 387-407.
10. Knowles, F. R., *Composite Steel and Concrete Construction*, Wiley, New York, 1973, pp. 83-88.
11. Roik, K., "Methods of Prestressing Continuous Composite Girders," *Proceedings of the Conference on Steel Bridges, Institution of Civil Engineers*, 24-26 June 1968. The British Constructional Steelwork Association Ltd., London, pp. 75-81.
12. Roik, K., "Highway Bridge over the Lied River at Unna." *Der Stahlbau*, no. 8, 1961, pp. 231-241.
13. Ibid.
14. Goldberg, D., et al., "Special Report, Precast Prestressed Concrete Bridge Deck Panels," *PCI Journal*, March-April 1987, pp. 27-45.
15. Kita, H., Takenaka, J., and Nakai, H., "Durability of Composite Girders Using Prestressed Precast Concrete Slabs Under Movable Vehicle," The First East Asian Conference on Structural Engineringand Construction Construction, Bangkok, January 15-17, 1986, pp. 681-692.
16. Takenada, H., Kishida, H., and Nakai, H., "A Study on New Composite Girder Using Prestressed Precast Concrete Slab by PPCS Method," *Stahlbau*, no. 6, 1986, pp. 165-174.
17. Biswas, M., "Precast Bridge Deck Design Systems," *PCI Journal*, March-April 1986, pp. 40-94.
18. Biswas, M., "On Modular Full Depth Bridge Deck Rehabilitation," *Journal of Transportation Engineering*, vol. 112, no. 1, January 1986, pp. 105-120.
19. Dischinger, op. cit.
20. Hoyden, A., "Bridge Building Under Special Consideration," Der Bauingenieur, no. 6, 1955, (in German).
21. Wilmes, "Construction of the New Motorway Bridge at Montabaur (Germany)," *Acier-Stahl-Steel*, no. 3,1957, pp. 106-108.
22. Wnek, H., "Recent Building of the West Roadway for Montabaur Highway Bridge," Der Stahlbau, no. 6, June 1954, pp. 129-134, (in German).
23. Sattler, K., "Result of Measurement on the Composite Montabaur Bridge," Die Bautechnik, vol. 34, 1957, p. 329, (in German).
24. Beyer, E., "Experience and Cable Investigation of Cable Stressed Composite Construction, (Section of Approach Structure of the Nord Bridge at Düsseldorf)," Der Stahlbau, vol. 26, no. 7, July 1957, pp. 177-183, (in German).
25. Anon. "Engineering Developments in the U.S.S.R. Prestressed Steel Bridge," *Civil Engineering and Public Works Review*, vol. 55, no. 653, December 1960, p. 1571.
26. Streletskii, N. N., "Reinforced Concrete Bridges," Edition Transport, Moscow, 1965, pp. 52-54 (in Russian).
27. Iwamoto, K., On the Continuous Composite Girder, Highway Research Board, National Academy of Science—National Research Council, Washington, D.C.
28. Popov, G. D., "Regulation of Forces at Bridge Structures," *Metal Constructions, Proceedings of MISI Institute*, no. 43, Moscow, Gosdortechizdat, 1962 (in Russian).

29. Meljnikov, N. P., *Metal Constructions—Present State and Development*, Stroiizdat, Moscow, 1983, pp. 362–363 (in Russian).
30. Seim, C., "Steel Beats Concrete for Idaho Bridge," *Civil Engineering/ASCE*, August 1983, pp. 28–32.
31. Anon. "Bridge Design Cuts Weight," *Engineering News-Record*, July 22, 1982, p. 13.
32. Anon. "Steel Design Sweeps Bid," *Engineering News-Record*, September 2, 1982, p. 12.
33. Anon. "Prestressed Steel Bridge a Winner," *Engineering News-Record*, December 1983, pp. 48–49.

Chapter 7
Box Girder Bridges

7.1 INTRODUCTION

The steel box girder is now, in its various forms, the most commonly used element for the deck structure for all but the shortest-span bridges. Almost as soon as the span and loading become such that a second web is required, the box girder may be used. Now, with the box girder, much larger spans are possible for a given amount of steel than was the case previously. The depth of the box girder is determined by a number of factors, one of which may be the requirement to have minimal construction depth. Other factors are the required vertical rigidity and the minimum quantity of the metal. These last conditions may be achieved by prestressing of the box girder.

Prestressing of steel box bridges of simple and continuous spans was proposed in 1949 by Dischinger in his articles considering highway and railway bridges.[1]

7.2 TYPICAL PRESTRESSED BOX GIRDER BRIDGES DESIGNED BY DISCHINGER

7.2.1 Highway Bridges

1. Prestressed Simple-Span Box Girder Bridge Having Span of 390 ft (119 m), Fig. 7.1

Figure 7.2 shows the longitudinal section with the positions of the polygonal cables. In cross section the roadway width is 59 ft (18m), the box height is 14.75 ft, and the thickness of the reinforced concrete composite deck is 9.5 in. (Fig. 7.3).

The bridge is prestressed by 24 polygonally shaped cables, having a total cross-sectional area of 124 in.[2]. During analysis three cases were considered, as shown in Table 7.1

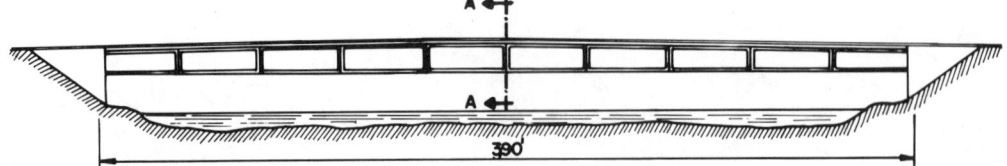

Fig. 7.1   Elevation of prestressed box girder highway bridge of simple span.

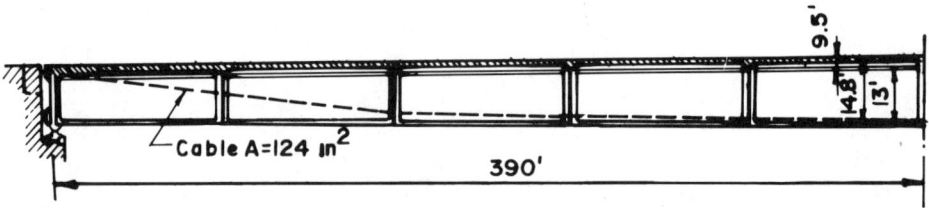

Fig. 7.2   Longitudinal section.

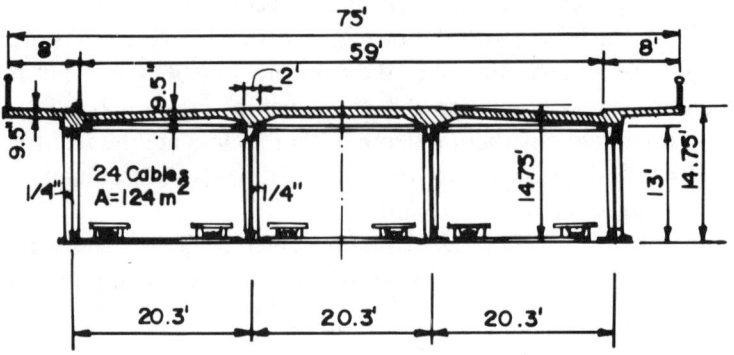

Fig. 7.3   Cross section.

Table 7.1

Case	Max. Simple Span l	Deflection $\dfrac{\Delta}{l}$
1. Concrete deck is not composite and bridge is not prestressed	283 ft (86.3 m)	$\dfrac{l}{850}$
2. Composite deck, but bridge is not prestressed	354 ft (108 m)	$\dfrac{l}{1130}$
3. Composite deck and prestressed bridge	390 ft (119 m)	—

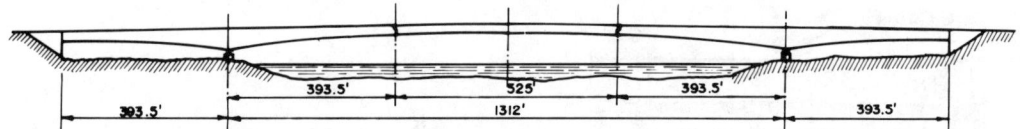

Fig. 7.4   Elevation of continuous prestressed highway box girder bridge.

2. Continuous Prestressed Highway Box Girder Bridge Having Three Spans of 394 + 1312 + 394 ft (120 + 400 + 120 m), Fig. 7.4

The configuration of the bottom flange of box girders in each span is parabolic and the height of the box at each middle pier is 48 ft and at the midspan is 19.7 ft. Figure 7.5 indicates that the bottom flanges of the box sections at the cantilevered spans are strengthened by the reinforced concrete slab of changeable thickness. Figure 7.6 shows the suspended part of the middle span having length of 525 ft (160 m).

Prestressing of the box sections was performed by high-strength steel cables having 5/16 in. diameter. Prestressing decreases parts of the dead-load stresses, and the composite action of the concrete deck reduces the cross sections of the steel members. For the cantilevered parts of the bridge which are under negative moments, the reinforced-concrete deck is stressed under tension and is strengthened by prestressed cables.

Typical cross sections of the bridge are shown in Fig. 7.7. Deflection of the suspended span under live load is

$$\frac{\Delta}{l} = \frac{1}{770}$$

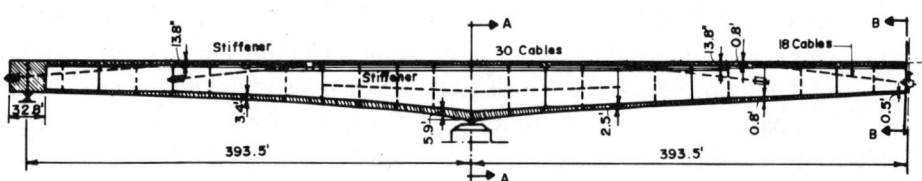

Fig. 7.5   Cantilevered part of the bridge. Longitudinal cross section.

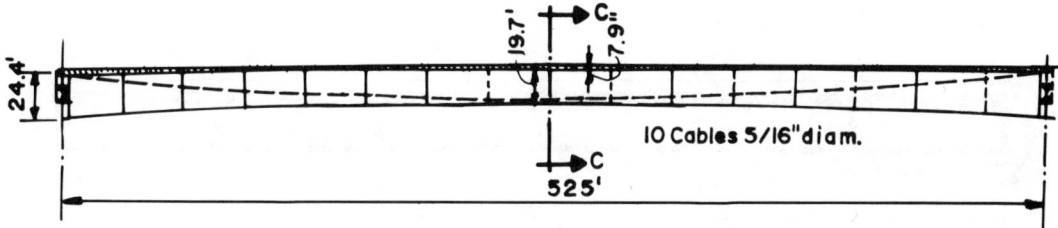

Fig. 7.6   Suspended box girder of the middle span.

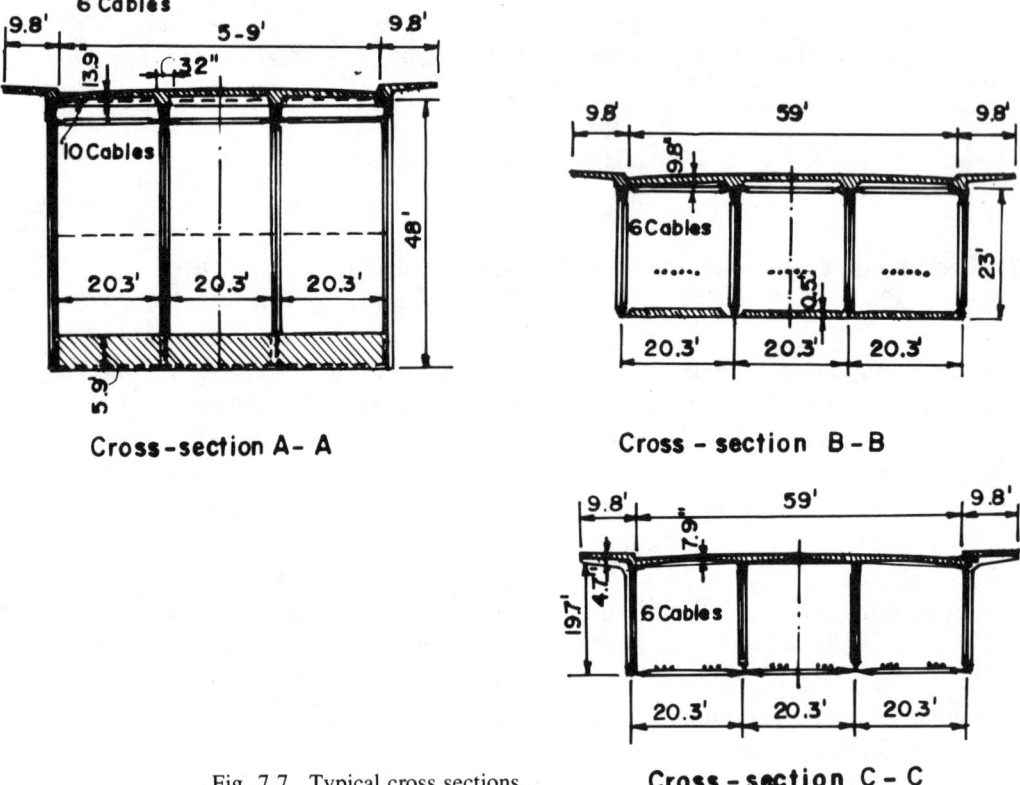

Fig. 7.7   Typical cross sections.

7.2.2 Railway Bridges

1. Prestressed Railway Bridge Having Single Track and Span, l = 157.5 ft (45 m), Fig. 7.8

In cross section, Fig. 7.9, the steel box has a depth of 7.4 ft. The bridge is prestressed by four cables of polygonal configuration having a total cross section of 124 in.[2] The reinforced concrete slab, acting compositely with the steel box, and holding ballast, has a thickness of 1.64 ft and total width of 13.1 ft. The calculated deflection of the bridge under live load is Δ = 0.152 ft.

$$\frac{\Delta}{l} = \frac{0.152}{157.5} = \frac{1}{1036} < \frac{1}{900}$$

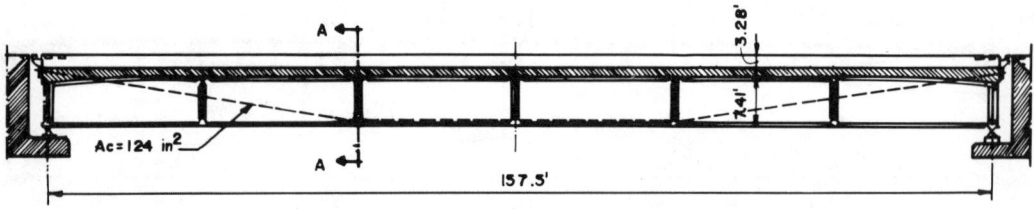

Fig. 7.8   Elevation of single-track prestressed box-type railway bridge.

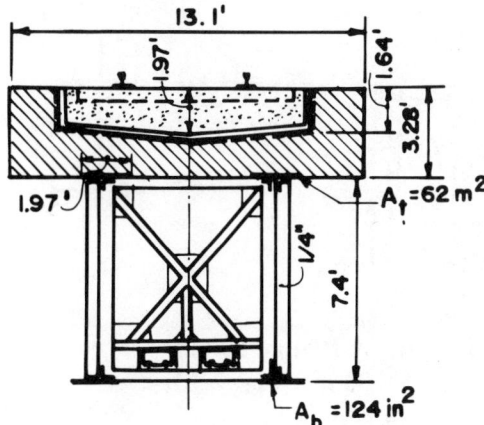

Fig. 7.9   Cross section A-A.

2. Prestressed Railway Bridge Supporting Two Tracks and Span $l = 328$ ft (100 m), Fig. 7.10

In cross section, Fig. 7.11, steel boxes have a height of 14.8 ft, and the width of both boxes is 26 ft. Spacing between the center lines of the tracks is 13.1 ft. Boxes are prestressed by 16 high-strength steel cables, each having a cross section of 109 in.^{2}. An anchorage connection detail is shown in Fig. 7.12.

The calculated deflection of the bridge under live load is

$$\frac{\Delta}{l} = \frac{1}{720}$$

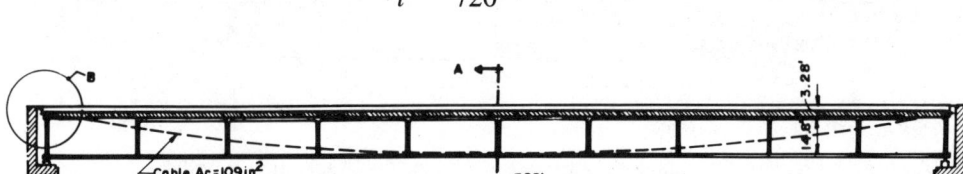

Fig. 7.10   Elevation of double-track prestressed box-type railway bridge.

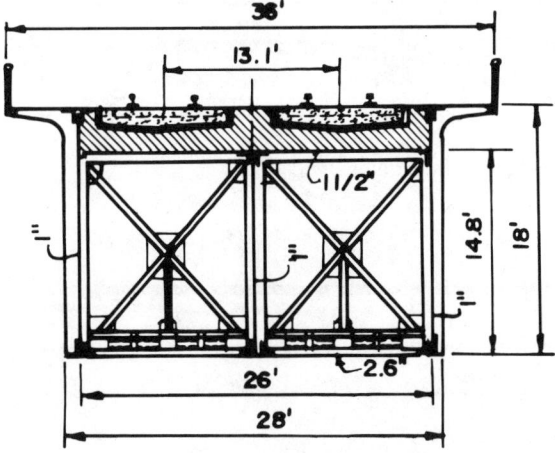

Fig. 7.11   Cross section A-A.

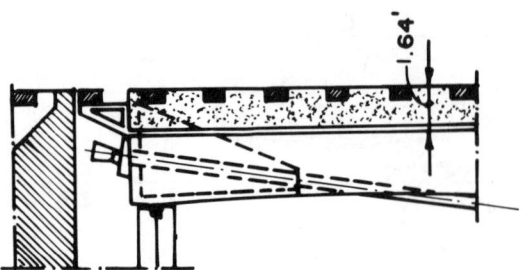

Fig. 7.12 Anchorage connection detail.

3. Continuous Prestressed Two-Track Railway Bridge Having Three Spans of 328 + 984 + 328 ft (100 + 300 + 100 m), Fig. 7.13.

The configuration of the bottom flange of the box girder in each span is curvilinear, and the height of the box girder above the middle supports is 36 ft and at mid-span is 32.8 ft.

Figure 7.14 indicates that the bottom flanges of the box sections at the cantilevered spans are strengthened by the reinforced concrete slab of variable thickness. Figure 7.15 shows the suspended part of the middle span having a length of 558 ft. Prestressing of the box section was performed by the 36 high-strength steel cables, each having cross section 26 in.[2]. Typical cross sections of the box girders are shown in Fig. 7.16.

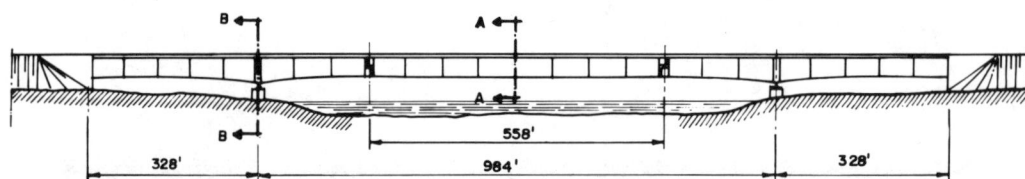

Fig. 7.13 Continuous prestressed box-type two-track railway bridge.

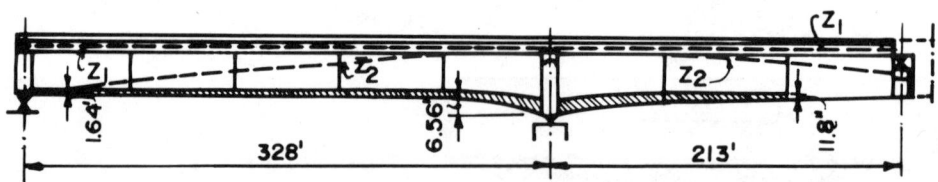

Fig. 7.14 Cantilevered parts. Longitudinal cross section.

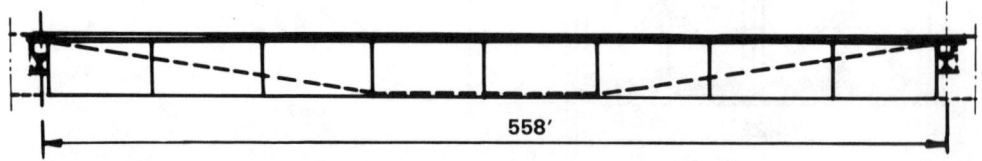

Fig. 7.15 Suspended box girder of the middle span.

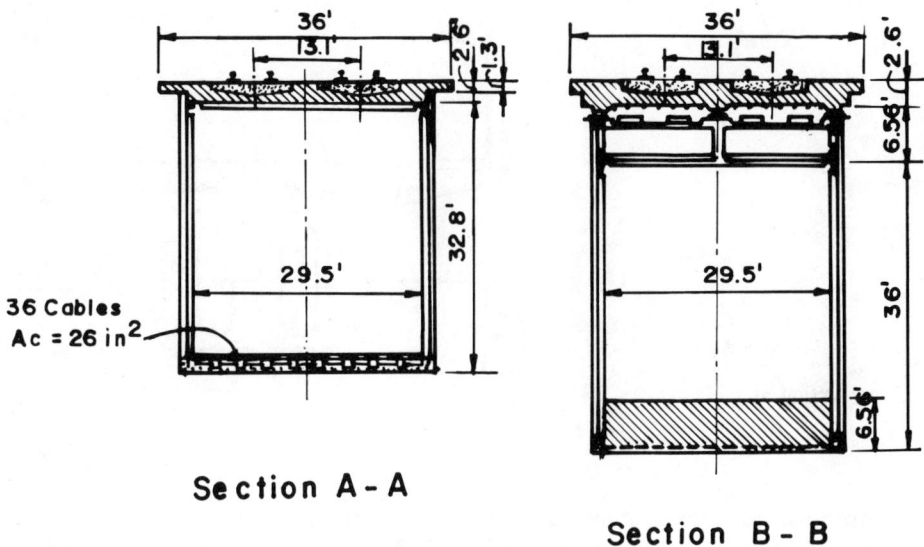

Fig. 7.16   Cross sections.

7.3 SCHKOPAU BRIDGE, GERMANY

Figure 7.17 shows a elevation of a prestressed box-type bridge, erected in 1963, supporting a double track for a streetcar, over the Saale River at Schkopau, having a span of 266 ft (81 m).[2] In cross section, Fig. 7.18, this bridge has a

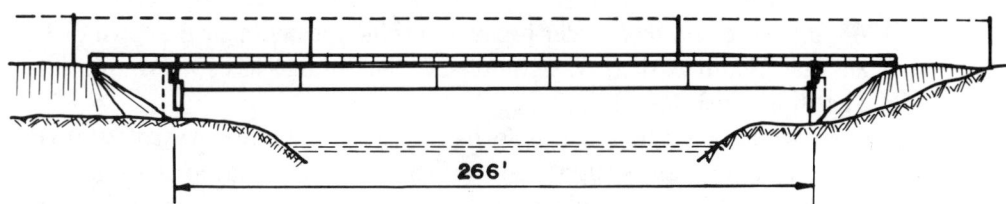

Fig. 7.17   Elevation of Schkopau Bridge.

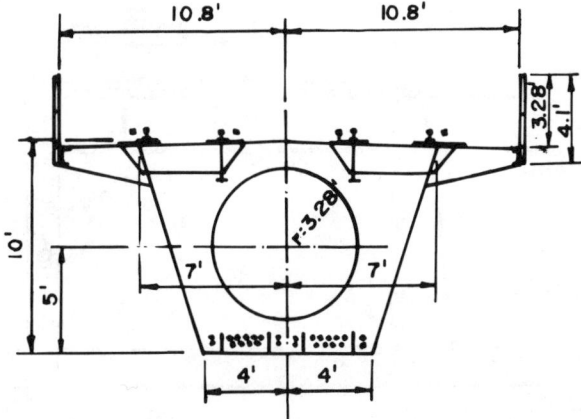

Fig. 7.18   Cross section of mid-span.

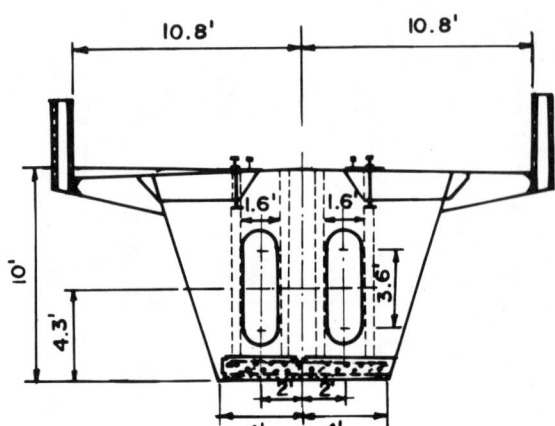

Fig. 7.19 Cross section at the tendon's anchorage.

box girder of trapezoidal configuration, with a height of 10 ft and widths at the top of 14 ft and at the bottom flange of 8 ft. Figure 7.19 indicates the arrangement of the 26 high-strength steel prestressing cables, each having diameter of 1 in. and placed above the bottom flange of the box. Figure 7.21 shows the arrangement of the tendons and anchorages in the plan as well as anchorage connections to the vertical stiffening diaphragms inside the box girder.

7.4 BRITS BOX GIRDER BRIDGE, SOUTH AFRICA

In 1964, a prestressed box girder highway bridge located near the town of Brits in Transvaal, South Africa, was completed. The bridge has four simple spans, 4 × 81 ft 6 in. = 326 ft (Fig. 7.22)[3,4]

In cross section the bridge has a 36-ft-wide deck and a roadway of 30 ft (Fig. 7.23). Each single span is supported by two steel boxes spaced at a distance of 20 ft. Box girders are strengthened by prestressing and are composite with a

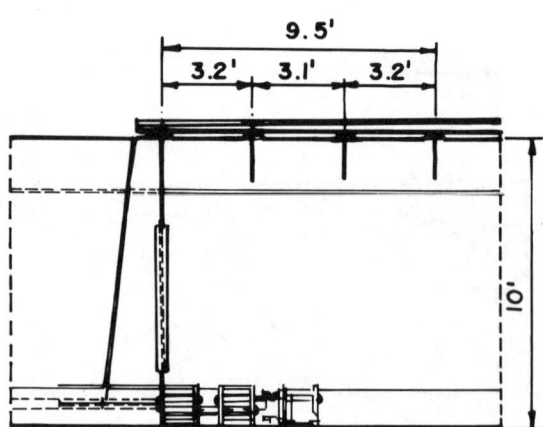

Fig. 7.20 Longitudinal cross section.

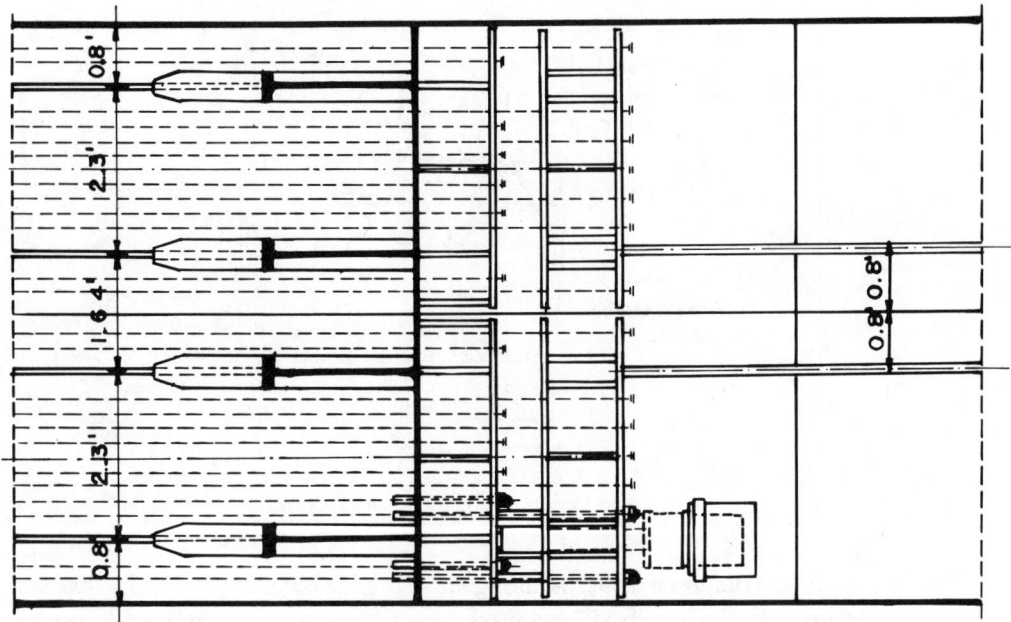

Fig. 7.21   Tendon anchorages in the plan.

cast-in-place concrete deck. Each shop-fabricated box has a troughed compression top flange and two webs.

The form of the box girder is that of a box with an externally attached side

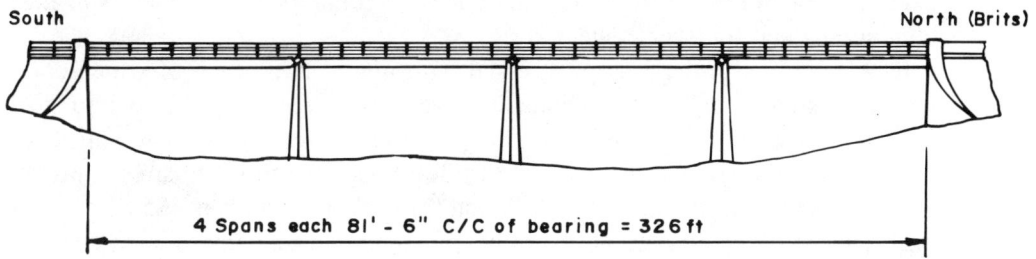

Fig. 7.22   Elevation of Brits Bridge.

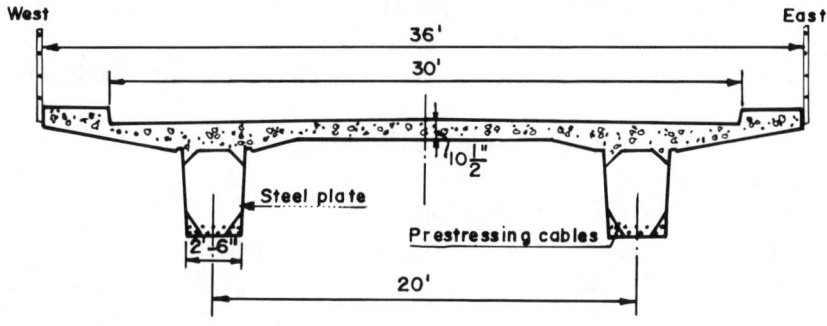

Fig. 7.23   Cross section of Brits Bridge.

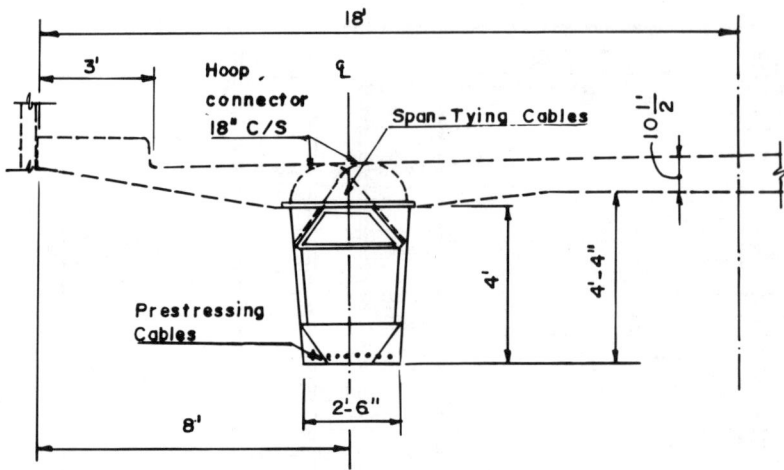

Fig. 7.24   Details of box girder in cross section.

piece to form triangular troughs along its top edges. Shop-attached hoop connectors provide the required structural union between the reinforced concrete deck slab and the girder top flange.

Internally, lengths of steel plate welded onto the bottom corners stiffen the box girder. At the top, the plates that form the top flange have the shape of an inverted trapezoid; the outer two faces comprise the faces of the external troughs. These top and bottom elements contribute materially to the strength and rigidity of each box girder (Fig. 7.24). The trough and the hoop connectors give an excellent bond between the steel and concrete with ample safety factors against longitudinal and transverse shear.

Each steel box girder was prestressed with nine cables, each of which comprised 12 strands of 0.276-in-diameter wire. The wires were placed into steel tubes and fixed onto the girder ends. The Freyssinet system was used to prestress the cables in the shop to a total force of 360 tons. The ultimate stress in these cables under the dead load of the deck and the design live load is 405 tons. After the beams had been erected, all the cables and their end anchorages were grouted.

Each pair of box girders was erected as a simple span and the reinforced concrete deck was cast in place in forms supported by scaffolding. The individual units in each line of girders butt together at their ends. However, when the deck was completed, the whole structure was tied together as a continuous unit by a single cable threaded through a steel duct under the hoops in the deck atop each girder.

The designer describes the completed structure as a "mechanically conditioned, three-hinged, four-span continuous beam bridge." The deck is fixed at one abutment and is free to move at the other.

Major advantages claimed for the new design include savings in weight and material costs, ease of erection, and speed of construction. For comparable spans and loadings, five or six conventional steel or prestressed concrete beams would be needed in each span to do the job carried out by the two box girders, according to the designers.

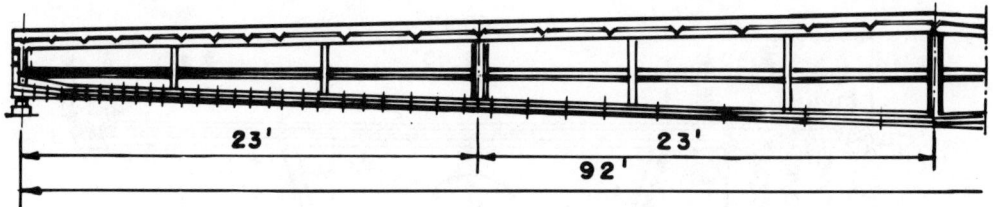

Fig. 7.25   Elevation of Willstress Bridge.

7.5 WILLSTRESS SYSTEM BRIDGE

A composite prestressed box girder bridge of the Willstress system was erected and tested as a prototype in Willebrock, Belgium, in 1960 by the firm Graver S.A.[5] This is a highway bridge having a span of 91 ft 10 in. and a changeable depth of the box girder which is 4 ft 9 in. (1.45 m) in the middle of the span and 2 ft 8 in. (0.82 m) at each support. The configuration of the bottom chord is parabolic, (Fig. 7.25). In cross section, the width of the concrete deck is 13 ft 1 in. (4 m) having a thickness of 4 3/4 in. (12 cm) which is supported by a steel box girder of trapezoidal cross section and variable depths (Figs. 7.26 and 7.27).

The special feature of the welded-steel box girder, whose plates are only 1/4 in. (7 mm) thick, lies in the fact that the angle formed by the wall and the bottom chord remains constant for the whole length of the bridge. In consequence, not only is the depth of the girder variable, but the width of the bottom chord is also, as are the cantilevers of the upper slab. The walls of the box girder are reinforced by longitudinal and vertical stiffeners.

The construction of this bridge consisted of two stages. The first stage was performed in the shop and consisted in the following steps:

1. Fabricating the welded box girder.
2. Fixing 270 high-strength bolts on the bottom chord and 7 longitudinal pre-stressing cables.

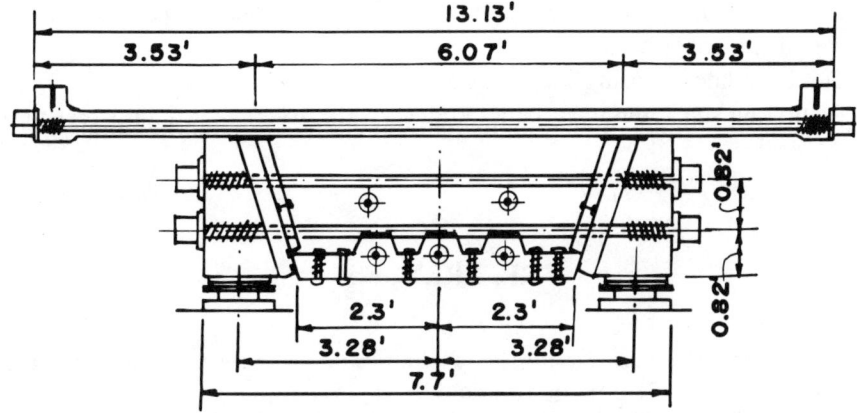

Fig. 7.26   Cross section at support.

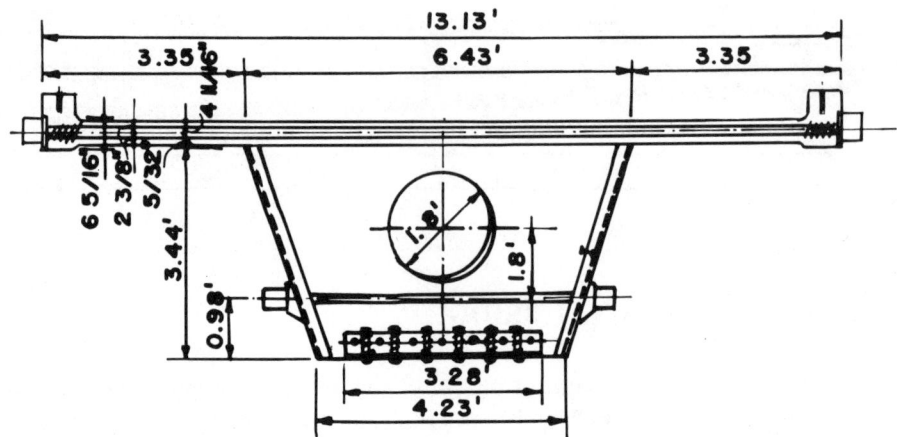

Fig. 7.27 Cross sections at mid-span. Prestressed composite beam of Willstress system.

3. Casting the lower sill and the conical anchors for the prestress cable.
4. After 14 days, putting the lower slab into partial prestress—5 cables of 7—
   on the concrete alone. High-strength bolts placed in ovaling holes provided
   in the bottom chord against displacements of the ends of the sill due to the
   contraction of the concrete.
5. Screwing the five lower sills tight with high-strength bolts.
6. Final prestressing by 2 cables.
7. After concreting the deck, prestressing it in the transverse direction.
8. Tensioning the prestress cables.

The second stage was performed on the site and consisted of the following:

1. Erecting the girder on its two end supports and on a temporary bent at the
   mid-span.
2. Placing the transverse prestress cables.
3. Casting the deck slab.
4. Five days later, putting this slab into partial prestress to counteract shrinkage of the concrete.
5. Ten days later, putting into final prestress.
6. Removing the temporary bent producing the longitudinal prestressing of
   the deck.

REFERENCES

1. Dischinger, F., "Composite Steel Bridges Prestressed by High Strength Cables," Der Bauingenieur, nos. 11 and 12, 1949, (in German).
2. Seidel, H., "Project of Two Lane Highway Bridge Over the Saale at Schkopau," International

Special Convention "Prestressed Steel Constructions," from 10 to 13 September 1963, 1964, no. 30, (in German).

3. Anon. Wissenschaftliche Veroffentlichungen aus der Fakultät für Bauwesen der Techische Universität Dresden, 1964, pp. 24, 27, (in German).

4. Anon. "Revolutionary Design for New Transvaal Bridge," reprinted from *Construction in Southern Africa*, June 1984, pp. 1–7.

5. Anon., "Willstress System Experimental Bridge," Acier, Stahl, Steel, no. 2, 1961, pp. 80–84.

Chapter 8
Trusses

8.1 INTRODUCTION

The prestressing of trusses may be achieved either by the use of tendons or, for continuous trusses, by vertical support displacements. The most developed prestressing truss method is the use of tendons of high-strength materials. The tendons should be located in such a way that under the tensioning of the most heavily loaded members, forces will originate which are in a reverse sign to forces originated due to loading. The scheme of the tendon locations defines the character of its prestressing and truss performance.[1,2,3]

The prestressing truss effectiveness depends on the design of the truss, its tendons, and the sequence of prestressing. From the location of the tendons and the effect upon their structural performance, the prestressed trusses may be divided into two basic types:

1. Trusses in which the tendons are located within the limits of those members most stressed
2. Trusses in which the tendons are located throughout or within that span section and are used to prestress several or all of the truss members

8.1.1 Prestressing of Single Truss Members

In trusses having single members acting in tension under loading, the tendons are installed along the members to prestress them by compression (Fig. 8.1). These members are prestressed during fabrication or during preassembly at the erection site. Such types of truss are more complicated in design and require a greater number of tendon anchorings. Such prestressed trusses are only effective for those large spans and loads where each prestressed member is an individual prefabricated unit. Such trusses are designed as statically determined systems.

The choice of prestressed cross sections is determined by the methods used for the design of those tension members which require prestressing. The savings in steel in such trusses may reach 10–15%.

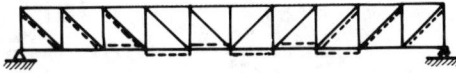

Fig. 8.1   The prestressed single members of a truss.

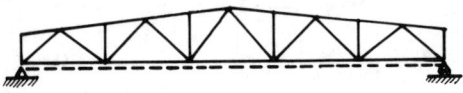

Fig. 8.2   The placement of tendons along a truss span.

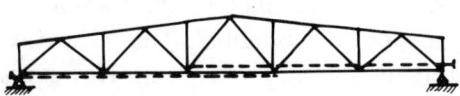

Fig. 8.3   Two or more tendons along a truss span.

8.1.2 Prestressing Along a Part or the Whole Truss Span

The prestressing of these types of trusses provides greater possibilities in terms of the tendon configurations and schemes for their placements. The simplest scheme is obtained when a tendon is located along the chord in tension and extended through a number of panels (Fig. 8.2). In this case, the tendon creates prestressing in those panels along which it is placed; however, other truss members are not prestressed. For large spans, when the difference of the panel forces are substantial, it is convenient to install two or more tendons (Fig. 8.3). The savings in steel in such trusses may reach 10–15%.

When a tendon is polygonally shaped, it is possible to create prestressing at the chord and web members of the truss (Fig. 8.4).

When a tendon is located outside of a truss, it is possible to obtain an economy in steel of 25–30% (Fig. 8.5). However, such exterior tendons cannot always be used, considering the design conditions and truss configuration. Aside from this, a tendon which is not connected to the bottom chord does not ensure a loss of stability. To eliminate a possible loss of truss stability during prestressing, it is necessary to perform prestressing during erection after the installation of bracings between trusses, which stiffen the truss structure against the loss of stability.

The prestressing of tendons may be achieved by constructing a space structure from connected trusses, as in Fig. 8.6(a), or from stable trusses having three-chord systems, as in Fig. 8.6(b).

The effective performance of a truss depends mainly on the conveniently chosen configuration, the chord slopes, the truss system, and the prestressing inten-

Fig. 8.4   A truss having a polygonal tendon.

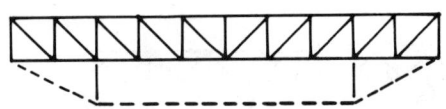

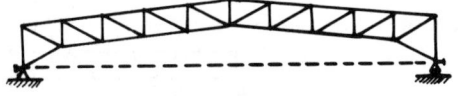

Fig. 8.5   Trusses having externally located tendons.

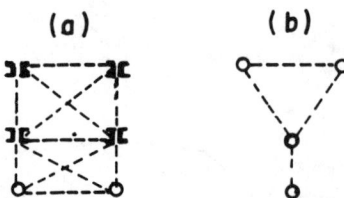

Fig. 8.6   Trusses with outside tendons.

sity. The magnitude of a single-stage prestressing is limited by the stability of its compressed members and, therefore, cannot be substantial. The most economical truss design may be obtained during multistage prestressing. However, it may not always be achieved considering the construction conditions.

During single-stage prestressing, it is rational to perform work in the following sequence. A truss with a tendon is installed and loaded by a partial or total dead load which produces forces of nonprestressed structures. Following that, a tendon is prestressed which not only may cancel those truss member forces originating under dead load, but also may create forces in a reverse sign. Further, the remaining part of the dead and live load is further applied to the truss to enable it to reach its limiting forces. Such a work sequence permits, during a single-stage loading, the achievement, to some degree, of two-stage prestressing.

In continuous trusses, straight tendons should be located along that part of the chord stressed in tension (Fig. 8.7). For a continuous truss, polygonal tendons or outside tendons can also be used (Fig. 8.8).

8.1.3 The Tendon Arrangements

It is more convenient to prestress the tendons before the trusses are erected on the site. Throughout their length, tendons are connected to the chord by diaphragms, spaced at intervals of $40r_{min}$ to $50r_{min}$, where r_{min} indicates the least radius of inertia of the chord cross section. This is to ensure the stability of the chord during prestressing.

Fig. 8.7   Tendon locations in a continuous truss.

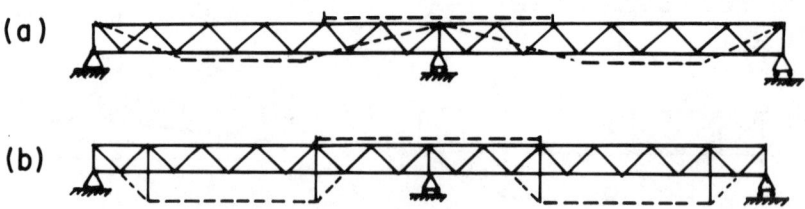

Fig. 8.8   A continuous truss. (a) Polygonal tendons. (b) Outside tendons.

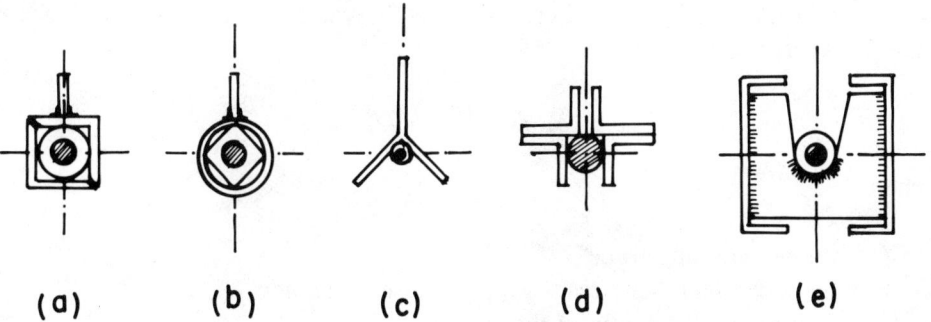

Fig. 8.9 The location of a single tendon in light trusses.

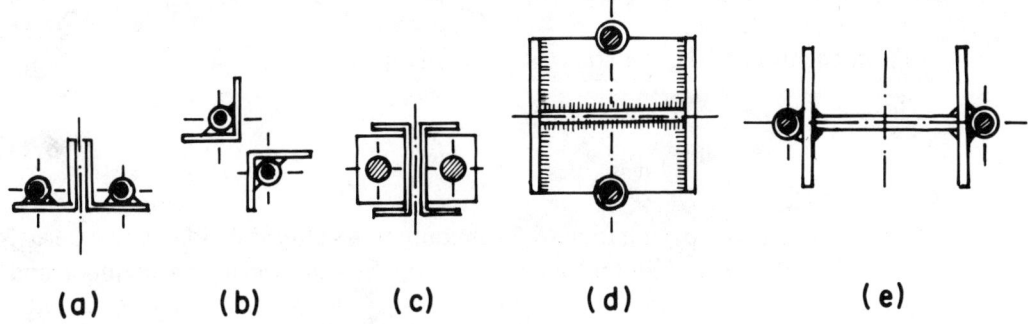

Fig. 8.10 The location of tendons in heavy trusses.

The number of tendons depend on the chord cross-sectional shapes. It is more convenient to have a single tendon such as that shown in Fig. 8.9. When two tendons are necessary, they should be placed symmetrically with respect to the center of gravity of the chord cross section, as in Fig. 8.10.

8.2 TRUSS ANALYSIS

Static truss analysis may be performed in two cases, namely, individually prestressed truss members and those trusses with tendons that prestress several members simultaneously.[4,5,6]

8.2.1 Prestressing of Individual Members

In the first stage, the truss is analyzed under a given loading without considering prestressing. In the second stage, those members acting in tension are designed considering their prestressing.

The cross section of the prestressed members in tension is designed by applying the intensity of the calculated force as follows. Assuming that the member is not loaded by external forces (own weight and live load), the general expressions between the prestress and cross-sectional areas of the member and tendon may be written as follows

$$X = -Af = A_c f_c \qquad (8.1)$$

where

X = the prestressing force
A = the cross-sectional area for a prestressed member
A_c = a cross-sectional area of the tendon
f = the compressive stress in the cross-sectional area of a prestressed member
f_c = the compressive normal stress in the tendon

From equation (8.1), the stresses in the member and tendon are:

$$f = -\frac{X}{A} \qquad f_c = \frac{X}{A_c} \qquad (8.2)$$

The stresses given by equation (8.2) are only those stresses without an external load. By applying an external axial load, F, the stresses in the member and tendon become

$$f = -\frac{X}{A} + \frac{F_m}{A} \qquad f_c = \frac{X}{A_c} + \frac{F_c}{A_c} \qquad (8.3)$$

where

F_m = the axial force in a member due to external load
F_c = the axial force in a tendon due to external load

The following condition should be satisfied

$$F = F_m + F_c \qquad (8.4)$$

The elongations of the prestressed member and tendon should be equal. According to Hooke's law, we have

$$\frac{F_m l}{EA} = \frac{F_c l}{E_c A_c} \qquad (8.5)$$

where

l = the length of a connected prestressed member and tendon
E, E_c = the moduli of elasticity of a member and tendon, respectively

By designating with $\alpha = EA/E_cA_c$, we obtain from equations (8.4) and (8.5) those forces F and F_c which are taken by the member and tendon.

$$\frac{EA}{E_cA_c} = \frac{F_m}{F_c} = \alpha \qquad (8.6)$$

and

$$F_m = \alpha F_c = \alpha(F - F_m)$$

Therefore

$$F_m = F\frac{\alpha}{1 + \alpha} \qquad (8.7)$$

and

$$F_c = F - F\frac{\alpha}{1 + \alpha} = F\frac{1}{1 + \alpha} \qquad (8.8)$$

Due to the external tensile force F, a tensile stress originates in the member which reduces the magnitude of force F. In the tendon, the stresses do not change signs, but rather increase through the tensile force F. Since the member and tendon elongations should be equal, we obtain

$$\frac{(F - \Delta X)l}{EA} = \frac{\Delta Xl}{E_cA_c} \qquad (8.9)$$

where $\Delta X =$ an increase in the tendon force X due to the external load.

From equation (8.9), we have

$$F - \Delta X = \Delta X\frac{EA}{E_cA_c} = \Delta X\alpha$$

from which we obtain

$$\Delta X = F\frac{1}{1 + \alpha} \qquad (8.10)$$

By using a summary of those member and tendon forces which could be expressed by the force F and prestressing force X, we obtain

$$F_m = F - X - \Delta X \qquad F_c = X + \Delta X$$

then the total corresponding stresses should satisfy the condition

$$f = \frac{F - X - \Delta X}{A} \leq f_{all} \qquad (8.11)$$

$$f = \frac{X + \Delta X}{A_c} \leq f_{c, all} \qquad (8.12)$$

where

$f_{all} = f_y/\nu$ = the allowable stress of the prestressed member
$f_{c, all} = f_t/\nu_t$ = the allowable tendon stress
f_y = the yield limit of steel for the prestressed members
f_t = the limiting prestressed tendon tensions
ν = the safety coefficient of the prestressed member
ν_t = the safety coefficient of the tendon

By substituting the expression for X from equation (8.10) into equations (8.11) and (8.12) and considering the relaxation of the tendon's steel, we obtain the following expression for the determination of the prestressed member and tendon stresses.

$$f = -\frac{\psi X}{A} + \frac{F\alpha}{(1 + \alpha)A} \leq f_{all} \qquad (8.13)$$

$$f_c = \frac{\psi X}{A_c} + \frac{F}{(1 + \alpha)A_c} \leq f_{c, all} \qquad (8.14)$$

The prestressed member, apart from the compressive stresses, may eventually be under buckling, which should be checked as follows:

$$\frac{\omega X}{A} < f_{all} \qquad (8.15)$$

where

ω = the buckling coefficient depending upon the slenderness of the member and type of steel
ψ = the coefficient considering the loss of tendon stress due to relaxation and creep in the tendon

A check of the stress expressed by equation (8.15) can be performed when the structure is prestressed under such conditions that the member is loaded by external loading.

To obtain the minimum amount of material, the cross sections of the prestressed members and tendons should be chosen to satisfy conditions $f = f_{all}$ and $f_c = f_{c, all}$.

By designating

$$\alpha_0 = \frac{E_c}{E} \tag{8.16}$$

and substituting this value into equation (8.6), where

$$\alpha = \frac{EA}{E_c A_c} \tag{8.17}$$

From equations (8.4), (8.13), (8.15), and (8.17), expressions for the calculation of the required cross sections of the prestressed members, tendons, and prestressing forces may be obtained after certain transformations, as follows:

$$A = \frac{F}{f_{all}} \frac{1 - \alpha_0 \dfrac{f_{all}}{f_{c,all}} \left(1 + \dfrac{\psi}{\omega} \right)}{\left(1 - \alpha_0 \dfrac{f_{all}}{f_{c,all}} \right) \left(1 + \dfrac{\psi}{\omega} \right)} \tag{8.18}$$

$$A_c = \frac{F - f_{all} A}{f_{c,all}} \tag{8.19}$$

$$X = \frac{1}{\psi} \left(\frac{FA}{\alpha_0 A_c + A} - f_{all} A \right) \tag{8.20}$$

8.2.2 Prestressing of Individual Members by the Optimization Method

1. Introduction

The following method concerning optimum prestressing of individual members of the truss was developed by Speransky.[7] During the design of separate prestressed steel members of the truss, the following coefficients are used:

a. Coefficient of overloading under prestressing force is $n_1 > 1$ considering the possibility of increase the actual prestressing over design prestressing, and the coefficient $n_2 < 1$, considering the reduction of the actual prestressing force, the loss due to the relaxation and yielding of the anchorage.
b. Coefficient of overloading, $n_1 = 1.1$, is used in the following two cases:
   • during checking of the member under prestressing, without external loading;
   • during checking of the member when the stresses under external loading coincide by sign with the prestressing.
c. The coefficient $n_2 = 0.9$ is used during checking of the member under

external loading, when the stresses are greater and have the reverse sign of the prestressing.

d. In the case of safe and direct control of the prestressing, $n_1 = n_2 = 1$.

e. During design of the member having a tendon of steel cable, and safe and direct control is assured, the values of these coefficients are $n_1 = 1.05$ and $n_2 = 0.95$.

2. Basic Design Assumptions

The rigid member and tendon working together are a statically indeterminate system. For their analysis it is necessary to assume the distribution of the material between rigid member and tendon as follows:

$$K = \frac{A_t}{\Sigma A} \quad (1 - K) = \frac{A_m}{\Sigma A} \quad \Sigma A = A_t + A_m \tag{8.21}$$

where A_t and A_m are the cross-sectional areas of the tendon and member, respectively.

The required total cross-sectional area is

$$\Sigma A = \frac{F_{tot}}{f_m \left[(1 - K) + K\frac{f_t}{f_m} \right]} \tag{8.22}$$

where

F_{tot} = total force acting in the member

f_t, f_m = allowable stresses of the tendon and member, respectively

$$A_t = K\Sigma A \quad A_m = (1 - K)\Sigma A \tag{8.23}$$

The prestressing force is

$$Z = \varphi f_m A_m \tag{8.24}$$

where φ is the coefficient of the longitudinal bending of the member, used in the range 0.9–0.95.

The force in the tendon acting under total design force under external loads is

$$X_t = \frac{F_{tot} A_t \dfrac{E_t}{E}}{A_t \dfrac{E_t}{E} + A_m} \tag{8.25}$$

where E_t and E are the moduli of elasticity of the tendon and member, respectively. The tendon stress is

$$f_t = \frac{Zn_1 + X_t}{A_t} < f_{\text{all}} \qquad (8.26)$$

The force in the cross section of the member under loading is

$$F_m = F_{\text{tot}} - (Zn_2 + X_t) \qquad (8.27)$$

where n_1 and n_2 are the coefficients of overloading and underloading under prestressing of the tendon. Checking of stress in the member under loading,

$$f_m = \frac{F_m}{A_m} \qquad (8.28)$$

With the above design method, optimal use of metal or cost of the prestressed member requires repetition of design procedure. However, it is possible to eliminate repetitive design by the following method. Considering the total design-carrying capacity of the member and tendon, we may write

$$Zn_1 + \Delta F_{\text{tot}} = f_t A_t \qquad (8.29)$$

$$-Zn_2 + (F_{\text{tot}} - \Delta F_{\text{tot}}) = f_m A_m \qquad (8.30)$$

Equation (8.29) relates to the tendon and equation (8.30) to the member. By solving the system of equations (8.29) and (8.30), we obtain the following formulas for the design of a prestressed truss member

$$A_t = \frac{n_1 \varphi \beta \alpha^2 f_m}{\alpha(\beta f_t + n_1 \varphi f_m) - F_{\text{tot}}} \qquad (8.31)$$

$$A_m = \alpha - \frac{A_t}{\beta} \qquad (8.32)$$

$$\alpha = \frac{F_{\text{tot}}}{f_m(1 + n_2\varphi)} \qquad (8.33)$$

$$\Delta F_{\text{tot}} = \frac{F_{\text{tot}} A_t}{A_t + \beta A_m} \qquad (8.34)$$

$$Z = \varphi f_m A_m \qquad (8.35)$$

where

ΔF_{tot} = part of the total force under external loading, acting in the tendon
α = cross-section of the member, reduced to the rigid material
$\beta = E/E_t$ = ratio of moduli of elasticity of the member and tendon

Knowing the total design force acting in the member and choosing the material, we may obtain, using formulas (8.31) and (8.35), cross-sectional areas of the member and tendon and also prestressing force.

From expression (8.32) it follows that it is possible to realize the prestressing only when $A_m > 0$, or

$$\alpha - \frac{A_t}{\beta} > 0 \qquad (8.36)$$

By the introduction the denominations for ratios of design stresses of the tendon and member $K = f_t/f_m$, and after certain transformations of formula (8.36), we obtain the condition of possible prestressing of the member

$$K > \frac{F_{tot}}{\alpha \beta f_m} \qquad (8.37)$$

and after substitution value of α, we have

$$K > \frac{1 + \varphi n_2}{\beta} \qquad (8.38)$$

At $\beta = 1$, $\varphi = 1$, and $n_2 = 0.9$, the minimum value of ratio $K \geq 1.9$. At smaller φ the ratio K also diminishes.

We intend to evaluate how the effect of prestressing of the members influences the change of the design stress ratio K and the coefficient of longitudinal bending φ, or more exactly, flexibility of the member. For this purpose we introduce into formula (8.31) the value K, and after certain transformations we obtain the following expression for the cross-sectional area of the tendon

$$A_t = \frac{n_1 \varphi \beta F_{tot}}{f_m(1 + n_2 \varphi)\left[\beta K + (n_1 - n_2)\varphi - 1\right]} \qquad (8.39)$$

8.2.3 Numerical Example

Find the cross-sectional area of a prestressed truss member that is under tensile force $F_{tot} = 380$ kips. The allowable stress of the member is $f_m = 30$ ksi. The high-strength steel tendon bar has the allowable stress $f_t = 135$ ksi. The coefficients are: $K = 5$, $\varphi = 0.9$, $\beta = 1$, $n_1 = 1.1$, and $n_2 = 0.9$.

Solution

a. The required cross-sectional area of the tendon after formula (8.39) is

$$A_t = \frac{n_1 \varphi \beta F_{tot}}{f_m(1 + n_2\varphi)[\beta K + (n_1 - n_2)\varphi - 1]}$$

$$= \frac{1.1 \times 0.9 \times 1 \times 380}{30(1 + 0.9 \times 0.9)[1 \times 5 + (1.1 - 0.9) \times 0.9 - 1]}$$

$$= \frac{376.2}{226.97} = 1.66 \text{ in.}^2$$

b. The required cross-sectional area of the member after formulas (8.32) and (8.33) is

$$A_m = \alpha - \frac{A_t}{\beta} = \frac{F_{tot}}{f_m(1 + n_2\varphi)} - \frac{A_t}{\beta}$$

$$= \frac{380}{30(1 + 0.9 \times 0.9)} - \frac{1.66}{1} = 7.00 - 1.66 = 5.34 \text{ in.}^2$$

c. Use the cross section of the member as two channels $2[7 \times 21/8 @ 9.8$ lb/ft, having cross section $A_m = 2 \times 2.85 = 5.7$ in.^{2}, and the cross section of the tendon from two high-strength steel bars, each having diameter $D = 1\frac{1}{8}$ in. Therefore,

$$A_t = 2 \times \frac{\pi D^2}{4} = 2 \times \frac{3.14 \times 1.125^2}{4} = 1.98 \text{ in.}^2$$

The force due to the initial prestressing of the tendon may be found from expression (8.35).

$$Z = \varphi f_m A_m = 0.9 \times 30 \times 5.7 = 154 \text{ kips}$$

To check the correct choice of the cross section of the member, we substitute the found values of Z, F_{tot} and into equations (8.29) and (8.30), namely

$$Zn_1 + \Delta F_{tot} = 154 \times 1.1 + 97.97 = 169.4 + 97.97 = 267.37$$

$$= f_t A_t = 135 \times 1.98 = 267.3 \text{ kips}$$

Also

$$-Zn_2 + (F_{tot} - \Delta F_{tot}) = -154 \times 0.9 + 380 - 97.97$$

$$= -138.6 + 380 - 97.97 = 143.43 < f_m A_m$$

$$= 30 \times 5.70 = 171.0 \text{ kips}$$

d. The reduction of the total cross-sectional area of the member due to the prestressing is:

$$\frac{F_{tot}/f_m - (A_m + A_t)}{F_{tot}/f_m} \times 100 = \frac{380/30 - (5.7 + 1.98)}{380/30} \times 100$$

$$= \frac{4.99 \times 100}{12.67} = 39.38\%$$

e. To secure the value of the coefficient $\varphi = 0.9$, it is necessary to install transverse diaphragms between both channels considering lateral flexibility of one channel. For $f_m = 30$ kips, we found from Fig. 4.18 the value of the flexibility coefficient $\lambda = 46$. However, considering that the diaphragm has a somewhat larger diameter of the hole than the diameter of the prestressing bar, we propose the actual flexibility as $0.8\lambda = 0.8 \times 46 = 37$.

Therefore, the transverse diaphragms should be spaced at $a = 37r_y$, where r_y is the minimum radius of gyration of a single channel.

8.3 A TRUSS HAVING A BOTTOM CHORD STRENGTHENED BY TENDONS

8.3.1 General

Prestressed trusses are considered as statically indeterminate systems where an increase of the tendon force is taken as an additional unknown. Applying the usual prestressing method with a straight tendon, the magnitude of the tendon force depends on the assumed cross section of a truss member, its external load, and the tendon cross section (Fig. 8.11).

The first condition for the prestressed member i is

$$F_{i,X} \leq \frac{F_{all}A_i}{\omega} \tag{8.40}$$

where

$F_{i,X}$ = a force in member i under a prestressing force X
F_{all} = the allowable stress in a member

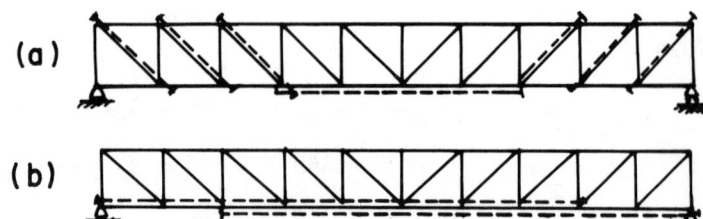

(a)

(b)

Fig. 8.11 Prestressed trusses. (a) Prestressed single diagonals and bottom chord. (b) Prestressing of a bottom chord with two tendons.

ω = the buckling coefficient
A_i = the cross-sectional area of a prestressed member, i

The second condition for a prestressed member is

$$X + \Delta X \leq f_t A_t \tag{8.41}$$

where

f_t = the allowable tendon stress
A_t = the cross-sectional tendon area
X = the magnitude of an assumed prestress force
ΔX = an increase in the tendon force under external load

ΔX can be calculated by using the Maxwell-Mohr principle.

For a simple span truss having a single tendon parallel to its bottom chord, the magnitude of ΔX is

$$\Delta X = \frac{\displaystyle\sum_i \frac{F_{i,X=1} \, F_{i,q} \times l_i}{EA_i}}{\displaystyle\sum_i \frac{(F_{i,v=1})^2 l_i}{EA_i} + \frac{l_t}{E_t A_t}} \tag{8.42}$$

where

$F_{i,X=1}$ = a force in member i due to a prestress force $X = 1$
$F_{i,q}$ = a force in member i due to an external load q
l_i = the length of the member
EA_i = the stiffness of the member
l_t = the length of the tendon
E_t = the modulus of elasticity of the tendon
A_t = the cross-sectional area of the tendom

The values of $F_{i,X=1}$ and $F_{i,q}$ are determined by conventional methods.

8.3.2 A Truss Design

Members of the truss prestressed by the tendon should be analyzed under different loading cases as follows.

a. Forces in the members under external loading should be investigated considering a nonprestressed truss. Also, a following check of the members should be conducted.

b. For compressed members on which the tensile force due to prestressing acts.

In the case when

$$F_{i,q} > F_{i,X=1}(X + \Delta X)$$

we have

$$F_{i,q} - F_{i,X=1}(X + \Delta X) \leq \frac{f_{\text{all}}A_g}{\omega} \tag{8.43}$$

$$\nu F_{i,q} + F_{i,X=1}(X + \nu \Delta X) \leq \frac{f_F A_g}{\omega} \tag{8.44}$$

and for the case when

$$F_{i,q} < F_{i,X=1}(X + \Delta X)$$

we have

$$F_{i,X=1}(X + \Delta X) - F_{i,q} \leq f_{\text{all}}A_n \tag{8.45}$$

$$F_{i,X=1}(X + \Delta X) - \nu F_{i,q} \leq f_y A_n \tag{8.46}$$

c. Compressed members, in which an additional compression force due to prestressing also exists.

$$F_{i,q} + F_{i,X=1}(X + \Delta X) \leq \frac{f_{\text{all}}A_g}{\omega} \tag{8.47}$$

$$\nu F_{i,q} + F_{i,X=1}(X + \nu \Delta X) \leq \frac{F_y A_g}{\omega} \tag{8.48}$$

d. Members under tension which are also under an additional prestressing tensile force

$$F_{i,q} + F_{i,X=1}(X + \Delta X) \leq f_{\text{all}}A_n \tag{8.49}$$

$$\nu F_{i,q} + F_{i,X=1}(X + \nu \Delta X) \leq f_y A_n \tag{8.50}$$

e. Members under tension which are also under compression due to prestressing. In the case when

$$F_{i,q} > F_{i,X=1}(X + \Delta X)$$

we have

$$F_{i,q} - F_{i,X=1}(X + \Delta X) \leq f_{\text{all}}A_n \tag{8.51}$$

$$\nu F_{i,q} - F_{i,X=1}(X + \nu \Delta X) f_{\text{all}} A_n \qquad (8.52)$$

In the case when

$$F_{i,q} < F_{i,X=1}(X + \Delta X)$$

we have

$$F_{i,X=1}(X + \Delta X) - F_{i,q} \leq \frac{F_{\text{all}} A_g}{\omega} \qquad (8.53)$$

$$F_{i,X=1}(X + \nu \Delta X) - \nu F_{i,q} \leq \frac{F_{\text{all}} A_g}{\omega} \qquad (8.54)$$

where

$F_{i,q}$ = a force in member i which is under an external load q (without the influence of tendon)
$F_{i,X=1}$ = a force in member i which is under a prestressing force $X = 1$
X = a prestressing force
ΔX = an additional tendon force due to an external load (after prestressing)
f_{all} = an allowable stress in the truss member
f_y = the steel yield limit for the truss members
ν = the safety coefficient for the truss members
ω = the buckling coefficient
A_g = the gross cross-sectional area of a truss member
A_n = the net cross-sectional area of a truss member

The carrying capacity of a tendon is found from Equation (8.41).

In practice, a simple span truss may be prestressed by two straight tendons (Fig. 8.12). In zone II, where the maximum tensile force acts, two tendons are used having single tendons close to the bearings. By a similar tendon arrangement, a large pressure will occur in the member at its bearings due to the tendon force.

Further, a reduction of the member cross sections will be obtained. The design follows in two stages, namely for zone I and zone II. In zone I, a force X acts and in zone II, a force $2X$, with such distribution that both tendons, AB and A′B′, are prestressed with equal intensities X.

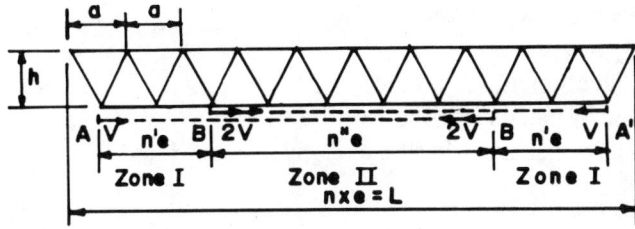

Fig. 8.12 A truss having two prestressed tendons.

In checking the bottom chord of a prestressed truss, equations (8.32) and (8.33) should be used. For zone II, we use

$$F_{i,q}^{II} - 2(X + \Delta X) \leq f_{all} A_n^I \tag{8.55}$$

$$2(X + \Delta X) \leq f_t A_t^{II} \tag{8.56}$$

$$\nu F_{i,q}^{II} - 2(X + \nu \Delta X) \leq f_y A_n^{II} \tag{8.57}$$

From equation (8.57) the safety coefficient is

$$\nu = \frac{2X + f_y A_n^{II}}{F_{i,q}^{II} - 2\Delta X} \tag{8.58}$$

From the equal extension condition of the tendon and bottom chord, we have

$$\frac{\sum_{I} F_{i,q}^{I} - n^I \Delta X}{E A_n^I} - \frac{\sum_{II} F_{i,q} - 2n^{II} \Delta X}{E A_n^{II}} = \frac{(n^I + n^{II}) \Delta X}{E_t A_t^I}$$

And the additional tensile tendon force, due to prestressing of the truss under the action of a load q, may be found from the above expressions

$$\Delta X = \frac{A_n^{1/I} \sum_{I} F_{i,q}^{I} + A_n^{1/II} \sum_{II} F_{i,q}^{II}}{\dfrac{n^I}{A_n^I} + \dfrac{2n^{II}}{A_n^{II}} + \dfrac{n^I + n^{II}}{\alpha_0 A_t^I}} \tag{8.59}$$

where $\sum_{I} F_{i,q}^{I}$ is calculated for a single truss.

The actual tendon stresses may be checked by the following equation

$$f = \frac{2(X + \nu_t \Delta X)}{\nu_t A_t^{II}} \leq f_t \tag{8.60}$$

where

$$\begin{aligned}
F_{i,q} ={}& \text{the tensile force in a bottom chord member } i \text{ at zone I under an}\\
& \text{external load } q \text{ acting upon the truss without considering the}\\
& \text{participation of the tendon}\\
F_{i,q}^{II} ={}& \text{the above for zone II}\\
\nu_t ={}& \text{the safety coefficient for a tendon}\\
n^I ={}& \text{the number of panels in zone I of the truss}\\
A_n^I \text{ and } A_n^{II} ={}& \text{the cross-sectional areas for the tendons in zones I and II, where}\\
& A_t^{II} = 2A_t^I
\end{aligned}$$

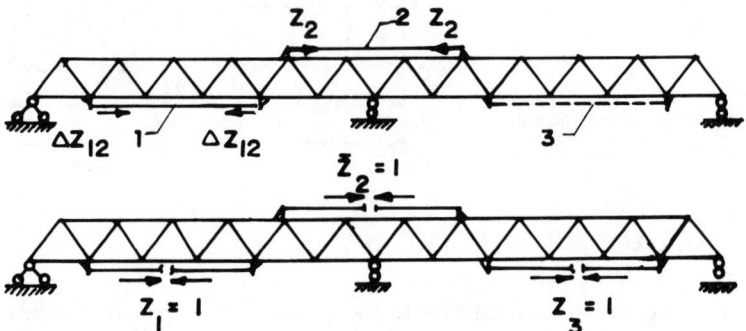

Fig. 8.13   Schemes of prestressed continuous trusses.

8.3.3 Design of Prestressed Continuous Trusses

In continuous trusses, prestressing of one part of the truss influences the remaining parts; therefore, partial prestressing affects whole system. Tensioning
of the cables will introduce forces in the truss members, which are determined
by the usual methods of structural analysis. Therefore, during this installation
of cables and their prestressing, it is necessary to consider the change of tensioning in the cables installed earlier.

Considering, for instance, the tensioning of cable 2 in. Fig. 8.13, by force Z_2
at the earlier installed cable 1, the tensioning change of the value ΔZ_{12} is defined
by the equation

$$\delta_{11}\Delta Z_{12} + \delta_{12}Z_3 = 0 \qquad (8.61)$$

where δ_{11} and δ_{12} are displacements caused in the basic system (statically determinate or indeterminate) under unit forces, applied at the cut of the cables 1 and
2.

By the tensioning of cable 3 by force Z_3, tensioning changes at cables 1 and
2 (ΔZ_{13} and ΔZ_{23}) will be determined from equations

$$\delta_{11}\Delta Z_{13} + \delta_{12}\Delta Z_{23} + \delta_{13}Z_3 = 0 \qquad (8.62)$$

$$\delta_{21}\Delta Z_{13} + \delta_{22}\Delta Z_{23} + \delta_{23}Z_3 = 0 \qquad (8.63)$$

By combining successively the similar equations, we may determine the losses
at each cable caused by the tensioning of the cables, and take them into account
during our choice of controlled forces applied during the tensioning of each cable.

To determine the forces at the members of the basic structure and cables installed into the structure under loading acting on the already prestressed system,
it is necessary to consider forces in the cables as the additional redundants and
to apply the conventional design methods for a statically indeterminate system.

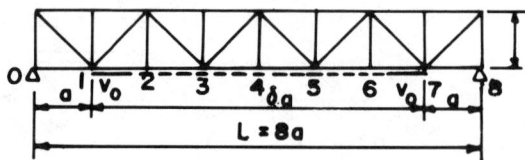

Fig. 8.14 A truss having a tendon shorter than that of its span.

8.4 DEFLECTION OF THE TRUSS

The determination of a deflection due to tendon force X_0 will be shown in an example of one truss which is prestressed by a tendon located at its bottom chord, and having a length less than that of its span (Fig. 8.14). The deflection will be determined by using Maxwell-Mohr rule, which is

$$f = \sum \frac{S_{i,i}S_{i,x}l_i}{EA_i} \qquad (8.64)$$

where

$\quad f$ = the deflection of a truss at the joint under consideration
$\quad S_{i,i}$ = a force in i, due to unit force $P = 1$ at its joint, for which the deflection
$\qquad$ is required
$\quad S_{i,x}$ = a force in member i due to X
$\quad l_i$ = the length of member i
$\quad A_i$ = a cross section of member i
$\quad E$ = the modulus of elasticity of the member

An example is given in Table 8.1, where the forces $S_{i,i}$ and $S_{i,x}$ which are in the truss members due to loading are shown at joint 4 under force $P_4 = 1$ in Fig. 8.14. The bottom chord along the whole span has a constant cross section. The maximum deflection at the middle joint of the bottom chord is

$$f = -4\frac{a^2X}{hEA} - 2x\frac{2a^2X}{hEA} = -8\frac{a^2X}{hEA} \qquad (8.65)$$

The total deflection of the truss is the sum of the deflection of the truss due to the influence of the tendon force and the deflection due to external loading, which act as concentrated forces at the joints.

Table 8.1 Forces in the Truss

Member	Forces	1-2	2-3	3-4	4-5	5-6	5-7
Force $S_{i,x}$ in member	Due to force in tendon	X	$-X$	$-X$	$-X$	$-X$	$-X$
Force $S_{i,i}$ in member	Due to unit force	$\dfrac{a}{h}$	$\dfrac{a}{h}$	$\dfrac{2a}{h}$	$\dfrac{2a}{h}$	$\dfrac{a}{h}$	$\dfrac{a}{h}$

To calculate the optimum carrying capacity, the tendon is located underneath of the bottom chord. In this case for many members of the truss, the required cross-sectional areas may be reduced. Irrespective of slight stiffness of truss members and irrespective of large loading, the total deflection of one prestressed truss is generally smaller than the deflection of the equal but not prestressed truss.

8.5 TYPICAL TRUSS BRIDGES

8.5.1 Pedestrian Bridge, Germany

Figure 8.15 shows a prestressed truss for a pedestrian bridge having a span of 98 ft and a triangular shape in cross-section.[7] The members of this truss have a tubular cross section, and prestressed cables are located inside their bottom chord.

8.5.2 Highway Bridge, Germany

Figure 8.16 shows an elevation and cross section of the bottom chord of a prestressed highway bridge having a 262-ft span.[8] A comparison of the cross-sections of prestressed and conventional structures are:

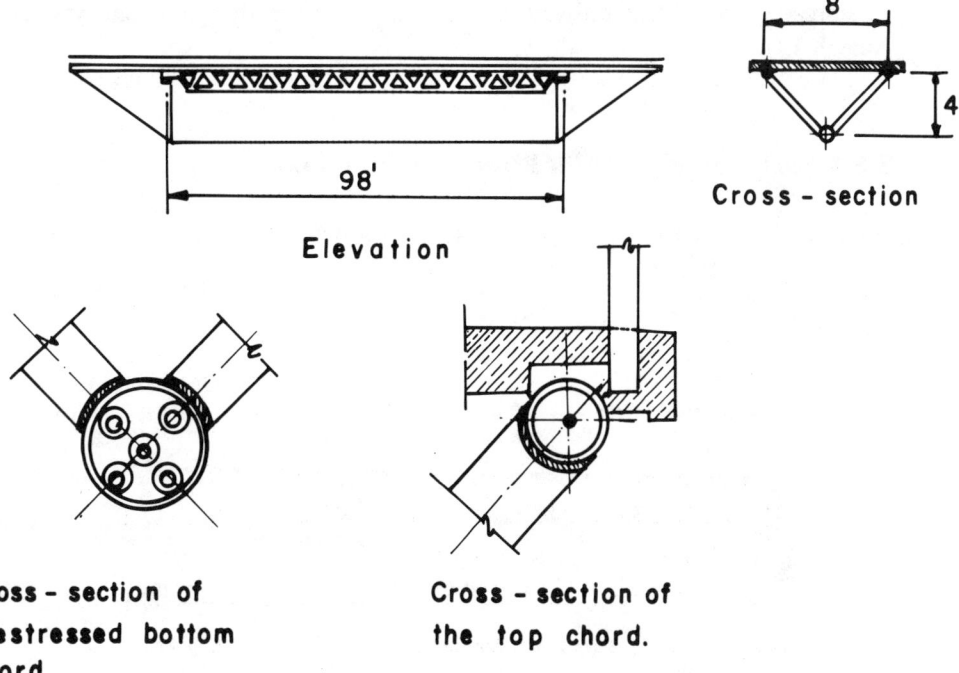

Fig. 8.15   A prestressed pedestrian bridge. Elevation and cross section.

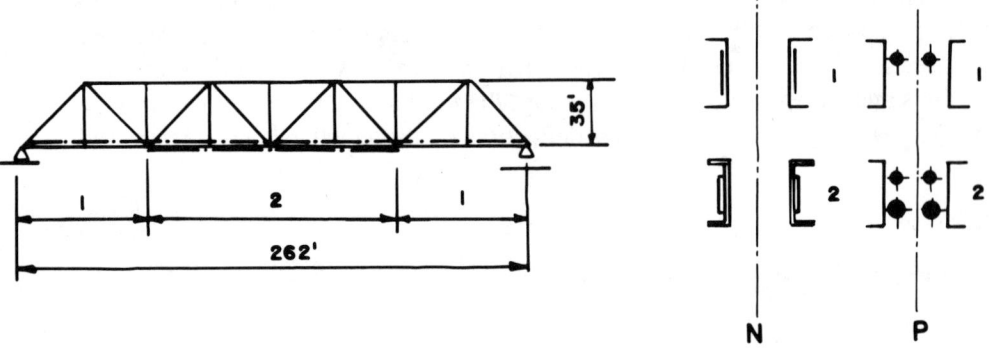

Fig. 8.16   A prestressed truss bridge.

$$\text{Conventional (C)}: A_1 = 96 \text{ in.}^2$$

$$A_2 = 214 \text{ in.}^2$$

$$\text{Prestressed (P)}: A_1 = 46 \text{ in.}^2 \quad \text{Cables: } A_{1C} = 13.5 \text{ in.}^2$$

$$A_2 = 46 \text{ in.}^2 \qquad A_{2C} = 44 \text{ in.}^2$$

8.5.3 Bridge over the Aare River, Switzerland

The prestressing of this railway bridge was performed by polygonal-type cables (Fig. 8.17).[9]

8.5.4 Bridge over the Orlik River, Czechoslovakia

Figure 8.18 shows a bridge prestressed by polygonal cables and having a suspended 482-ft mid-span.[10]

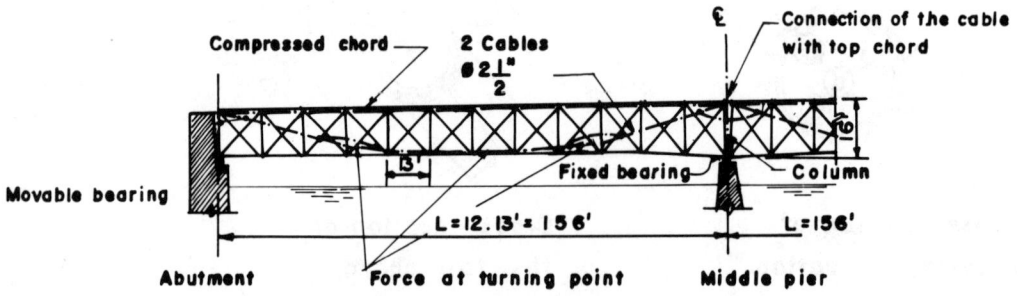

Fig. 8.17   Prestressing of a truss by polygonal cables.

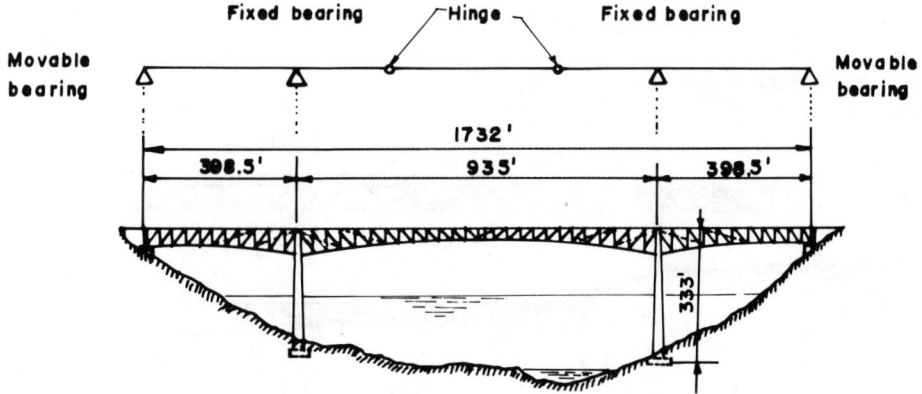

Fig. 8.18   Elevation of a prestressed bridge.

8.5.5 Pipeline Prestressed Frame Bridge

Figure 8.19 shows a bridge over the Elbe Rive, having a span of 328 ft, for the crossing of a gas pipeline.[11] A prestressed portal strut substantially reduces the bending moment at the mid-span.

8.5.6 Prestressing of a Pipeline Bridge by the Regulation of Its Bending Moments

In the pipeline bridge shown in Fig. 8.20, prestressing is achieved by the action of concrete blocks at the cantilevered sections of the supporting welded truss of a triangular cross section.[12]

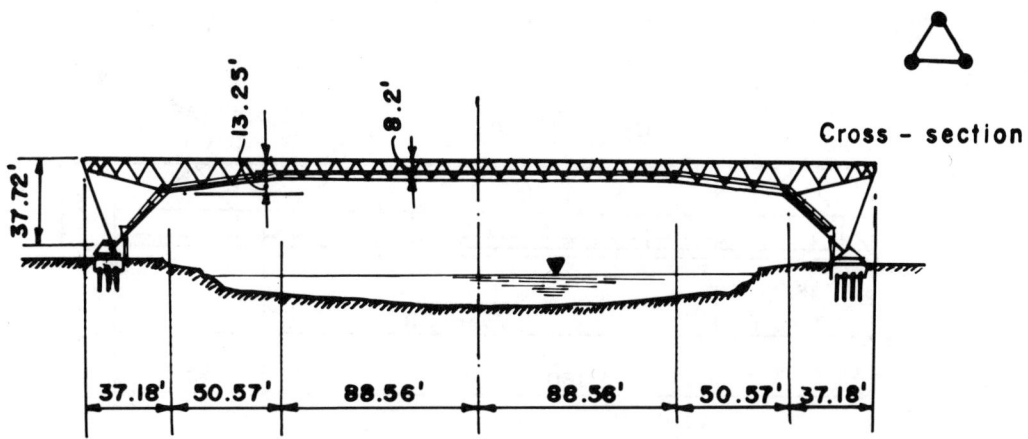

Cross - section

Fig. 8.19   A prestressed frame—supporting pipeline.

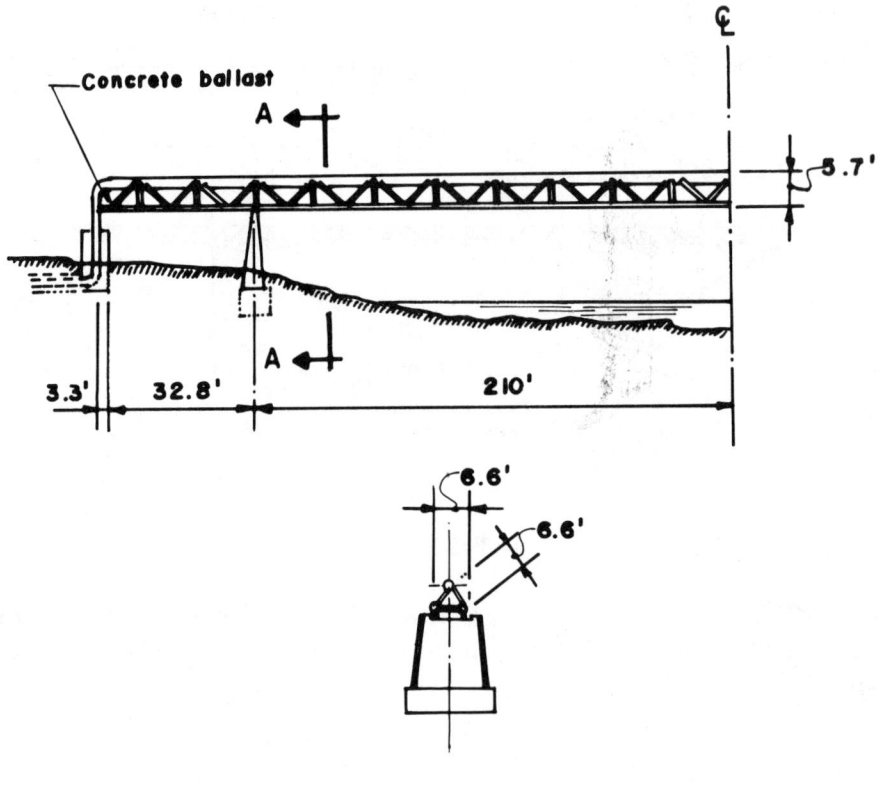

Cross - section A - A

Fig. 8.20   The regulation of bending moments in a pipeline bridge.

8.6 NUMERICAL PROBLEM: PRESTRESSED TRUSS

Design a truss with its bottom chord prestressed and under a uniformly distributed load of $q = 2500$ lb/ft. The tendons are at the axis of the bottom chord
connected at locations between sections 2 and 5 (Fig. 8.21). Moduli of elasticity

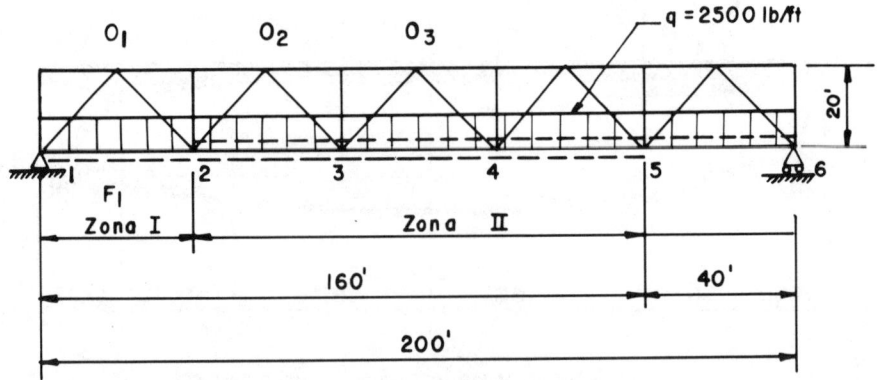

Fig. 8.21   Elevation of the truss.

are: for the steel $E = 30 \times 10^6$ psi, for the high-strength wire tendon $E_t = 29 \times 10^6$ psi. For the tendon wires are used having diameter $d = 0.196$ in. and cross-sectional area $= 0.0302$ in.^{2}.

1. *Axial tensile forces at the bottom of the nonprestressed bottom chord are as follows:*
Panel 1-2

$$\text{Reaction: } A = \frac{2{,}500 \times 200}{2} = 2{,}500 \times 10^2 \text{lb}$$

$$M_{0_1} = 2{,}500 \times 10^2 \times 20 - 2{,}500 \times \frac{20^2}{2} - F_1 \times 20 = 0$$

$$F_1 = \frac{1}{20}\left[2{,}500 \times 20\left(10^2 - \frac{20}{2}\right)\right] = 2{,}500 \times 90 = 225{,}000 \text{ lb}$$

Panel 2-3

$$M_{0_2}: F_2 = \frac{1}{20}\left[2{,}500 \times 60\left(10^2 - \frac{60}{2}\right)\right]$$

$$F_2 = 2500 \times 3 \times 70 = 525{,}000 \text{ lb}$$

Panel 3-4

$$M_{0_3}: F_3 = \frac{1}{20}\left[2{,}500 \times 100\left(10^2 - \frac{100}{2}\right)\right]$$

$$F_3 = 2{,}500 \times 5 \times 50 = 625{,}000 \text{ lb}$$

2. The cross section of the bottom chord is used $2[12 \times 5]$, having cross-sectional area as $A_1 = A_2 = 2 \times 9.26 = 18.52$ in.^{2}.
Tendons
   3. a. Zone I: $_1A_t = 45 \times 0.0302 = 1.359$ in.^{2}
      b. Zone II: $_2A_t = 2 \times 45 \times 0.0302 = 2.718$ in.^{2}
4. The increase of the prestressing in the tendon force will be found by equalizing the elongation of the bottom chord to the elongation of the tendon, namely:
   a. Elongation of the bottom chord in zone I

$$\Delta l_1 = \frac{(F_1 - n_1 \Delta X)a}{EA_1}$$

   b. Zone II

$$\Delta l_2 = \frac{[(2F_2 + F_3) - 2n_2\Delta X]a}{EA_2}$$

c. Elongation of the tendon

$$\Delta l_t = \frac{n_1 a \Delta X}{E_t A_t} + \frac{2 n_2 a \Delta X}{2 E_t A_t} = \frac{(n_1 + n_2) \, a \Delta X}{E_t A_t}$$

where n_1, n_2 = number of panels in zone I and II, respectively.
  d. $\Delta l_1 + \Delta l_2 = \Delta l_t$

$$\frac{F_1 - n \Delta X}{E A_1} + \frac{(2 F_2 + F_3) - 2 n_2 \Delta X}{E A_2} = \frac{(n_1 + n_2) \Delta X}{E_t A_t}$$

$$\Delta X = \frac{\dfrac{F_1}{A_1} + \dfrac{(2 F_2 + F_3)}{A_2}}{\dfrac{1}{A_1} + \dfrac{2 n_2}{A_2} + \dfrac{(n_1 + n_2)}{A_t} \times \dfrac{E}{E_t}}$$

$$\Delta X = \frac{\dfrac{225 \times 10^3}{18.52} + \dfrac{(2 \times 525 + 625) 10^3}{18.52}}{\dfrac{1}{18.52} + \dfrac{2 \times 3}{18.52} + \dfrac{(1 + 3)}{1.359} \times \dfrac{30 \times 10^6}{29 \times 10^6}}$$

$$\Delta X = \frac{12{,}149.03 + 90{,}442.76}{0.054 + 0.324 + 3.045}$$

$$\Delta X = \frac{102{,}591.79}{3.423} = 29{,}971.30 \text{ lb}$$

5. The permissible force in one tendon consisting of 45 wires having $d = 0.196$ in.

$$45 \times 0.0302 = 1.359 \text{ in.}^2$$

$$125{,}000 \times 1.359 = 169{,}875 \text{ lb}$$

6. $X_0 = X + \Delta X = 169{,}875$ lbs

$$X = 169{,}875 - 29{,}971 = 139{,}904 \text{ lb}$$

Use 139,000 lb.
  7. The stress in the tendon is

$$\sigma_t = \frac{X + \Delta X}{A_t} = \frac{169{,}875}{1.359} = 125{,}000 \text{ psi}$$

8. The stress in member 3-4 is

$$\sigma = \frac{F_{3,4} - 2(x + \Delta X)}{A_2} = \frac{625,000 - 2 \times 169,875}{18.52}$$

$$= \frac{625,000 - 339,750}{18.52} = \frac{285,250}{18.52} = 15,402.26 \text{ psi}$$

9. The stress in member 3-4 during prestressing

$$\sigma = \frac{2X}{A_2} = \frac{2 \times 139,904}{18.52} = 15,108.42 \text{ psi}$$

REFERENCES

1. Belenya, E., *Prestressed Load-Bearing Metal Structures*, Mir Publishers, Moscow, 1977, pp. 229–240.
2. Ferjencik, P., and Tochacek, M., "Prestressing of Steel Structures, Theory and Construction Practice," Edition Wilhelm Ernst und Sohn, Berlin 1975, pp. 192–223 (in German).
3. Belenya, E., *Metal Constructions*, Stroiizdat, Moscow, 1976, pp. 278–281 (in Russian).
4. Brodka, J., and Klobukowski, J., "Prestressed Steel Constructions," Edition Wilhelm Ernst und Sohn, Berlin, 1969, pp. 22–24; 41–44 (in German).
5. Streletzkii, N.S. et al, "Metal Construction," Edition of Literatury po Stroiteljstvu, Moscow, 1965, pp. 58–64 (in Russian).
6. Brodka, J., Jerka-Kulawinska, K., Kwasniewski, M., "Prestressed Steel Girder, Statical Calculation," Verlags Gesellschaft Rudolf Muller, Cologne-Braunsfeld, 1968, pp. 65–74 (in German).
7. Speransky, B.A., *Prestressed Metal Trusses*, Edition of Literature on Construction, Moscow, 1970, pp. 88–95 (in Russian).
8. Fritz, B., "Application of prestressing at steel constructions," Zeitschrift für Verein deutsche ingenieur, 1965, no. 22 (in German).
9. Muller, Th., "Rebuilding of highway bridge over the Aare River at Aarwangln," Schweizerische Bauzeitung, 1968, no. 11 (in German).
10. Zeman, J., "Alternative design of our largest steel bridge," Inz. Stavby, no. 6, 1955 (in Czechoslovakian).
11. Hoyer, W. et al, "Handbook for Steel Constructions," vol. III, Verlags für Bauwesen, Berlin, 1974, p. 228 (in German).
12. Ferjencik, P., and Tochacek, M., "Prestressing of Steel Structures, Theory and Construction Practice," Edition Wilhelm Ernst und Sohn, Berlin, 1975, p. 65 (in German).

Chapter 9
Prestressed Arch Bridges

9.1 INTRODUCTION

The prestressing of arch-type steel bridges was proposed by Dischinger in 1949, and in his articles he discussed the detailed prestressing design of such structures as follows.[1]

1. *Through-type arch bridge:* This highway bridge has a central span of 1033 ft (315 m) and its arch rises 115 ft (35 m) (Fig. 9.1). In cross section the bridge has a roadway 56.4 ft (17.20 m) wide and a reinforced concrete composite deck slab $9\frac{1}{2}$ in. thick. The deck slab is supported by six plate girders, each one 11.5 ft (3.5 m) high (Fig. 9.2).

The deck slab is prestressed underneath by 18 high-strength steel cables having a total cross section of 108.5 in.[2]. Also the superstructure is prestressed by 30 cables, having a total cross-section of 225 in.[2], located above the bottom flanges of the plate girders.

2. *Deck-type arch bridge:* This shallow two-hinged prestressed deck arch highway bridge has a span of 820 ft (250 m) and a rise of 49 ft (15 m) (Fig. 9.3). In cross section the width of the roadway is 49 ft (15 m) and the reinforced concrete composite deck slab is 8.7 in. thick.

The deck is supported by four plate girder arches having changeable height (Fig. 9.4). Each arch is prestressed by the high-strength steel cables placed underneath the deck slab. Total arch load due to the own weight and live load is 23900 tons, but part of this load—13,000 tons—is taken by the prestressing cables and the remaining 10,900 tons is taken by the steel arches.

The cables are prestressed at both ends of the bridges by the action of heavy suspended blocks which are supported by the inclined cantilevered beams that are hinged to the arch foundations, as shown in Fig. 9.3.

9.2 DORSTEN BRIDGE, GERMANY

In 1954 a prestressed through arch bridge was built over the Wesel-Dotteln-Canal at Dorsten, having a span of 214 ft (65.2 m) (Fig. 9.5).[2] In cross section

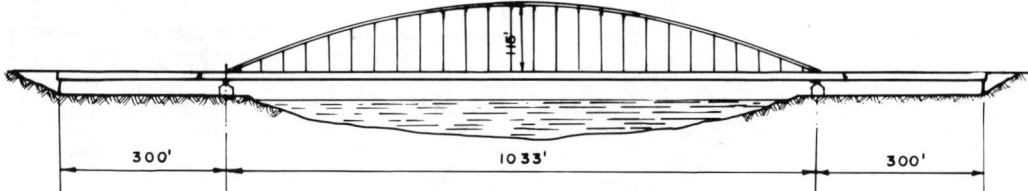

Fig. 9.1 Elevation of through-type arch bridge.

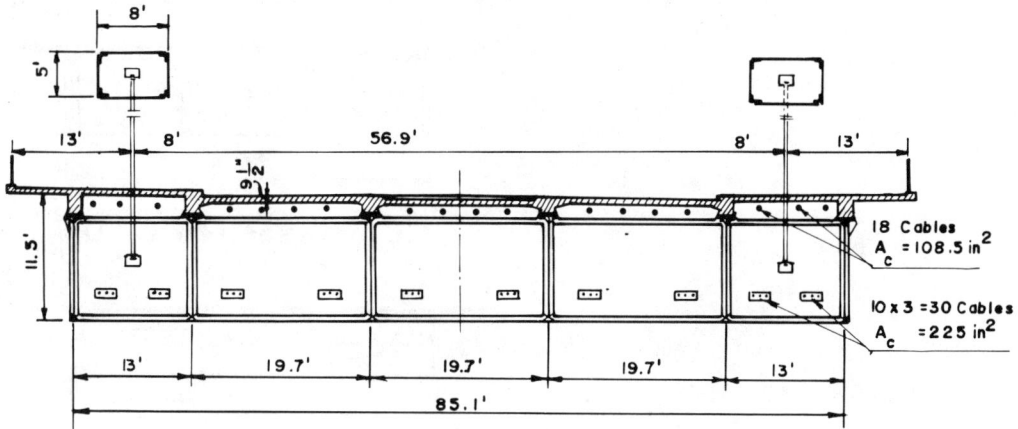

Fig. 9.2 Cross section of the bridge.

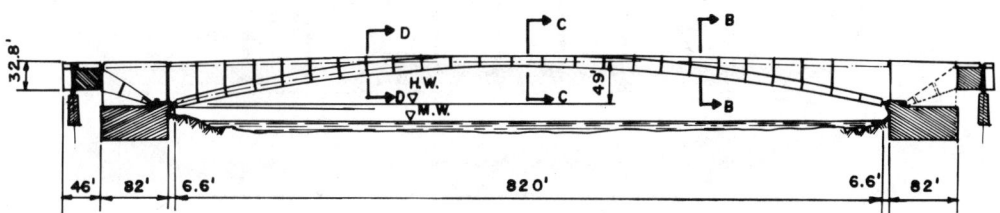

Fig. 9.3 Elevation of the deck arch bridge.

the bridge arches are spaced at 31.8 ft (9.7 m); the roadway width is 26.2 ft (8 m) and two sidewalks each have a width of 13.44 ft (4.1 m) (Fig. 9.6).

The concrete deck slab is prestressed in the longitudinal and transverse directions by the Freyssinet system. The bridge was designed without an upper wind bracing system between the arches. Hangers work together with cross beams as open frames.

The total horizontal tension, which is around 800 tons per arch, is taken by the four high-strength steel poststressed cables, each having a diameter of 3 in. (79 mm).

Details of the anchorage for the prestressing cable are shown in Fig. 9.7.

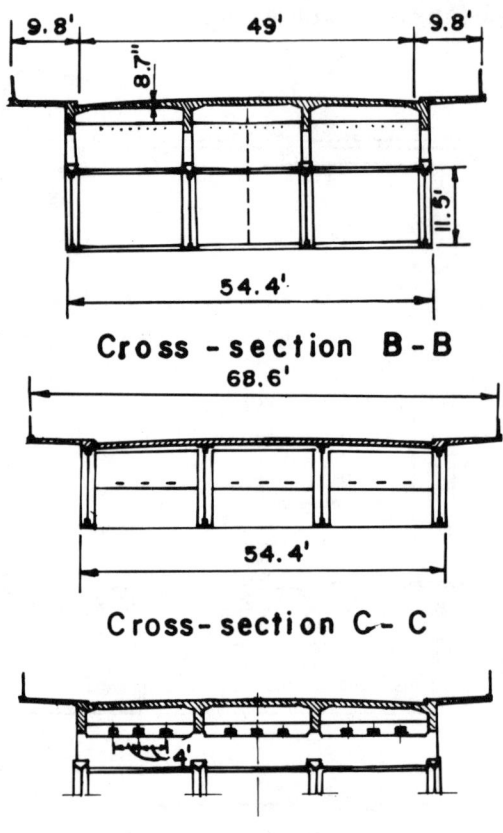

Cross - section B - B

Cross - section C - C

Cross - section D - D

Fig. 9.4 Cross sections of the deck.

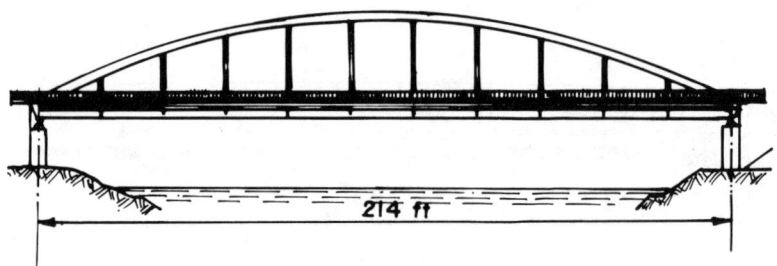

Fig. 9.5 Elevation of Dorsten arch bridge.

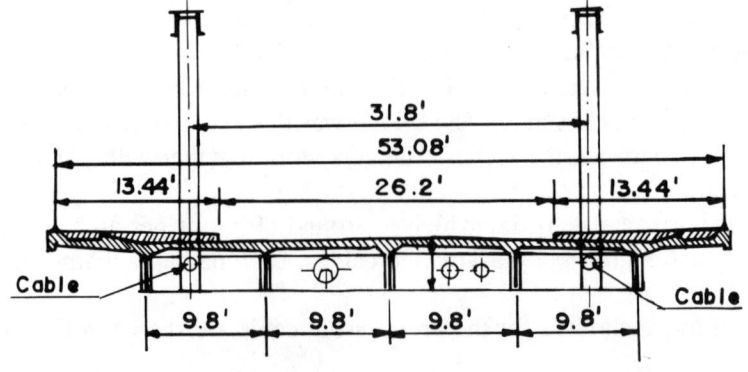

Fig. 9.6 Cross section of Dorsten Bridge.

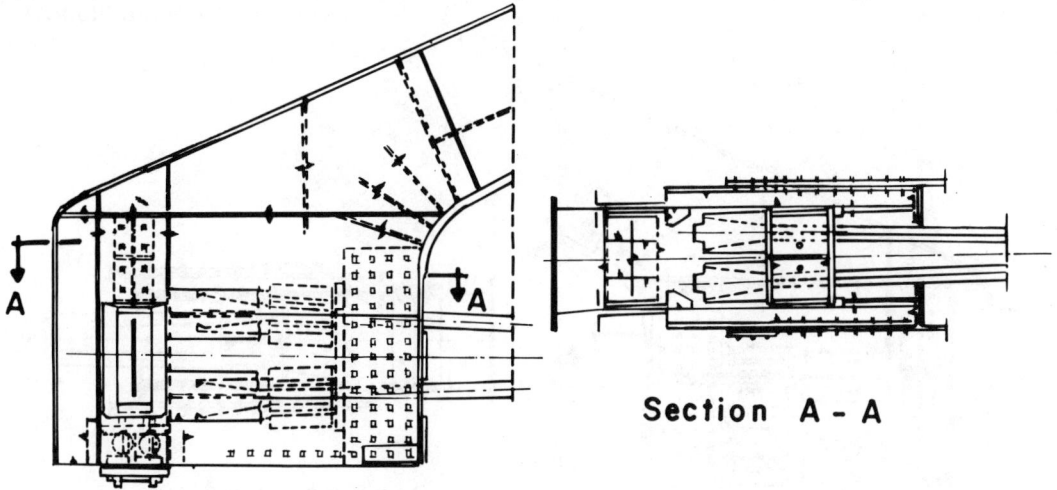

Fig. 9.7   Details of a prestressed cable anchorage.

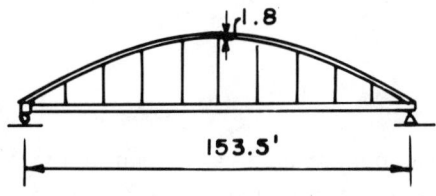

Fig. 9.8   Elevation of Gahmen arch bridge.

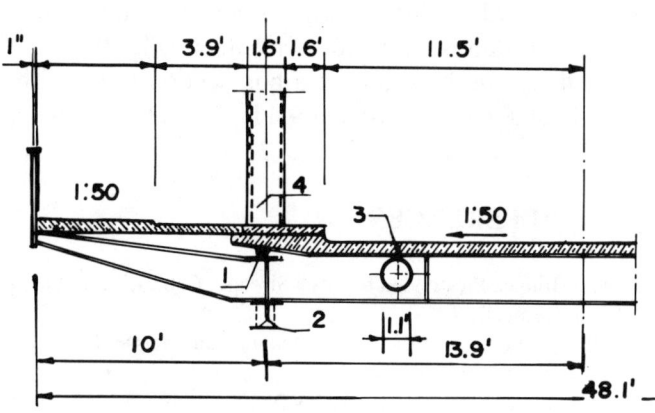

Fig. 9.9   Cross section of Gahmen bridge.

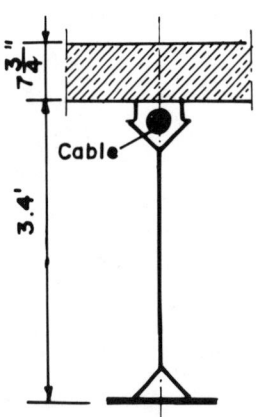

Fig. 9.10   Cross section of the main plate girder.

297

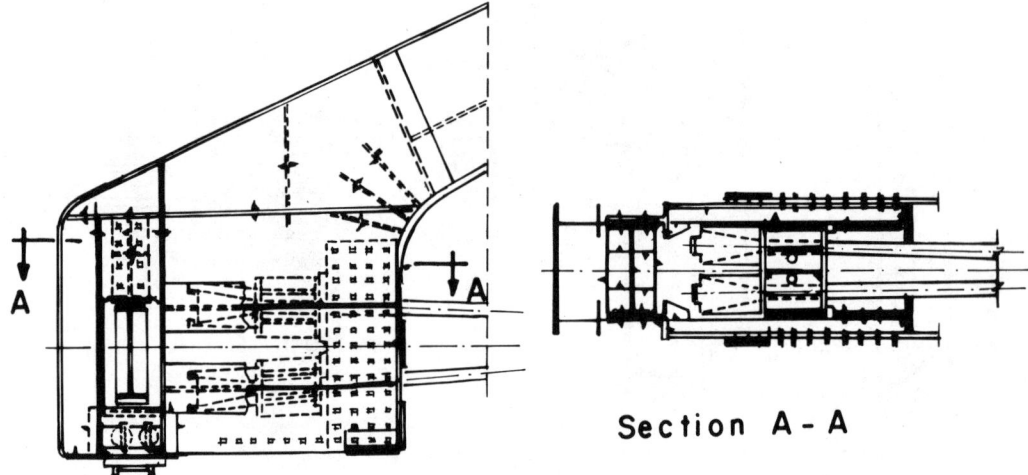

Fig. 9.11   Details of a prestressed cable anchorage.

9.3 GAHMEN BRIDGE, GERMANY

This prestressed through arch bridge of the Langer system has a span of 153.5 ft (46.8 m) (Fig. 9.8).[3] In cross section the spacing of two arches is 27.9 ft (8.5 m) and the width of the roadway is 23 ft (7.0 m). Each sidewalk width is 12.5 ft (3.8 m). The reinforced concrete slab is $7\frac{3}{4}$ in. (0.20 m) thick (Fig. 9.9).

Each main girder has a top flange in the shape of a box section to carry the prestressing high-strength steel cable (Figs. 9.10 and 9.11).

REFERENCES

1. Dischinger, F., "Composite Steel Bridges Prestressed by High Strength Cables," Der Bauingenieur, nos. 11 and 12, 1949 (in German).
2. Bauch, Fr., and Joos, I., "Highway Bridge Over the Wesel-Dotteln Canal in Dorsten," Der Stahlbau, no. 11, 1955, pp. 241–244 (in German).
3. Fritz, B., "Application of Prestressing at Steel Structures," VDI-Zeitschrift, vol. 22, 1956, p. 1279 (in German).

Chapter 10
Cable Truss Bridges

10.1 BRIEF HISTORY

The stiffening of suspension bridges by means of cables is not an entirely new idea. In the past century the famous bridge builder John A. Roebling applied this basic approach to some extent by installing diagonal stays on his suspension bridges in ordinary stiffening trusses. Roebling realized the importance of the adequate stiffening of suspension bridges against wind forces, and his bridges stood up while those built by his contemporaries were wrecked by the wind. After his time the lessons learned from the failures of early suspension bridges were more or less forgotten. It was still recognized that a suspension bridge had to be designed to be safe against horizontal wind loading, but it was also believed that wind loading would not require much vertical strength of the deck beyond that required by traffic demands.

For a long-span suspension bridge, the cables are usually so heavy that the traffic loads become a smaller proportion of total weight than for short-span bridges. This condition resulted in the design of relatively more flexible stiffening trusses and girders for long spans than for short ones. This trend toward the more flexible stiffening girders, together with an insufficient understanding of aerodynamic forces brought, on the Tacoma suspension bridge disaster in 1940.

The collapse of the Tacoma bridge made evident the vital part unforeseen aerodynamical effects can play in structures designed primarily for static loads. Since then the attention of bridge designers has focused on providing an adequately stiffened system that will be effective not only in distributing deck loading, but also in stiffening against aerodynamical forces.

However, it is well known that the satisfaction of both the conditions outlined above resulted in a conventional suspension bridge truss that is relatively uneconomical, since it carries no dead load and yet accounts for about 50% of the entire steelwork weight.

The trend to secure stability against wind forces and still provides an economical system led to a general interest in the possibility of using only systems of

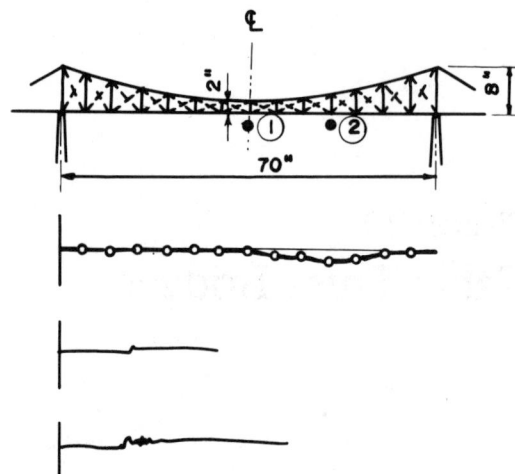

Fig. 10.1   Model of three-member hanger system
proposed by Maney.

cables to aid stiffening in future large structures. Researchers have proposed the
use of single or double diagonal hangers instead of the ordinary vertical ones to
obtain vertical stiffness in suspension bridges. Studies were made concurrently
in 1941 by Maney, who proposed a system of double diagonals between suspenders in each panel (Fig. 10.1),[1] and by the staff of Modjeski and Masters,
who proposed the design of a triangular suspended rope system which seems to
offer damping characteristics of a high order, (Fig. 10.2).[2] To reduce the amount
of steel in stiffening trusses Ostashevskii proposed cable trusses having inclined
suspenders (Fig. 10.3).[3]

During World War II, model tests carried out by the U.S. Engineering Board
at Fort Belvoir and at the John A. Roebling Sons Co. at Trenton confirmed the
theoretical predictions of Maney regarding the behavior of the cable-stiffened
truss (Fig. 10.4), as described by Grove.[4] Further, Grove discussed in detail the
problem of a suspension bridge stiffened by a system of diagonal cables (Fig.
10.5).[5] Shearwood suggested a modification to substitute diagonals for the ver-

Fig. 10.2   Triangular suspended rope system.

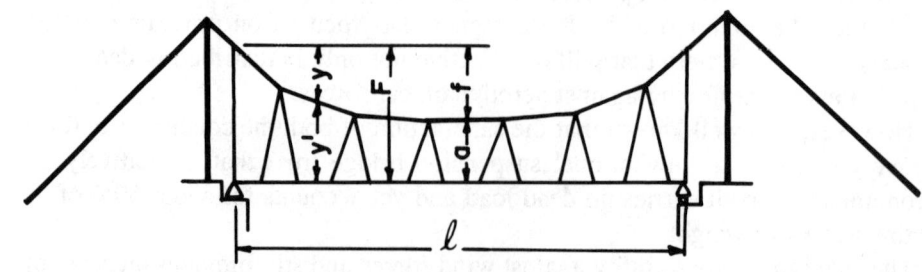

Fig. 10.3   Cable truss having inclined suspenders.

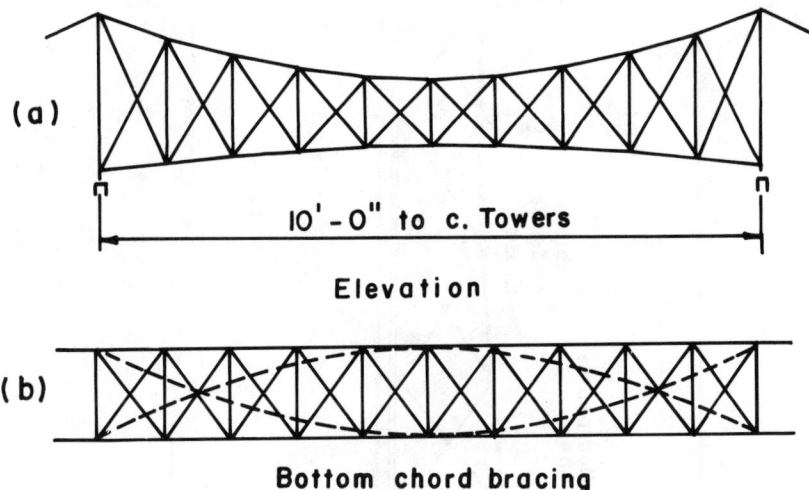

Fig. 10.4   Tests of a cable-stiffened suspension span.

tical hangers and stiffening truss of the conventional type with the cable and a stiff bottom chord (Fig. 10.6).[6]

The first large-scale practical application of these ideas was achieved by John A. Roebling's Sons Co., which in 1954 designed and built with marked economy the continuous five-span cable truss San Marcos Bridge in San Salvador, Central America. A detailed description of this bridge is given later in this chapter.

10.2 GENERAL

Cable truss bridge structures were proposed in 1940 by Ostashewsky.[7] They consist of the main carrying cable supported by pylons and diagonal cables, which create prestressing of the diagonals (Figure 10.7 (a)), or substituted by a prestressed bottom chord with or without a stiffening beam or nonstressed stiffening beam.

An intermediate solution is also possible, namely with the bottom chord from the cable and with stiffening beam connected by vertical suspenders to the joints of the diagonals, shown in Fig. 10.7(b). The thrust of the cables is taken by the anchorages.

The cable truss shown in Fig. 10.7(a) possesses great rigidity of its own and, due to this, may be applied without a suspended stiffening beam. Such a system was applied at the cable truss crossing over the Volga River during construction of the hydroelectric station there.

10.2.1 Structural Behavior of Cable Trusses

Cable trusses with a stiffening beam are systems having changeable static indeterminacy. The diagonals of the truss, consisting of flexible strands that may

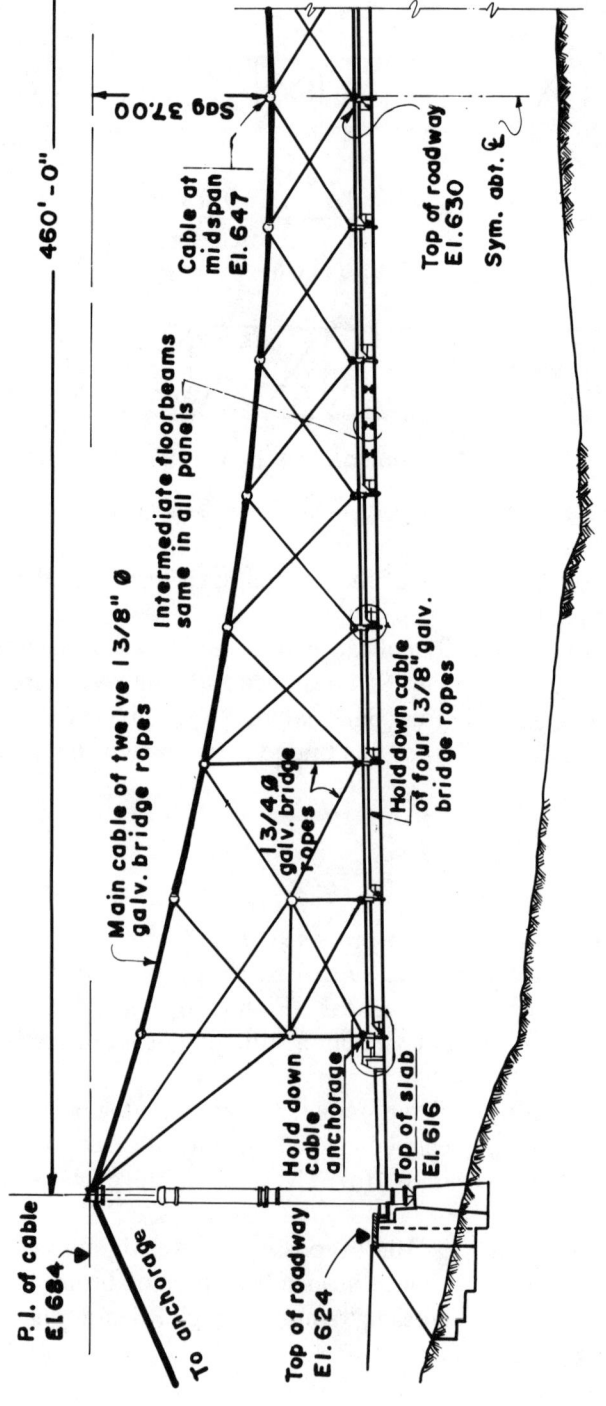

Fig. 10.5 Stiffened suspension bridge.

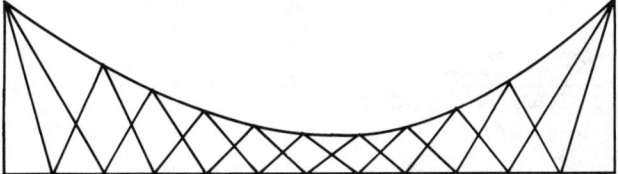

Fig. 10.6 Shearwood's proposal.

work only in tension and after a change in force sign, are excluded from the work, and the working scheme of the truss changes correspondingly.[8] To each loading there is a corresponding "working scheme," the truss scheme that contains only the working elements of the truss. The excluded diagonals are not considered.

From the possible cable truss schemes shown in Fig. 10.7, we will consider only schemes having a stiffening beam which satisfy structural requirements. The systems with the beams that do not work together with the main trusses (Fig. 10.7(a)) are used only in cases without strong requirements regarding deformations of the structure. In the schemes with a stiffening beam, exclusion from the work part of the diagonals at a definite combination of loadings does not influence the geometric configuration of the system at a noncomplete truss.

As theoretical and experimental investigations indicate, the number of diagonals excluded from the work depends mostly on the ratio of the intensities of line load, q, and dead load, g. Thus at $q/g = 1.8$ almost all diagonals are excluded which are necessary for a nonchangeable geometric configuration of the system, and at the ratio of $q/g = 0.09$ no diagonal is excluded from the work. Exclusion and inclusion of the diagonals are performed smoothly without dynamic impact even with a dynamic change in the moving of live load.

The tests indicate also that the change of the truss working scheme is not proportional to the change of loading and follows nonlinear loads, which leads to the nonlinear work of the whole system under loading.

The stiffening beam at a complete cable truss performs only under local loading as a continuous beam on elastic supports which are truss joints having spans equal to the panel length. At a noncomplete truss, namely when separate diagonals are excluded from the work, the participation of the stiffening beam in the

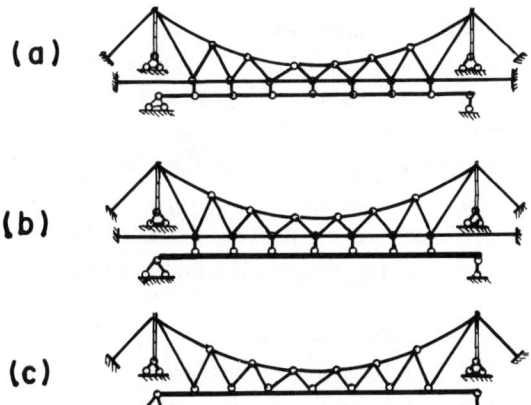

(a)

(b)

(c)

Fig. 10.7 Schemes of cable trusses. (a) Prestressed bottom chord without stiffening beam. (b) Same as (a) with stiffening beam. (c) Nonstressed stiffening beam.

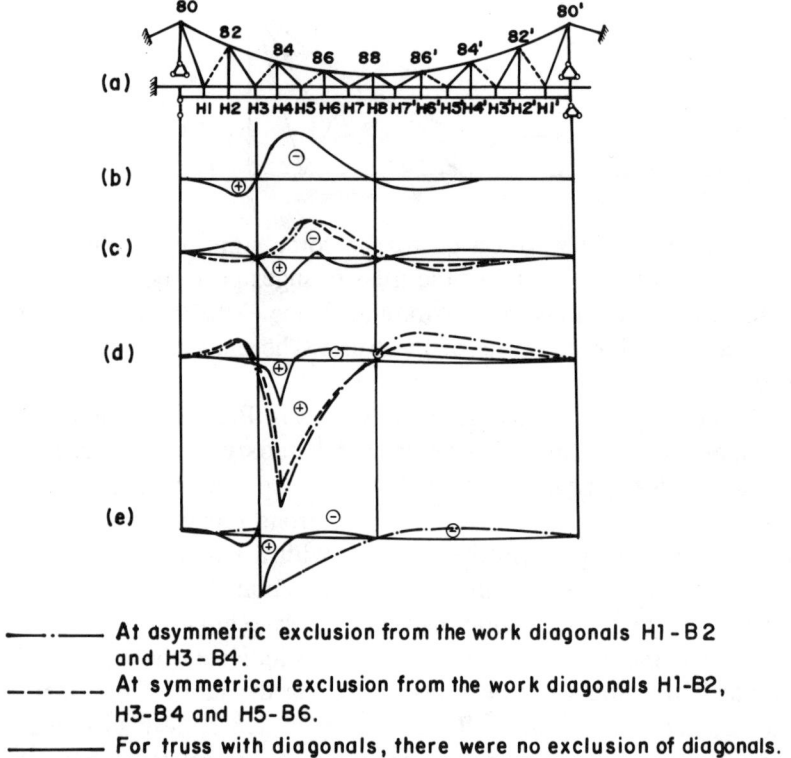

———·——— At asymmetric exclusion from the work diagonals H1-B2
and H3-B4.

——————— At symmetrical exclusion from the work diagonals H1-B2,
H3-B4 and H5-B6.

——————— For truss with diagonals, there were no exclusion of diagonals.

Fig. 10.8   Working scheme of a cable truss. (a) Basic system. (b) Influence line for the force in diagonal H3-B4. (c) Same force in suspender H4-B4. (d) Same force for bending moment in the stiffening beam at section at $\frac{1}{4}$ of span. (e) Same force for transverse force at section $\frac{3}{16}$ of span.

work of the structure increases. It starts to work in general bending together with the truss, ensuring nonchangeable geometry of the system at parts with noncomplete trusses. Bending moments in the stiffening beam increase abruptly, surpassing normal (at a complete truss) a few times. When separate diagonals are excluded from the work, their loading is transmitted to the neighboring elements of the truss, changing their forces not only according to their magnitude, but also by their sign (Fig. 10.8).

The deflections of the cable truss are virtually independent of the beam rigidity and are determined by the rigidity of the truss itself. The influence of the beam on the deformability of the system is expressed in a small degree and practically only during exclusion from the work part of diagonals.

The increase of deflection amplitude during exclusion from the work part of diagonals may reach 150–160%, but the deflection increase is only due to that part of the loading which causes the inclusion or exclusion of diagonals. The basic part of the loading acts on the truss having diagonals which are not excluded from the work.

The prestressing of diagonals is useful and necessary only in systems having a bottom chord of cables, as in Fig. 10.8(b). For a system having diagonals connected to the stiffening girder, the introduction in the truss of the internal forces on account of redistribution of forces among diagonals does not achieve the purpose, because this system does not eliminate the diagonals' inclusion or exclusion from the work.

Trusses having prestressed diagonals work similarly to those having nonprestressed ones. This property of cable trusses having rigid beams is explained by the insignificant deformation of prestressed diagonals in comparison with general deformations of a cable truss under loading.

The degree of static indeterminacy of a cable truss is determined by the formula

$$i = n - D \qquad (10.1)$$

where

n = number of joints of diagonal connections with bottom chord
D = number of diagonals which are excluded from the work

In the absence of fixed supports of the beam the degree of static indeterminacy is correspondingly diminished. The formula is correct at any inclination of the suspenders.

During the design of a cable truss it is necessary to consider the change of its working scheme under different loadings. Under constant uniformly distributed loads of the first stage, the stiffening beam may be analyzed as continuous on rigid supports, taking into account that during erection the settlements of the beam may be eliminated by raising the joints of the truss. Under constant loading of the second stage, or under loading applied to the truss after erection, the truss is designed as a cable truss taking into consideration the work of all diagonals.

Under live load the system is designed with separate changeable connections for each combination of loadings. In each case this requires, first, assuming a working scheme for the truss and, after full analysis, checking the correspondence of the obtained directions and displacements with the accepted working scheme. If the forces at some diagonals of the working scheme happen to be compressive or if displacements along the directions of some diagonals absent in the working scheme are moving the ends of these diagonals away from one another, then the accepted working scheme should be made more precise and the analysis should be repeated.

After analysis by second approximation, the results are again compared with the more precise working scheme. The correspondence of the signs of forces in the diagonals and the direction of the displacements of the joints are checked against the accepted working scheme.

For the preliminary determination of the forces in the diagonals at each position of the loading it is useful to apply approximate methods. In particular it is possible to consider as a first approximation a statically determinate cable truss system, obtained from the design by the introduction of hinges in joints, assuming conditionally that all members work equally in tension and compression, the so-called principle of solidification.

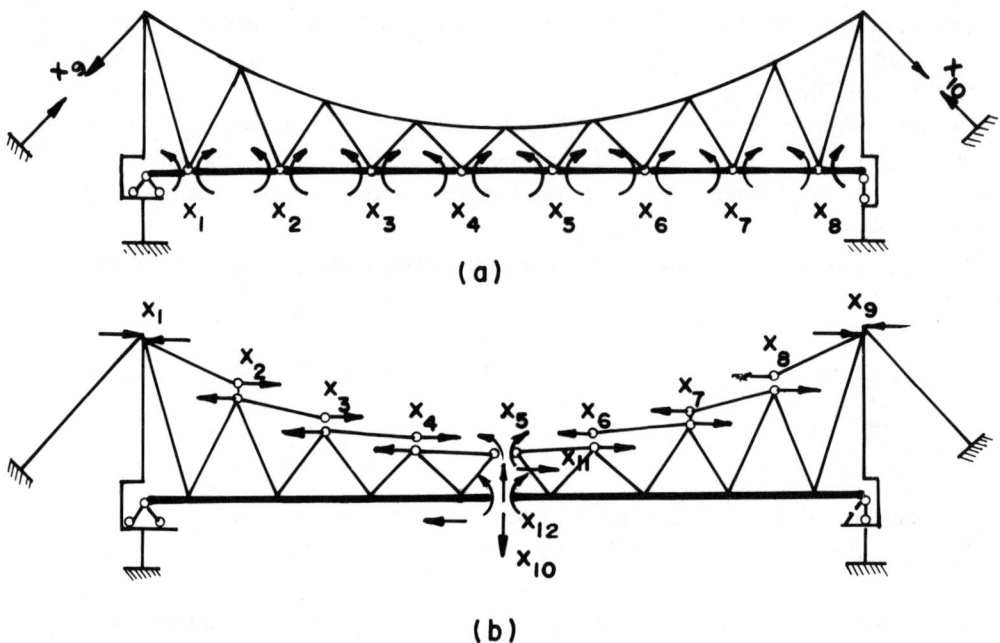

(a)

(b)

Fig. 10.9 Calculations for basic systems of cable trusses having diagonals. (a) By hand or with calculators. (b) With computers using corresponding programs.

During calculation by hand or with calculators, it is recommended that the designer choose a basic system such as that shown in Fig. 10.9(a), using as unknowns moments at beam joints along the axis. Such a basic system conforms closely in work to the design and does not require very exact calculations.

For calculations using typical computer programs, the basic system shown in Fig. 10.9(b) is used, considering the span as a cantilever. Such artificial methods decrease the difficulty of calculation and eliminate a great number of working schemes. The scheme of the truss is assumed to be nonchangeable, considering that diagonals are not excluded from the work at any load position of the loads, but the rigidity of flexible elements of the truss is assumed changeable (Fig. 10.10).

During tension the stiffness of the members corresponds to the designed, during compression conditionally used 2500 times smaller. As calculations show, such an assumption gives an error of no more than 0.004% with respect to the

Fig. 10.10 Diagrams showing the dependence of elastic deformations of flexible diagonals on the magnitude and sign of the forces. (a) The diagonal in the truss corresponding to the actual work. (b) Conditional with changeable values of diagonals' rigidities applied for the design.

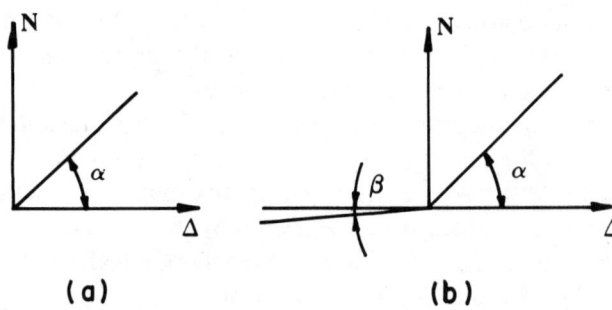

(a)   **(b)**

exact value, but simplification resulting from such artificial methods is great. Instead of providing new equations for the calculation of all displacements, it is sufficient only to exchange rigidities.

Due to a change of the static scheme of the truss, the dependence of forces in members under loading is nonlinear and the conventional method of influence lines is impossible to use. Instead, influence lines are constructed as "integral influence lines," which summarize the results of the separate calculations and express the force in the element depending on load position, or instead of $e(x)$ the following is determined

$$E(x) = \int_0^x e(x)\,dx \tag{10.2}$$

where $E(x)$ is the force in the member as a function of x—the distance of the end of the uniformly distributed loading q moving to the bridge from support (Fig. 10.11). Each ordinate of such a line represents the total force in the element of the truss at a given load position.

The construction of integral influence lines is shown in the simplified example of the cable truss bridge crossing the Volga River at Volgograd. If we assume that only diagonal D may be excluded, and we examine the truss work at two possible statical schemes—at the truss work with all diagonals and at the excluded diagonal—then the summary of these calculations gives the integral influence line.

The influence line for the second scheme of truss work is conditional and is constructed under the assumption of lowering the diagonal rigidity 2500 times with comparison with the actual rigidity (Fig. 10.9(b)). The construction of it is necessary for the determination of limits of sections under loadings, at which the diagonal is excluded from the work and changes the static scheme of the truss (points x_1 and x_2). Actually it is excluded not only one diagonal, but a few and instead two it is obtained more points of passing. Using integral influence lines, one may determine the calculated maximum and minimum forces and the corresponding positions of the loading.

10.2.2 Approximate Design of Cable Trusses

For approximate design, a cable truss is considered conditionally as a truss in which all the diagonals may work in compression and tension. The rigidity of the bottom chord beam and eccentricities of connections of suspenders are neglected. The axial deformations of the beam and pylons, and the deformations of the anchorage connections of the upper chord, are also neglected.

At the start forces are determined under a uniformly distributed dead load. The forces in the top chord and thrust of the system are determined as for the conventional suspension system, namely:

$$H = \frac{gl^2}{8f} \qquad N = \frac{H}{\cos\varphi} \qquad D = \frac{ga}{2\sin\alpha} \tag{10.3}$$

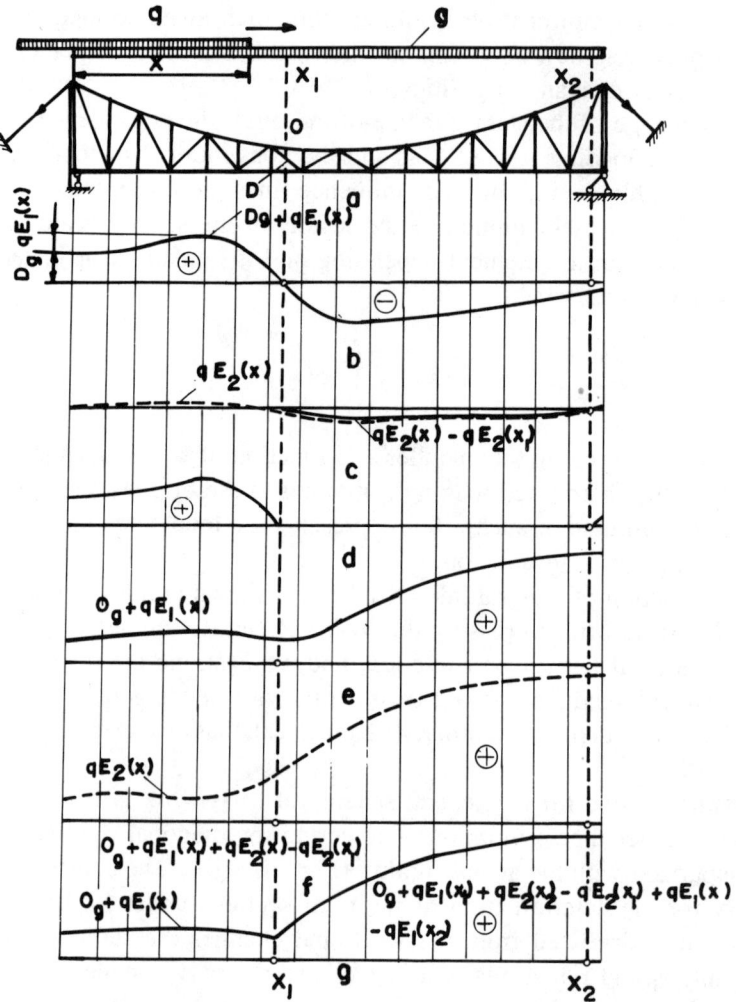

Fig. 10.11 Construction of integral lines of force in elements of a truss with excluded diagonals. (a) Scheme of truss with flexible diagonals D and uniformly distributed loading q_i; (b) "integral influence line" for diagonal D during truss work under first scheme; (c) same during truss work under second scheme (ordinates are increased 20 times); (d) resulting influence line; (e) "integral influence line" for element of chord O during truss work by first scheme; (f) same during truss work by second scheme; (g) resulting influence line.

where:

    g = intensity of dead load
    l = span of the cable truss
    f = sag of the upper chord
    N = axial force in the beam
    H = thrust of the cable truss
    a = length of the panels
    α = inclination angle of diagonal
    φ = inclination angle of tangent to top chord with respect to horizon
    D = force in diagonal.

Bending moments in the stiffening beam are determined as for the multispan continuous beam on the rigid supports. It is assumed that during erection settlements of joints due to cable elongation will be removed. The spans are equal to panel length a.

The design under live load is carried by successive approximations. At the first stage it is assumed that all diagonals participate in the work. By loading of the truss by some parts of the live load it is determined in which elements compressive stresses originate and, therefore, which elements are excluded from the work.

At the second stage the system with an incomplete truss is analyzed, taking into consideration the influence on the system of the forces in the diagonals excluded from the work. If it is apparent that at some of the remaining diagonals originate compressive forces, then these diagonals are excluded from the work and the design is again repeated.

Usually it is enough to have two or three approximations. For a changeable working scheme it is important to choose the critical position of the live load. When the live load is a distributed lane load the basic position is loading of the span and half of the span. For live loads as concentrated loads the basic positions are at mid-span and at the greater of the span.

The design of the truss at the second approximation, or by exclusion from the work part of the suspenders, which at the first stage were compressed, is carried out as follows. For the determination of the overloading due to the exclusion of diagonals, two forces are applied at the location of each excluded suspender. These forces (D_{DL} and D_{LL}—forces in the diagonals) are obtained during the first approximation under dead and live loads (Fig. 10.12). The additional forces

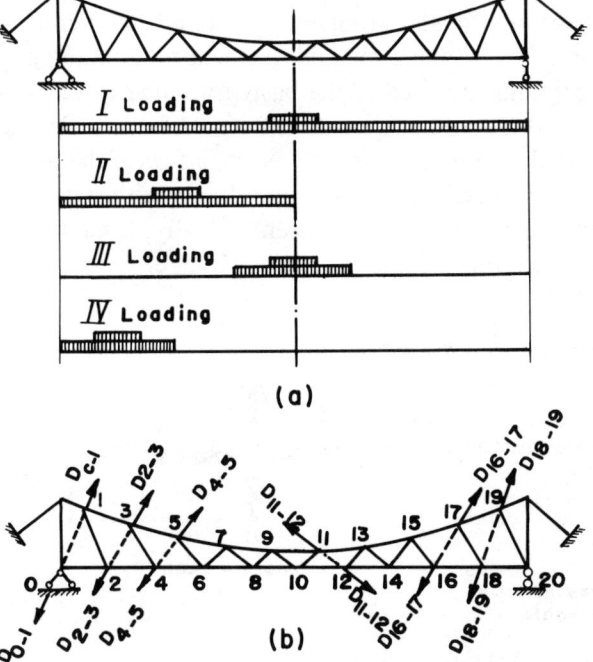

Fig. 10.12 Approximate design of cable truss. (a) Scheme of the truss and design loadings. (b) Work of the truss with partial exclusion of the diagonals.

cause additional loading of neighboring diagonals and additional bending of the stiffening beam.

The approximate design method provides results that are relatively close to the exact result at the flexible stiffening beam.

10.2.3 Structural Analysis

The prestressed steel cable truss concept has been employed for bridges, essentially different from conventional suspension bridges, to provide increased effective stiffness and aerodynamic stability. Such a bridge is a pair of pretensioned cable trusses consisting of an upper chord, a lower chord, and diagonals between them for the transfer of shearing forces (Fig. 10.13).

The lower chord of each truss is a cable running from one end of the bridge to the other, and tensioned against the steel or concrete deck. The diagonals which support the dead and live loads of the roadway are also subjected to initial tension and therefore can act as compression members.

A characteristic of such trusses is prestressed flexible members, which ensure that all members of the system are working in tension under any combination of external loadings. This condition is achieved only by the action of dead load; that is, tensional stresses in the elements of a cable-stiffened truss under dead load should be greater than its compressive stresses which could originate in these members when live loads are placed in the compression zone of the influence lines.

This study provides analysis of the prestressed cable trusses under a combination of external loadings and design of the cross-sectional members, and is treated in three parts. First, a prestressed cable truss under symmetric constant loading is analyzed.[9] For the solution of the problem, that is, all the members of the truss to be in tension only, the simplex method of linear programming is used.[10] Modified Jordan's exceptions are used as the basis for numerical calculations. In the second part, the prestressed cable truss subjected to variable and movable loadings is analyzed with a similar approach. Next, a few of the existing cable truss bridges are examined and conclusions are drawn with regard to the advantages of the prestressed steel cable truss system. Finally, a numerical example illustrates the loading case for a single cable truss bridge.

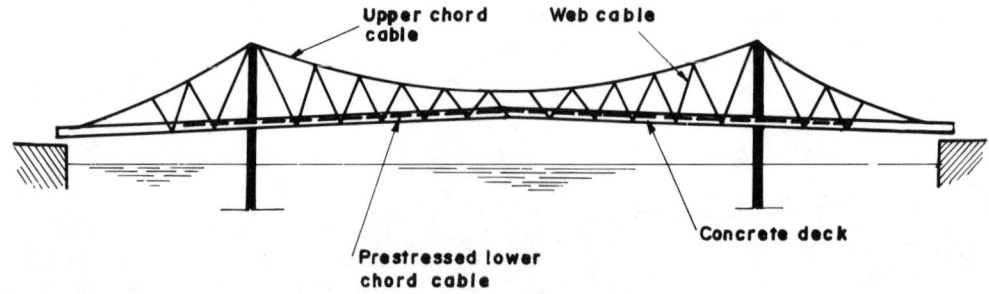

Fig. 10.13 Prestressed cable truss bridge.

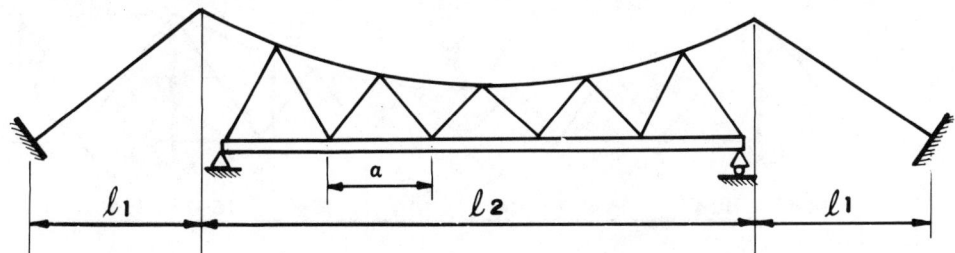

Fig. 10.14 Elevation of typical cable truss.

The basic requirement that all members of the truss (Fig. 10.14) are in tension under any combination of external loadings can be expressed as

$$W(A_1 - A_2) = \alpha P A_2 \quad \text{and} \quad \alpha = \frac{W}{P}\left(\frac{A_1}{A_2} - 1\right) \tag{10.4}$$

where

W = dead load per unit length
P = live load per unit length
A_1, A_2 = corresponding areas of the positive and negative parts of the influence lines for members under consideration
α = factor for tensional stresses

By designating $W/P = \beta$ and $A_1/A_2 = \lambda$ and substituting into equation (10.4)

$$\alpha = \beta(\lambda - 1) \tag{10.5}$$

Thus, to achieve the requirement of no compressive stress, the coefficient α should always be greater than 1.

10.3 ANALYSIS OF PRESTRESSED CABLE TRUSSES

The truss is analyzed under prestressing of external loads and unit forces. Those forces due to prestressing in all members of a truss to be tensile are achieved by selecting a suitable corresponding geometrical system.

10.3.1 Analysis of Prestressed Cable Trusses Under Symmetric Constant Loading

The following approach is considered for the analysis: (a) determine the forces in the system, (b) select the cross section, and (c) find the prestressing forces and the sequence of their application.

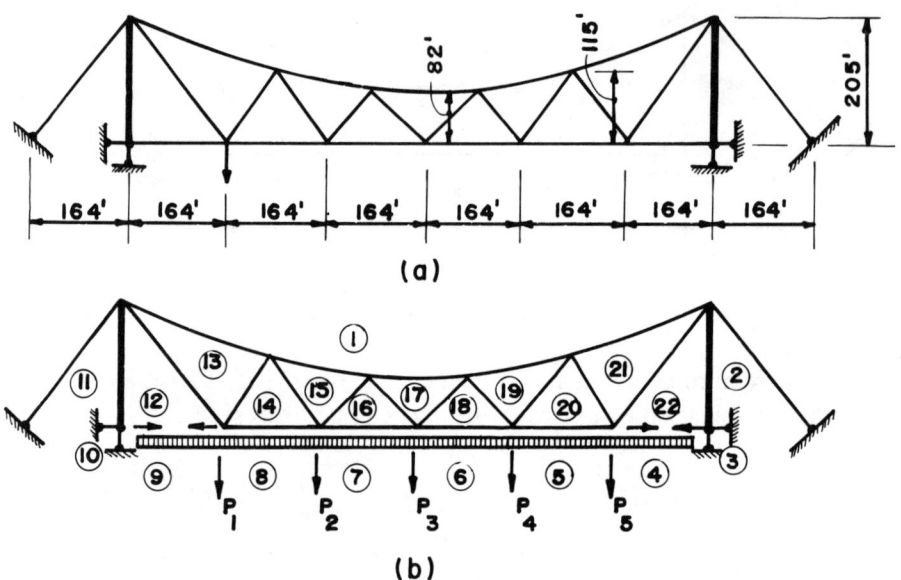

Fig. 10.15   Geometrical and loading schemes of cable truss bridge.

Assuming a geometrical scheme and the loading scheme (Fig. 10.15), the basic statically determinate system is chosen. The forces in its members are found under external loading and unit force in the redundant members. All members are considered capable of taking the tensile and compressive forces.

Once the forces in the basic system are determined, the performance conditions can be stated for each member. Since the truss consists of flexible members, the members will perform their function only if they are in tension. Thus, the performance condition of i members in the n times statically indeterminate system is

$$P_i = P_{pi} + P_{li}F_1 + P_{2i}F_2 + \cdots + P_{ni}F_n > 0 \qquad (10.6)$$

where

$$i = \text{the number of the member in the basic system}$$
$$P_{pi} = \text{the force in the member of the basic system under external loading}$$
$$P_{li} \text{ to } P_{ni} = \text{forces in the member under unit forces in the redundant members}$$
$$F_1 \text{ to } F_n = \text{total forces in the redundant members (including prestressing)}$$
$$n = \text{number of the redundant member}$$

Each redundant member should satisfy the condition

$$F_n \geq 0 \qquad (10.7)$$

The performance conditions should be considered only for those members which under any loading, including the forces in the redundant members, are in compression.

Equations (10.6) and (10.7) formulate the system under consideration, and usually the number of equations is greater than the number of unknown redundants. The number of unknown redundants in a simple-span cable structure with a triangular truss system and under symmetrical loading are 3 and 2, respectively. In the case of two unknowns,

$$
\begin{aligned}
P_1 &= P_{11}F_1 + P_{21}F_2 + P_{e1} > 0 \\
P_2 &= P_{12}F_1 + P_{22}F_2 + P_{e2} > 0 \\
&\vdots \qquad \vdots \qquad \vdots \qquad \vdots \\
P_k &= P_{1k}F_1 + P_{2k}F_2 + P_{ek} > 0
\end{aligned}
\tag{10.8}
$$

With the solution of the system of equations the values of total forces in redundant members are computed corresponding to all members to be in tension. The equations may be satisfied with infinite possible combinations of values for the forces in redundant members, or, in a statically determinate prestressed system, an infinite number of alternatives for distribution of internal forces is possible. Therefore, it is necessary to separate from the multiple alternatives the one at which the system will be most economical, that is, to find the optimum distribution of internal forces.

The simplex method of linear programming is used for the solution of the problem. The basic problem is formulated as

$$
Z = B_1x_1 + B_2x_2 + \cdots B_nx_n \tag{10.9}
$$

And given the system $m > n$ of linear inequalities

$$
a_{i1}x_1 + a_{i2}x_2 + a_{in}x_n < a_i \ (i = i \cdots m) \tag{10.10}
$$

which is written as

$$
y_i = -a_{i1}x_1 - a_{i2}x_2 \cdots -a_{in}x_n + a_i > 0 \ (i = 1 \cdots m) \tag{10.11}
$$

Thus to maximize (or minimize) (10.9) subjected to condition (10.10) is similar to finding the minimum volume (cost) of the prestressed cable system. Expression (10.9) is composed of the summation of the product of forces in all members by their length, including redundant members,

$$
Z = \sum_{i=1}^{i=m} P_i l_i = \sum_{i=1}^{i=m} (P_{pl} + P_{1i}F_1 + P_{2i}F_2 + \cdots P_{ni}F_n)l_i \tag{10.12}
$$

The required cross-sectional area of an element of the flexible system under full load will be

$$
A_i = \frac{P_i}{\sigma_i}
$$

where

σ_1 = design strength of steel
V_i = $\psi_i A_i l_i$-volume of steel section
C = $V_i c \gamma$-cost of the element
c = cost of unit weight of steel
γ = specific weight

Assuming the design strength of steel and the construction coefficient constant, expression (10.12) is

$$Z = \sigma_i C \psi \gamma \sum_{i=1}^{m} V_i \qquad (10.13)$$

Neglecting the free term, $\sum_{i-1}^{m} P_{pi} l_i$, since it does not depend on the changeable values, the function then takes the form

$$Z = \sum_{i=1}^{m} P_{1i} l_i F_1 + \sum_{i=1}^{m} P_{2i} l_i F_2 + \cdots \sum_{i=1}^{m} P_{ni} l_i F_n \qquad (10.14)$$

The simplex method consists of finding a solution between solutions of a system of linear inequalities (10.10). The basis for the numerical calculations of the simplex method are modified Jordan's exceptions. Therefore, using the simplex method we may determine the total forces in the redundant members corresponding to the minimum weight.

For problems involving two unknowns, a graphical method may be used to find the optimum total forces. In this case, it is necessary to construct the polygon in coordinates x_1-x_2 and to determine its top coordinates closest to the given straight line.

$$Z = B_1 x_1 + B_2 x_2 = 0 \qquad (10.15)$$

The sides of the polygon in this case will be straight lines,

$$a_{i1} x_1 + a_{i2} x_2 \le a_i \qquad (10.16)$$

or the performance conditions (10.8), and the straight line Z-function (10.14) having the equation

$$x_i = -\frac{B_2}{B_1} x_2 \qquad (10.17)$$

Once the total forces in the redundant members are computed, the design forces at the members of the system can be determined by the formula

$$P_i = P_{1i} F_1 + P_{2i} F_2 + \cdots + P_{ni} F_n + P_{ei} \qquad (10.18)$$

Then the cross sections are selected. The optimum distribution of forces is achieved by introducing prestressing.

In certain cases it is possible that the solution for the system of inequalities may not exist. This means that for a particular geometrical scheme and loading system it is not feasible for all members to be in tension. In this case, it may be necessary either to change the cable truss scheme or to introduce certain additional loading. The additional loading system is chosen to achieve maximum tension in those members which are not in tension. The magnitude of the additional loading can be introduced in the conditions (10.6) and (10.14) as additional unknown, F_{n+1}.

It is essential to note that by transferring the given statically indeterminate system into statically determinate, the solution for optimal distribution of forces under constant loading by the simplex method may lead to zero force in some members. Therefore, in this case members with zero forces may be chosen for minimum structural requirements.

However, in practice loadings are often of variable magnitude. Thereby forces in some members of the system may be greater at smaller loading. This is applicable to elements which are in compression under external loading.

For this reason, the cross sections of such elements should be chosen for the forces exerted by prestressing of redundant members. Since the prestressing forces in the redundant members are not known until all cross sections are chosen, it is possible first to choose the sections of such elements corresponding to the known total forces in redundant members and then to determine the corrected prestressing forces for members in compression under external loads.

Once the cross sections are chosen, the prestressing forces in the redundant members can be found using the known methods of structural mechanics to solve a statically indeterminate system to find the prestressing forces in the redundant members.

The prestressing forces in the redundant members can be found as a difference of the total forces,

$$F_{\text{tot}i} = F_i - F_{\text{pr}i} \qquad (10.19)$$

Further, it is necessary to determine the performance interval of the system or to find the minimum loading under which all members of the prestressed truss will be in tension. Also, it may be necessary to apply initial loading to counteract the compressive forces in some members of the system due to prestressing. Hence, the system will perform at certain intervals of loadings, the minimum and maximum value of loadings as the lower and upper limits of performance of the system, respectively. Here, it should be noted that the initial loading may be of varying intensity.

The required value of initial loading may be found considering the performance conditions of members compressed by prestressed redundant members. Also, it is necessary to find the loading value corresponding to each inequality, transfer the inequality into the equation, and then choose the maximum value.

Next, since sections of some elements are chosen corresponding to total forces in redundant members, contrary to prestressing forces, the section should be modified for this difference. However, any change in the section will lead to a corresponding change in prestressing force, and so on. Therefore, the required accuracy can be achieved by the method of successive approximations.

Each time, the prestressing forces should be calculated from the difference of the upper and lower values and not under the total loadings. Compared with the design of a conventional statically indeterminate system, where the initial ratio of stiffnesses influences the economy of the system, in this case there is no such influence as the final total forces in redundant members are found from the optimal distribution of internal forces without designing the statically indeterminate system.

Thus, the design procedure for an indeterminate cable truss under constant loading can be summarized as follows:

a. Choose a statically determinate basic system and determine forces in its elements under external loadings and unknown redundants.

b. Write the expressions for performance conditions of each member and expressions for volume of steel (or cost of truss).

c. Find the optimum values of total forces in redundant members.

d. Determine the design forces and select a section (final under constant loading and preliminary for the number of members under variable loading).

e. Determine the prestressing forces in the redundant members.

f. Determine the performance interval (under loading) of the system and modify the sections of members which are compressed by the external loads.

10.3.2 Design of Prestressed Cable Trusses Under Variable and Movable Loadings

The design problem is similar to the one in the previous section. Given the geometrical system and the variable or movable loading schemes (Fig. 10.15), the solution of the problem can be treated as follows.

First, a statically determinate basic system is chosen, and the forces in its members under the unit force and variable external loadings are determined. Next, in order to design the members, the maximum and minimum values of forces under movable loadings are to be calculated using influence lines.

Once the forces in the basic system are found, the performance conditions for each member can be stated. These conditions should be considered only for members which under any loading are under compression.

The performance conditions of i members in the n times statically indeterminate system is the same as equation (10.16):

$$P_i = P_{pi} + P_{1i}F_1 + P_{2i}F_2 + \cdots P_{ni}F_n > 0 \qquad (10.20)$$

In these inequalities, the variable value of forces under external loading and of the total forces in redundant members are to be substituted.

The number of system inequalities (10.20) is greater than the number of remaining unknowns. For single-span cable trusses having a triangular web, there are three unknowns, and the system inequalities composed of performance conditions of k members are:

$$P_1 = P_{p1} + P_{11}F_1 + P_{21}F_2 + P_{31}F_3 > 0$$
$$P_2 = P_{p2} + P_{12}F_2 + P_{22}F_2 + P_{32}F_3 > 0$$
$$\vdots \qquad \vdots \quad \vdots \quad \vdots \quad \vdots \quad \vdots \quad \vdots$$
$$P_k = P_{pk} + P_{1k}F_1 + P_{2k}F_2 + P_{3k}F_3 > 0$$

$$(10.21)$$

The free terms of the system inequalities are variable values; however, in order to achieve the condition that all members of the system under an arbitrary portion of external loads be in tension, it should be satisfied at all possible values.

If one considers the design under movable or variable loading as a series of designs under constant loading (taking each position of loading, changing its application scheme as immovable loading), then for each position of the loading the most favorable distribution of total forces in redundant members may be found corresponding to the minimum volume (cost) of steel. However, the total forces will be changing with a change in the position of load. Thus, in order to find those values, all inequalities (10.20), written for each position of changeable or movable loading, should be substituted in system (10.21).

Considering only P_{pi} values are changing for the different positions of external loading, at limits written for the same member, the system with variable free terms can be substituted by the one with the constant. For the design under movable load, the positive free terms are zero, since it is possible to have no load case in the truss.

The free terms of the inequalities (10.20) for symmetrical members of the truss, under movable and variable loadings, have the same limiting values. With these conditions of external loadings, the number of unknowns can be reduced to two.

The expression for the function, volume, or cost of a steel truss is the summary of volumes (costs) of all members of the truss, as in the case of constant loading,

$$Z = \sum_{i=1}^{m} P_{ii} = \sum_{i=1}^{m} (P_{pi} + P_{1i}F_1 + P_{2i}F_2 + \cdots P_{ni}F_n)l_i \quad (10.22)$$

The free term $\sum_{i=1}^{m} P_{pi}l_i$ in formula (10.22) which has varying values does not depend on the variables and thus may be neglected. Hence, the expression of function does not depend on the external loading position and is the same for constant, variable, or movable loadings.

In some instances, the inequality system (10.21) may not have a solution, which indicates that one or a few members are under compression in a combination of forces in redundant members. Thus, it is necessary to either change the geometrical scheme of the truss or apply additional loading. The magnitude of the loading should be taken as unknown and included in inequalities (10.21) and function (10.22).

For single-span trusses, the typical inequality, including additional loading, is

$$P_i = P_{pi} + P_{1i}F_1 + P_{2i}F_2 + P_{3i}F_3 + P_{4i}F_4 \geq 0 \qquad (10.23)$$

where P_{4i} is the force at i member of the basic system under unit additional loading and F_4 is the full weight of the additional loading.

After analyzing the performance of the system an additional loading scheme should be chosen to create maximum tensile forces in the members and with the solution of inequalities and analysis of total forces in redundant members, the sections can be chosen.

In order to choose the section, it is necessary to find the total forces. The total forces in redundant members for each loading position, found by the simplex method, are the limiting values at which any arbitrary position of external loads, all members are in tension. Actual total forces in redundant members always will be greater, and only at certain loading positions will the values of total forces be equal to those found by the simplex method. Therefore, to choose sections initially, the total forces in redundant members can be considered as constants and can be modified after determining the prestressing.

Once the sections are chosen for maximum design force, the design of the statically indeterminate system may be performed by the conventional methods of structural mechanics, and the prestressing can be found. The latter are variable values and depend on the position and value of external loading. Since only the prestressing may be constant, and the total forces at any arbitrary position of loading should not be less than that determined by the simplex method, then in the formula for prestressing,

$$F_{\text{tot}i} = F_i - F_{\text{pr}i} \qquad (10.24)$$

The minimum positive or maximum negative value of $F_{\text{pr}i}$ should be substituted. The condition providing the values of $F_{\text{tot}i}$ satisfy the limitations due to inequalities (10.21). The design under variable and movable loading, as a rule, shows the necessity of special additional loading.

Next the special additional loading required in the design under variable and movable loading may be accomplished by the weight of the structure itself. This additional loading should not be considered during prestressing so that it is applied to the system during its construction.

The modification achieved is as follows: After the prestressing forces are found, the maximum values of forces are determined in all elements and stresses are checked. The section of members in which design stresses differ from the allowable are modified. Next, the new prestressing forces are determined. The process is repeated until the required stresses in all members are achieved.

The cable truss concept has served well in providing economical stiffening for moderate, multiple-span bridges. The geometry of the system ensures that all members of the truss are working in tension under any combination of external loadings.

The analysis and design of such prestressed cable truss system, considering the internal force distribution corresponding to the minimum volume and cost of material, results in substantial savings in material. The truss analyzed under symmetric constant loading, and variable and movable loading cases, provide the optimal solution to achieve this requirement. The simplex method, used for this purpose, proves to be simple and effective in analyzing this problem.

Although this involves the solution to a number of equations and trial-and-error computation, the required accuracy can be achieved relatively easily by using the available computer programs.

10.4 TYPICAL CABLE TRUSS BRIDGES

The following are examples of typical cable truss highway bridges.

10.4.1 The San Marcos Bridge, El Salvador

The San Marcos Bridge, erected in El Salvador in 1957, is a typical cable truss structure having all its members made of cables (Fig. 10.16).[11, 12, 13, 14, 15] In cross section, the San Marcos Bridge is a two-lane highway bridge designed for AASHO H15-S12-44 traffic loading. There is one 3 ft 4 in. sidewalk on each side of the roadway (Fig. 10.17).

The main cables, spaced 30 ft apart, are of open-type construction. Each main cable is composed of 24 galvanized bridge strands of $1\frac{1}{2}$ in. diameter. Two additional strands added to the backstay portion of each main cable hold the end towers upright. Each lower chord cable consists of eight galvanized bridge strands of $1\frac{3}{8}$ in. diameter. One length of the diagonal cables extends from the main cable to the lower chord, where it passes over an inverted saddle, and back again to the main cable at another point. In general, the diagonals consist of one galvanized bridge rope, being $1\frac{3}{8}$ in. diameter, but where the loading requires it, there are two.

In the continuous truss, each main cable performs the function of an upper chord. Those cables parallel to the deck serve as their lower chord.

Under dead load, the main cables and diagonals carry tensile stresses. When compressive forces are exerted on these tension members by live loads or winds, they merely reduce the dead-load tensile stresses.

Lower-chord cables have been prestressed; when the compressive forces are applied, they merely act to reduce the prestress.

Another novel feature of the design is the anchorage of the prestressed lower-

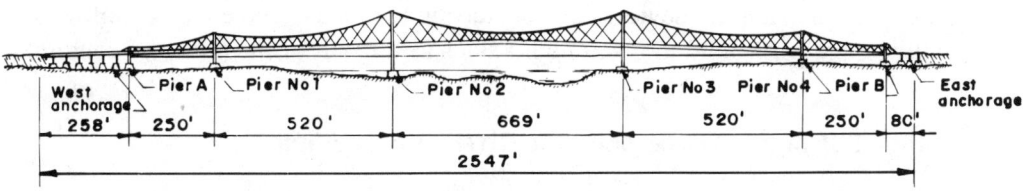

Fig. 10.16 Elevation of San Marcos Bridge, El Salvador. (Courtesy of the *Engineering News Record*)

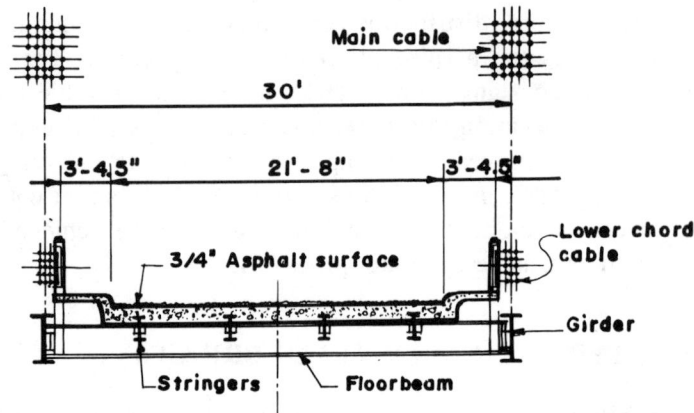

Fig. 10.17 A cross-sectional view of the San Marcos Bridge, El Salvador. (Courtesy of the *Engineering News Record*)

chord cables in the concrete floor of the bridge (Fig. 10.18). With provision made to permit the floor to slide on the suspended steel, the deck, in effect, is a prestressed slab 2160 ft (658.54 m) long. Those stresses and deflections due to live loads and temperature were determined with electric strain gages from a model built to a scale of 1:20. Studies indicated that if the cable truss were permitted to have a zero bending moment at its dead load, some diagonals would appear slack under its live load. A decision was made to prevent it.

During erection, therefore, bending moments were induced in the bridge at its dead load in the opposite sense to that produced by live load. This was accomplished by adjusting the diagonal system to a condition of zero bending moment under dead load plus one-third live load. On removal of this live load, the desired reverse bending moment existed in the truss at dead load. This illustrates what can be done with a cable truss system if the designer disapproves of slack diagonals.

The cable of a suspension bridge, which supports the dead load of the structure, makes an excellent upper-chord member for the truss. A main cable with an initial tension, say, of one million pounds, may be subjected to compressive stresses up to one million pounds before buckling takes place. Buckling in this case means simply that the cable becomes slack, and no structural damage results.

The diagonals, which support the dead and live loads on the roadway, are also subjected to initial tension and therefore can act as compression members. The lower-chord cables are tensioned by hydraulic jacks.

Since the tensions in the cables for the various critical loading combinations were obtained from the model, it was not difficult to choose cable sizes and proportion clamping devices for transferring the loads. The effective stiffness of the bridge was found to reach high values as a result of the prestress condition.

10.4.2 Bridge over the Zambeze River, Mozambique

The proposed continuous-type bridge with cable-stiffened truss over the Zambeze River, at Tete, Mozambique, has three central spans of 590 ft (180 m)

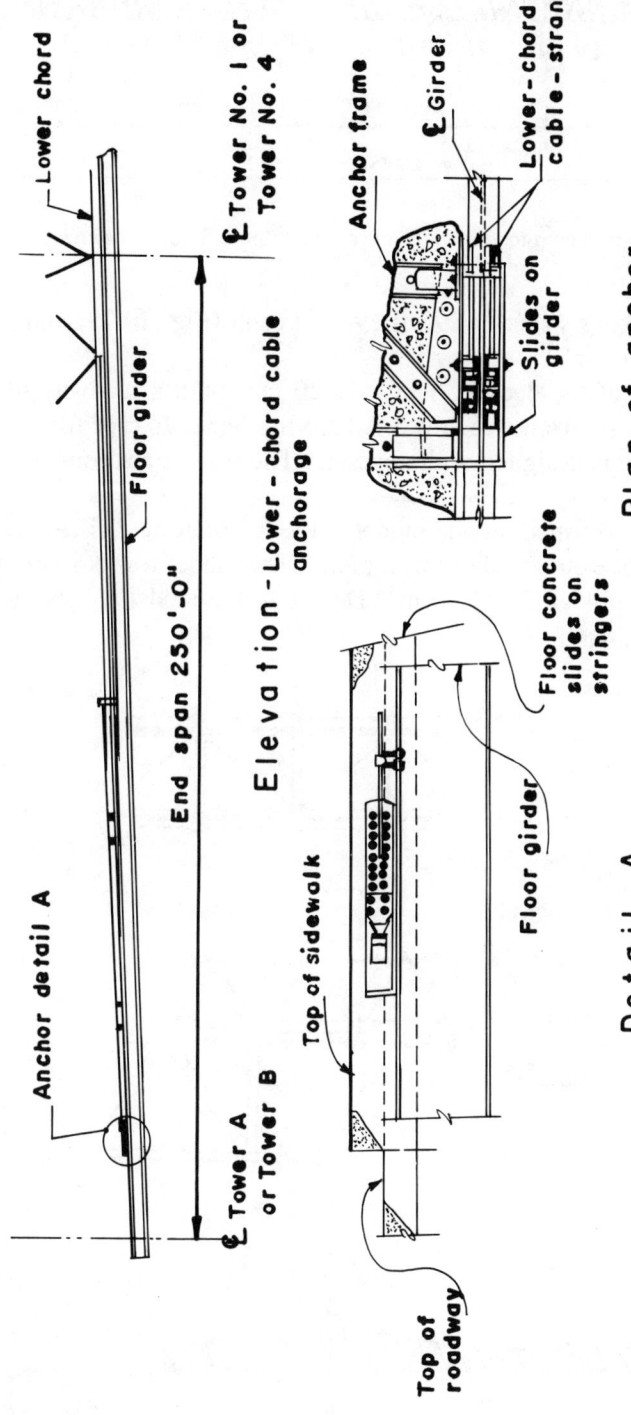

Fig. 10.18 The San Marcos Bridge. Lower-chord cables are prestressed and anchored to the concrete floor in the end spans. The floor is free to slide on the suspended steel.

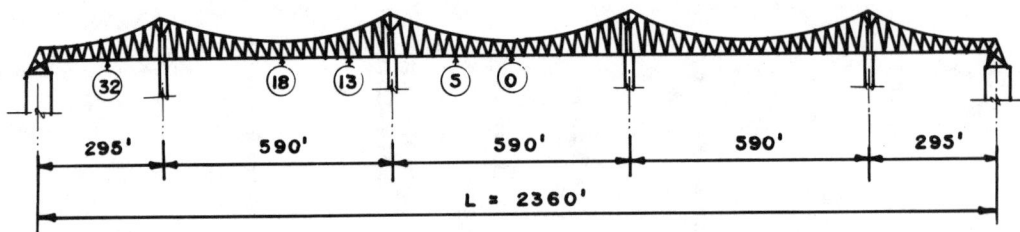

Fig. 10.19 Elevation of the prototype bridge over the Zambeze River, Mozambique.

each and two flanking spans of 295 ft (90 m) each (Fig. 10.19) and has been investigated on a model.[16]

The total width of the deck is 37 ft (11.20 m), with a roadway of 23.6 ft (7.20 m) and two sidewalks each 6.6 ft (2.0 m) wide. Except for the cables, the whole structure is designed with concrete. The trusses are constructed from prestressed cables.

The tests were performed on one model at the Laboratoire d'Essais de Matériaux et de Mécanique du Sol, de Mozambique. The model was 1:50 in scale and had a total length of 47.2 ft (14.40 m). Details of the model are shown in Fig. 10.20.

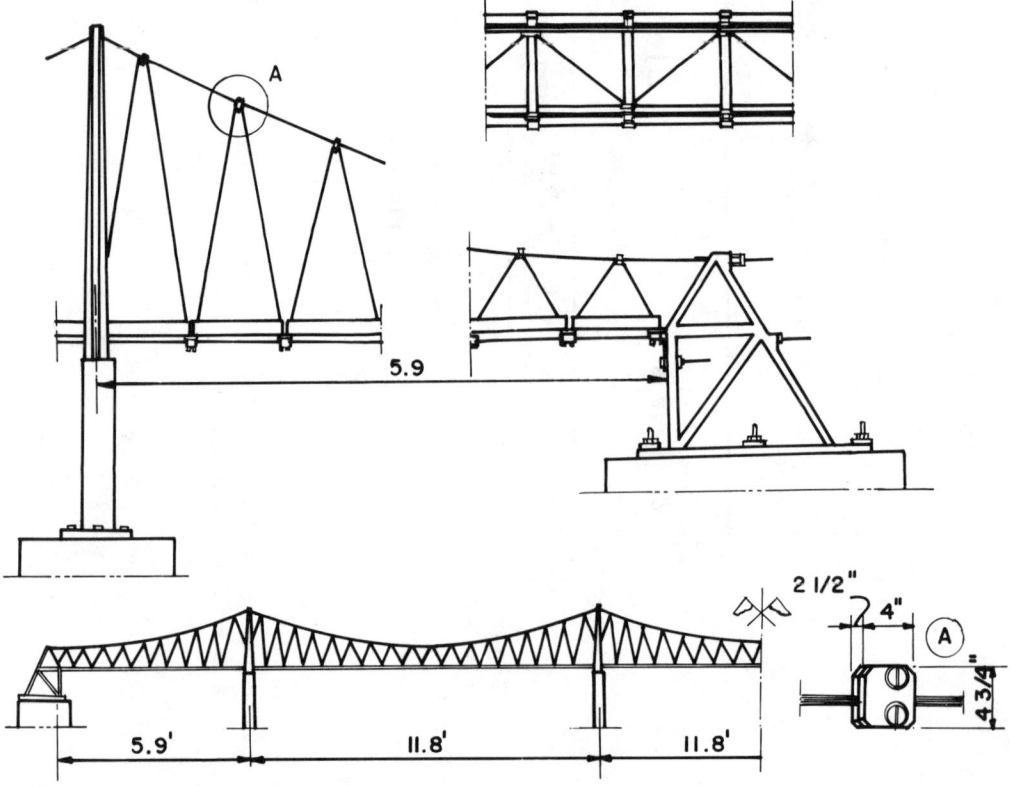

Fig. 10.20 A model of the bridge over the Zambeze River, Mozambique.

Fig. 10.21   Cable truss bridge over the Volga River, USSR, general view.

The values of the forces were determined by prestressed cables on a model and they corresponded to the forces in the prototype. The conditions for similitude were as follows:

Linear displacement $d_{\text{prot}} = 50 d_{\text{mod}}$

Angular displacement $\alpha_{\text{prot}} = \alpha_{\text{mod}}$

Tensions $\sigma_{\text{prot}} = \sigma_{\text{mod}}$

The model investigations confirmed the design assumptions.

10.4.3 Highway Bridge over the Volga River, USSR

This highway bridge has a central span of 1400 ft (470.0 m), each flanking span being 635 ft (193.60 m), and as the main carrying system, a cable truss (Fig. 10.21).[17] In cross section, the roadway width is 44.6 ft (13.6 m), and the two sidewalks are 7.38 ft (2.25 m) each. The main cables consist of parallel wires and a diagonal member of wire ropes. The stiffening girder is an orthotropic reinforced concrete slab connected to the steel truss, working together. The quantity of steel used for this bridge is 73 lb/ft^2.

10.5 INDUSTRIAL BRIDGES

10.5.1 Introduction

The past 50 years have shown great progress in the development of industrial or utility bridges to which belong pipeline bridges and material-handling bridges.[18,19,20] Pipeline bridges serve mainly for the transportation of oil, natural gas, and water and may be called "energy bridges." Material-handling bridges, or "conveyor bridges," serve mining, the pulp and paper industry, plants, and construction sites. Industrial or utility structures are erected across rivers, gorges, roads, and similar obstacles by manufacturers, contractors, and transportation and service companies. These are generally specialized structures, carrying heavy loads and simultaneously withstanding high wind pressures. For their construction, high-strength materials were introduced, new fabrication techniques were developed, and prestressing was applied. These structures are usually supported by cables or cable stays. Prestressing may also be achieved by the regulation of the bending-moment method in the structure.

By prestressing, it is possible to achieve an optimal solution for an industrial bridge crossing and achieve the following:

1. To provide safe stability conditions to resist vertical and lateral loadings
2. To reduce to a minimum stresses and deformations of the supporting system
3. To achieve the most economical solution, taking into account the steel quantity to be used and erection expenses

10.5.2 Suspended Pipeline Bridges

Pipelines are generally supported in a vertical plane by the suspension of a main cable in the vertical plane, and a separate structural system is provided to supply necessary lateral stability. This normally takes the form of a lateral parabola-shaped poststressed cable system on either side of the main structure. The lightness, flexibility, and shape of the pipe introduces the likelihood of aerodynamic instability which is required in a lateral cable system. It should be noted that in the case of a pipeline suspension bridge, the loading is such that longer spans are more economical, while the structure is more flexible.

For spans in the range of 160–980 ft (50–300 m), the most rational solutions used were pipelines supported by suspension cables or ropes. Typical schemes of such pipeline bridges are shown in Fig. 10.22.

Against wind action, there are "wind cables" in a transverse direction of a horizontal plane in the shape of either inclined stays or having a parabolic shape. To provide stability against wind action, the wind cables are poststressed after their erection. The suspension and wind cables are connected to anchor blocks. This system reduces any vertical and/or horizontal oscillations which may occur due to wind action.[21]

The pipe is supported by hangers which are connected to one or two suspen-

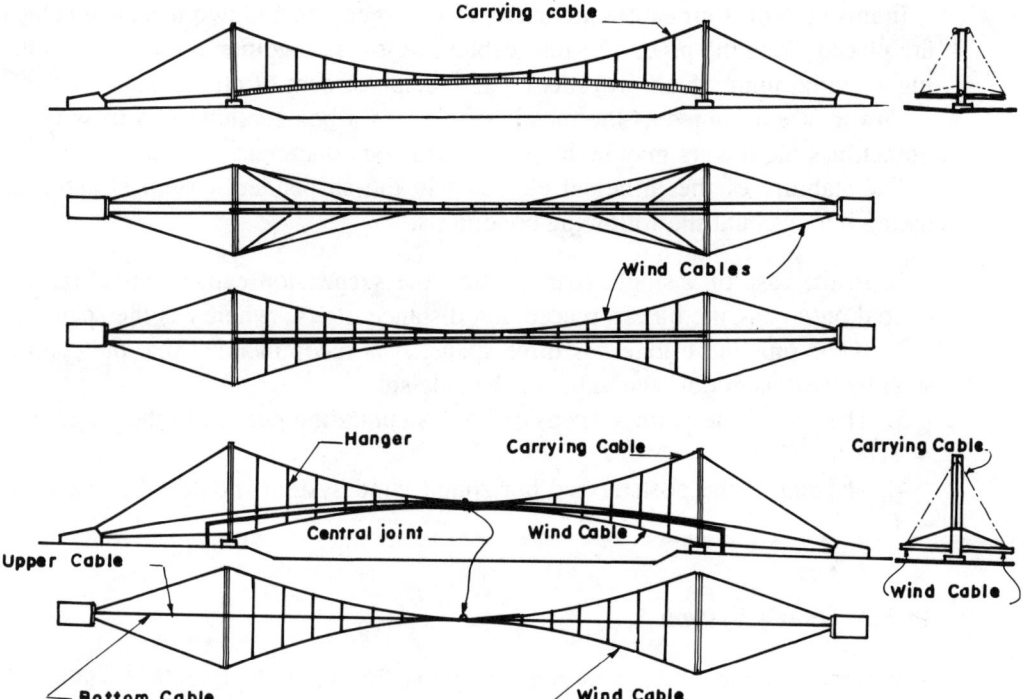

Fig. 10.22 Typical schemes of a pipeline bridge where the pipe is connected to the main cable by vertical suspenders.

sion cables. The cables are supported by two vertical towers and anchored to their massive anchor blocks.

To increase the stiffness of a pipeline supporting system, it is recommended that tension cables also be installed along the pipe. These cables, together with the carrying cables, form a closed spatial system. At the piers, these cables are supported either by triangular or V-shaped towers (Fig. 10.23).

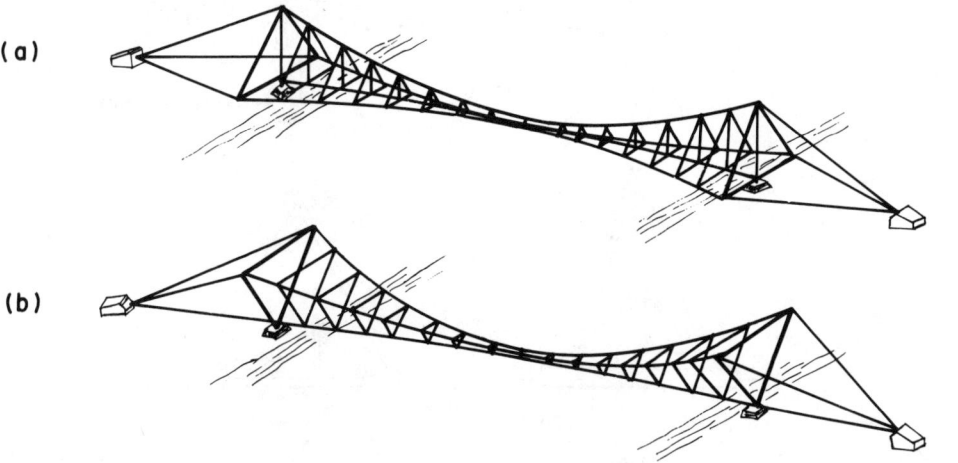

Fig. 10.23 Suspension systems with vertical and inclined cable planes.

In this system, four cables are used: Two suspension and two tensional cables are placed along the pipe. The four cables are joined together at mid-span forming a central joint which prevents relative displacement of all the cables.

Towers are designed as the metal structure of a box section or as truss type. Sometimes the towers may be built of reinforced concrete.

The stability of the principal elements in the suspended system is assured, taking into account the following conditions:

1. In the case of a single-span system, the suspension cables should be anchored onto concrete blocks placed at a distance of $l/4$, where l is the span.

2. If the pipeline bridge has three spans, it is recommended that the central span be used as double the value of the side spans.

3. The sag of the main suspension cables should be chosen in the range of $f = (1/8$ to $1/10)l$.

4. The sag of the poststressed horizontal wind system should be taken as $f_0 = l/14$.

10.5.3 Tesar's System

A different pipeline bridge system, designed by Tesar in Czechoslovakia, is shown in Fig. 10.24.[22] In this system a number of triangular rigid frames are connected to the suspension cable and to the poststressed wind cables. Inside of the external triangular frames there are smaller internal triangular frames carrying the pipeline and connected to the external frame by poststressed ties.

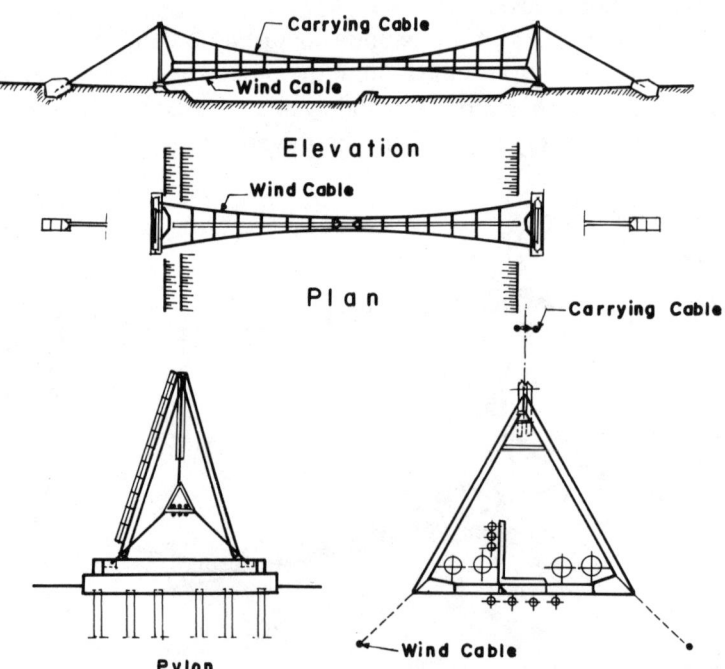

Fig. 10.24   Pipeline bridge designed by Tesar.

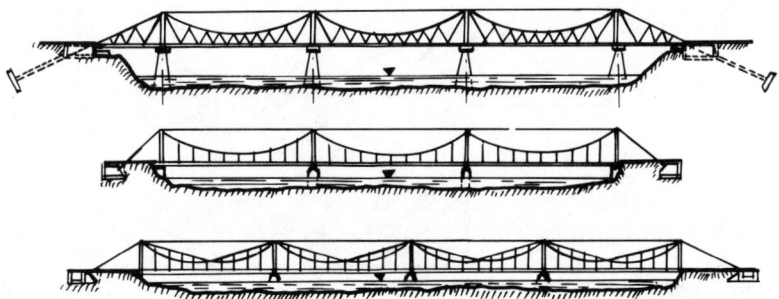

Fig. 10.25   A multiple-span pipeline bridge having a prestressed cable system.

This system has considerable space stiffness. Poststressing is applied in such a way that the structure possess a stiffness, as in the case of the large deflection in the carrying cable.

10.5.4 Multispan Bridges

Poststressing of the cables is also applied for multiple spans of suspension pipeline bridges (Fig. 10.25).[23,24] The pipeline is hung from the continuous main suspension cable by vertical suspenders. At either end these cables are attached to the anchor blocks. The pipeline follows a parabolic curve in every span. The outer poststressed wind-bracing cables follow more or less closely the curvature of the pipeline, so that they do not lie in a horizontal plane but exert a pull in a slightly downward direction to produce a stabilizing effect against any wind forces. In the same plane as the pipeline, two horizontal arrays of wire ropes are inserted and fitted with clips which grip the wind-bracing cables.

Also, the tops of all the towers are connected by additional continuous poststressed cable, which serves not only for erection, but is installed to provide stability against the wind acting in a longitudinal direction along the bridge axis.

10.5.5 Pipeline Cable-Stayed Bridges

The pipeline may be supported by a system of inclined stays extending between the towers and the pipe (Fig. 10.26).[25] Figures 10.26(a) and 10.26(b) show cable-stayed systems with 2×2 inclined stays at their central span, therefore providing four supports for the pipeline. However, it is possible to support a pipeline by a greater number of inclined cable stays.

In the cable-stayed pipeline system, Figure 10.26(a), there are horizontal components from inclined stays producing compressive axial force on the pipeline. To eliminate this axial force, it is possible to join the connections of the inclined stays together with that of the pipeline by horizontal cables, shown in Figure 10.26(b), or to provide the inclined cable system shown in Figure 10.26(c). In this case, all axial forces cancel each other out.

The pipeline is again stiffened against wind action by a system of horizontal

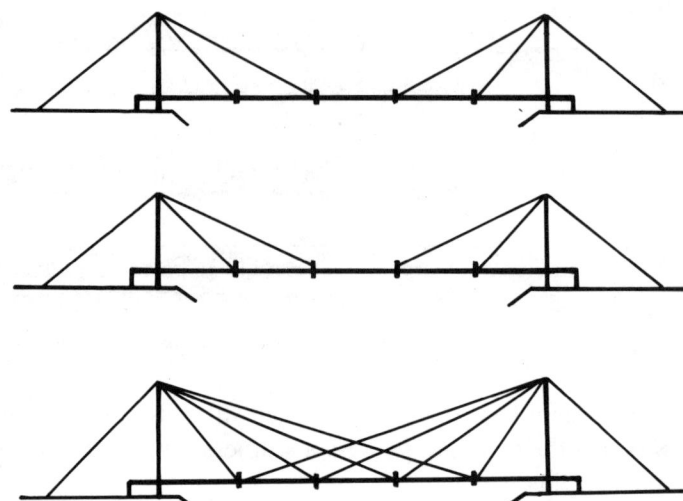

Fig. 10.26 Three types of cable-stayed pipeline bridges.

cable stays, as shown in Fig. 10.27. The stiffening systems of pipeline bridges consisting of cable stays in their vertical and horizontal planes are all post-stressed during and after erection.

This system is advantageous mainly for small and mid-spans, and their erection is relatively simple. However, the pipe itself should possess substantial stiffness, since it performs as a continuous beam on elastic supports.

For the movement of large quantities of construction aggregates and the crossing of obstructions, material-handling bridges, generally as poststressed suspensions or cable-stayed bridge systems, are designed. Their designs follow the basic concepts used for the design of pipeline bridges as described in previous sections. Such bridges appear to be more economical in comparison with other bridge systems used for the same purpose.

10.5.6 Cable Truss Pipeline and Material-Handling Bridges

For large-span pipeline crossings, the suspended cable truss possesses economical advantages in comparison to other systems. The main carrying cable is always under tension and may be fabricated from high-strength steel. For the erection there is no need for falsework, because after one panel is installed, it serves as support for installation of the next one. In the cable truss the stiffening girder requires much less steel than that for a conventional suspension cable system

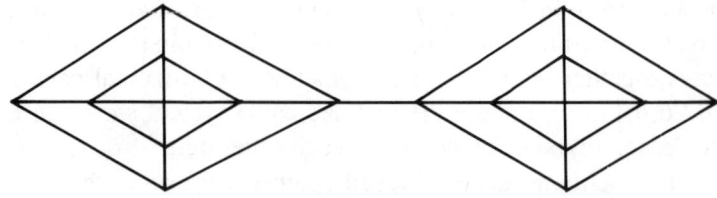

Fig. 10.27 Cable-stayed pipeline bridge. A typical example of wind bracing.

having vertical hangers. The cable-truss system possess great stiffness and may be applied without a special stiffening girder.

There are three basic types of cable trusses for pipelines, namely: (1) a suspended cable truss supporting pipeline system where the pipeline is used as the carrying bottom chord of the cable truss; (2) pipeline supported by the deck which is carried by the cable trusses, and (3) the rigid truss supporting pipeline carried by the cable trusses.

The cable truss system is also widely used for material-handling bridges. In the following sections, some typical examples of cable truss bridges are described.

10.5.7 A 1280-Foot (390.24 m) Gas Pipeline Bridge over the Amu-Darya River, USSR

In 1964, a prestressed cable truss bridge over the Amu Darya River, having a central span of 1280 ft (390.24 m), was completed (Fig. 10.28).[26,27] The deck consists of a one-lane roadway 11.5 ft (3.5 m) wide, of orthotropic plate, also supporting two gas pipes. The cable truss consists of nine panels, each being 141.7 ft (43.2 m), at the mid-length and 143.7 ft (43.8 m) at the end, (Fig. 10.29).

To eliminate compression, the cables were prestressed. For this purpose, a dead loading was first applied which was taken by the diagonals before connecting the bottom-chord cables with those of the diagonals. At this stage of erection, the flexible hanging system was formed. The diagonals obtained tension under the weight of a suspended deck and the tension was transferred onto the upper chord. The bottom cable trusses at this stage had temporary hinges, making the cantilevering system. This automatically ensured the proposed tensions in the diagonals (Fig. 10.29(a)).

The bottom chord was prestressed by tensioning at its ends with hydraulic jacks, by a force greater in magnitude than any possible compression stresses which might arise under certain undetermined conditions. After that, the bottom chord was connected with the diagonals, resulting in the stiffened cable truss. After elimination of the erection hinges and by its substitution by rigid connections, the cable-truss took its final shape.

The system is 11 times statically indeterminate and has been analyzed by computer.

For the top chord and diagonals, steel cables were used with parallel wires, $2\frac{5}{16}$ in. in diameter. They were galvanized by zinc. The bottom chords consisted of two cables, each having a diameter of $2\frac{5}{32}$ in.

The joints of the upper chord were constructed, as shown in Fig. 10.30. A cross section of the anchor cable consists of nine cables, each having a diameter of $2\frac{5}{16}$ in. The upper sections of the anchor cables have special installation for tensioning. The anchor cables were poststressed by hydraulic jacks. Its erection joints were connected by high-strength bolts.

The horizontal stiffness of the cable truss is ensured by wind cables consisting

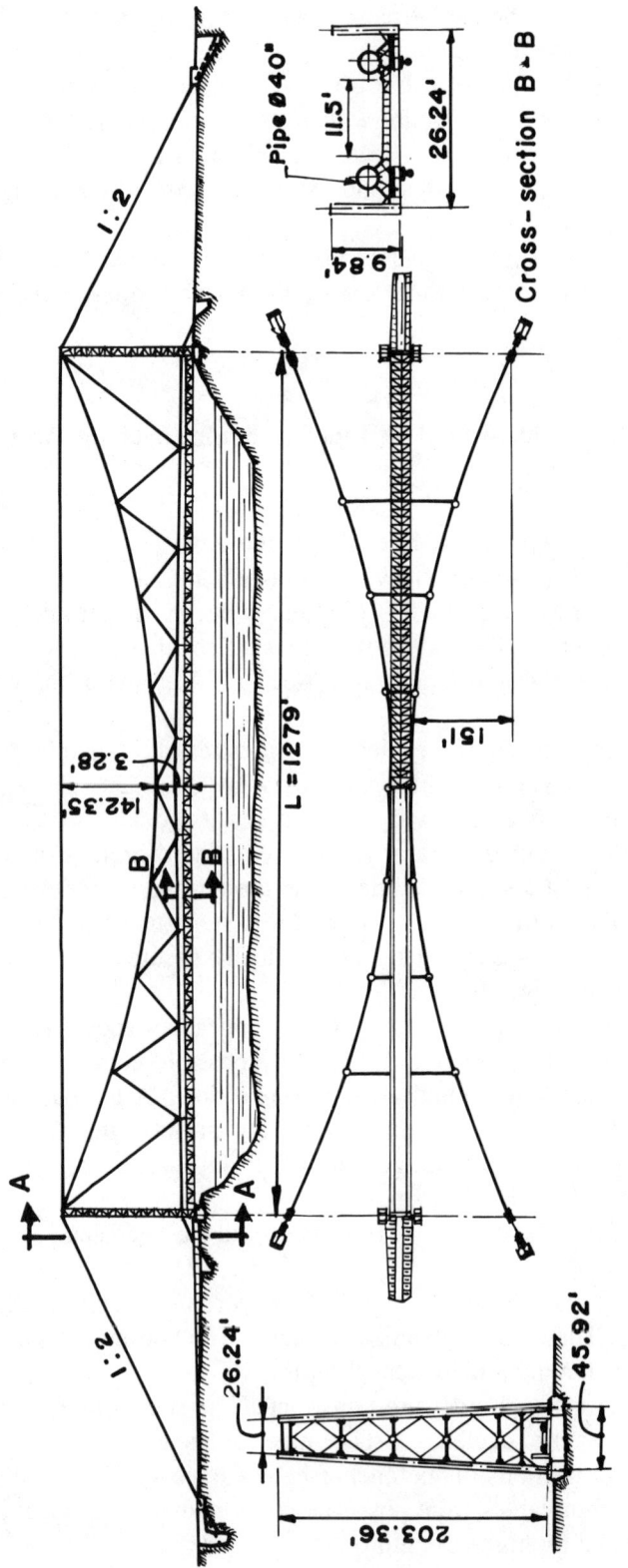

Fig. 10.28   Elevation, plan, and cross-sectional view of the bridge over the Amu-Darya River, USSR.

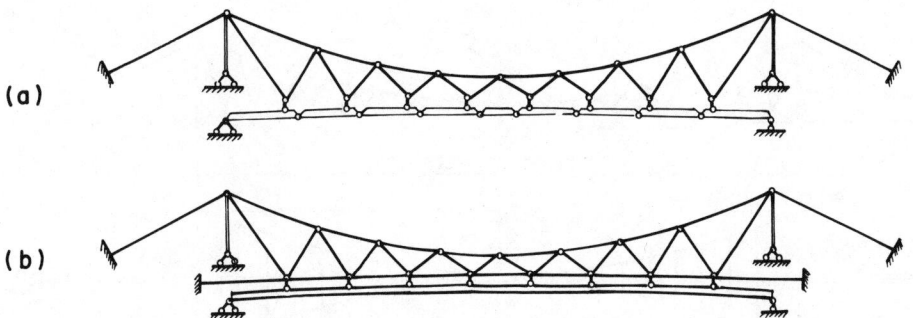

(a)

(b)

Fig. 10.29   The construction of a cable truss of the Amu-Darya River Bridge. (a) During erection. (b) Final position.

of two horizontal cables, which at mid-span are rigidly connected to the bottom chord. Steel pylons, each 203 ft (61.9 m) high, are constructed of welded structure.

The erection of the cable truss deck was performed from both river banks toward the center of the span.

10.5.8   A 2165-Foot (660.0 m) Gas Pipeline Bridge over the Amu-Darya River, USSR

This cable truss gas pipeline bridge, having a 2165-ft (660.0 m) span, was completed in 1974 over the Amu-Darya River (Fig. 10.31).[28,29] In cross section, the deck is supported by two cable trusses spaced at 9.2 ft (2.8 m) and carries

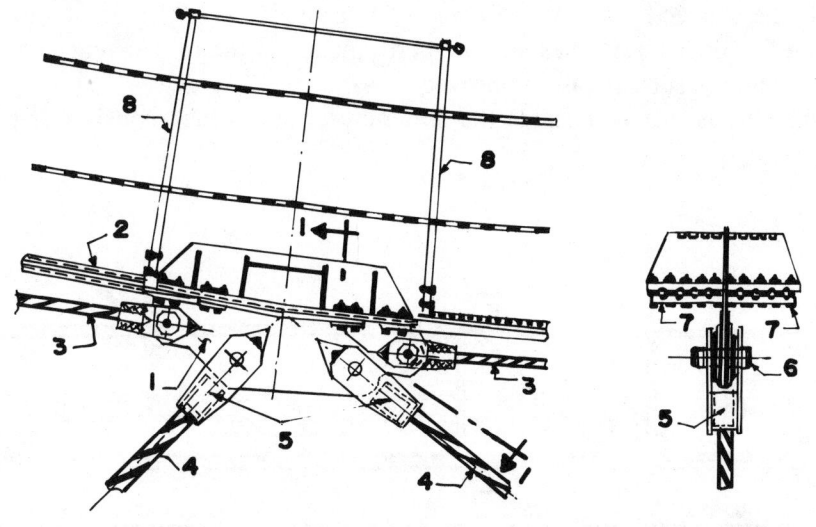

Cross – section I–I

Fig. 10.30   A joint of the upper chord of the Amu-Darya River Bridge. 1, Gusset plate; 2, upper chord; 3, spacing cable; 4, diagonal; 5, cable; 6, joint; 7, compression plate; 8, railing posts.

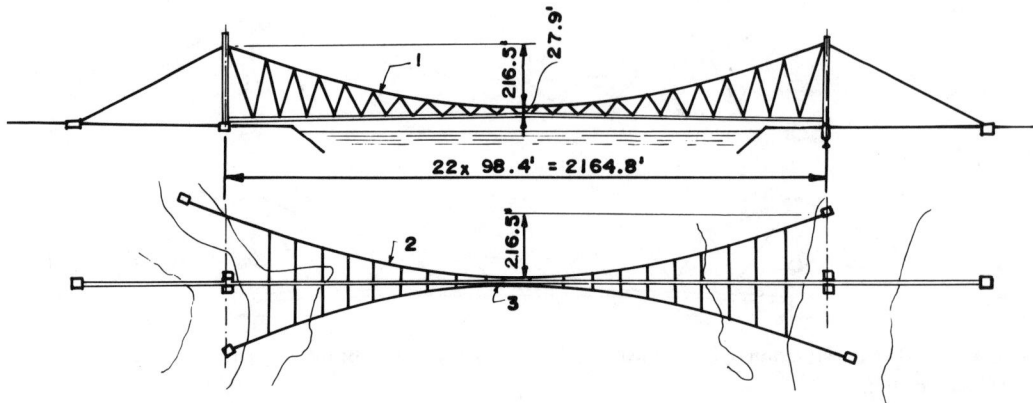

Fig. 10.31 Elevation and plan of the bridge over the Amu-Darya River. 1, Cable truss; 2, pre-stressed wind cable truss; 3, wind and main cable truss joints.

a gas pipeline have a diameter of 32 in. and four electrical cables, as well as a pedestrian sidewalk for maintenance purposes (Fig. 10.32).

For this crossing, a new cable truss system was used, being connected by inclined cables to the prestressed horizontal wind cables, performing as a space-type structure. The stiffness in the transverse direction was achieved by two prestressed horizontal wind cables of parabolic shape, having their ends fixed at each anchor block. Each wind cable consists of three galvanized strands, each having a diameter of $2\frac{3}{4}$ in. Each cable underwent a prestressing of 75 tons. Each main cable consists of six galvanized strands, each having a diameter of $2\frac{3}{4}$ in. The diagonals consist of seven galvanized wire ropes, each having a diameter of $1\frac{1}{2}$ in. Each of the two steel towers has an A shape and a height of 279 ft (85.0 m).

Due to the connection of the main and wind cables, the amplitude of vertical deflection is reduced 1.5 times and the performance of the bridge against aerodynamic action is substantially improved.

The pipeline is installed inside the stiffening girder, which consists of pipe

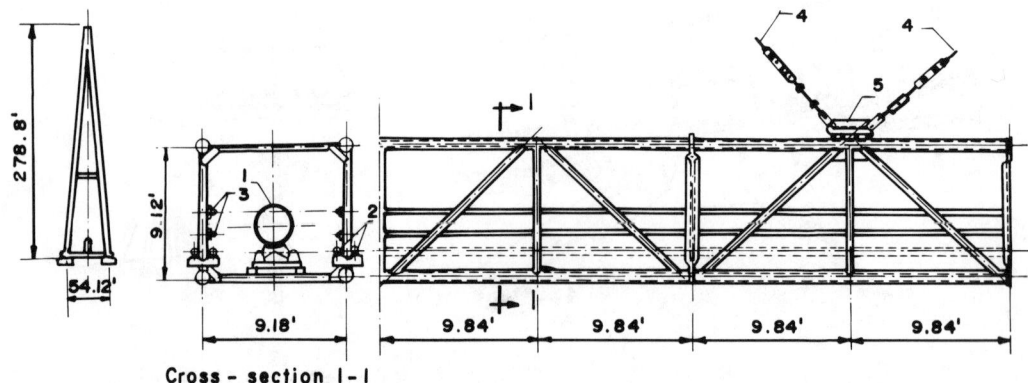

Cross – section I–I

Fig. 10.32 A cross-sectional view of the bridge over the Amu-Darya River. 1, Gas pipeline; 2, oil pipeline; 3, stiffening pipes; 4, diagonals; 5, gusset plate.

members, which reduce the wind pressure by 40%. Also, for the reduction of aerodynamic forces, sidewalks consist of grid-type steel plates.

It should be noted that the bending center of the pipeline, which takes the substantial part of the wind pressure, does not coincide with the bending center of the stiffening girder. Under wind pressure oscillations of the pipeline are reduced due to the frictional forces developed at the contact of pipeline with bearings.

This crossing was built over a period of 24 months, and erection of the cables and complete suspension system, including regulations, tests, and painting of the structure, was accomplished within 8 months. The total amount of steel used for bridge construction was 1600 tons, and for the cables, 600 tons.

10.5.9 Cable Truss Pipeline Bridge over the Dnieper River, USSR

This cable truss pipeline bridge crossing the Dnieper River has a span of 2362 ft (720.0 m) (Fig. 10.33).[30] In this bridge the main carrying cable truss performs together with the horizontal wind cables. A similar space system was first applied at the pipeline crossing over the Amu-Darya River and displayed an excellent performance in the worst weather conditions. Under nonuniform loadings, the frequencies of oscillations in the vertical and horizontal planes reduce each other and produce conditions of aerodynamic stability of the structure.

The suspended cable truss, having a span of 2362 ft (720.0 m), has 24 panels, each 98 ft (29.88 m) long. The sag of the cable is 230 ft (70.1 m), or approximately 1/10 that of the span. The height of the cable truss at mid-span is 26.6 ft (8.1 m), or 1/89 that of the span, and at the towers, 266 ft (81.1 m). The spacing of the cable trusses is 9.18 ft (2.8 m). The horizontal wind cables have parabolic configurations and sags of 230 ft (70.1 m).

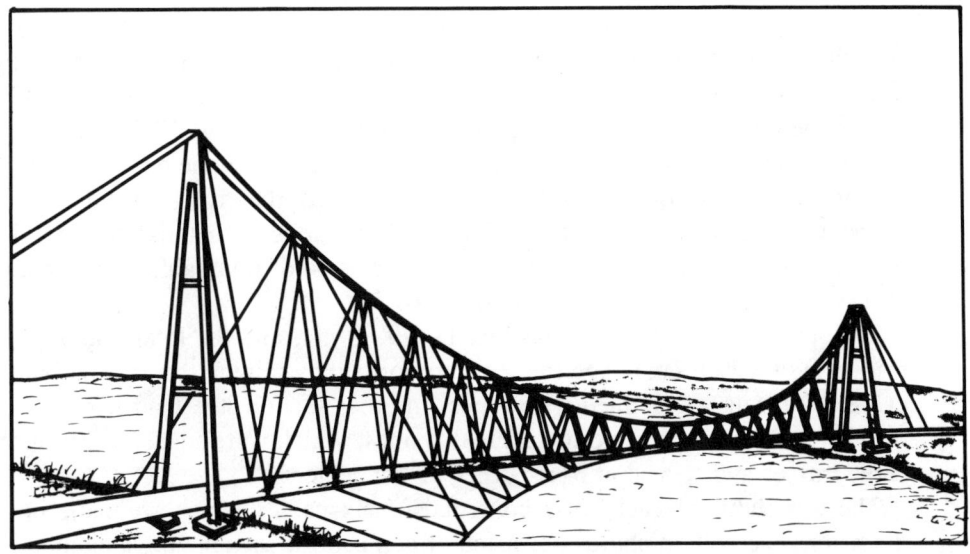

Fig. 10.33 General view of pipeline bridge over Dnieper River, USSR.

The main carrying cable of the top chord at each cable truss consists of six strands, each having a diameter of $2\frac{7}{8}$ in. and breaking stress is not less than 10–12 kips.

The first diagonal cable of the truss descends from the tower top to the first joint, along with the stiffening girder, which simultaneously performs as the bottom chord of the cable truss.

In the vertical plane, the stiffening girder has a curved configuration with the camber at mid-span of 9.8 ft (2.98 m), about 1/240 that of the span. At both ends, the stiffening girder has movable bearings in the longitudinal direction. The longitudinal forces from the pipe acting on the stiffening girder are transferred onto the anchor-type bearings through the diagonals and carrying cables of the cable truss and also by chords of wind trusses which are connected to the stiffening girders at their mid-span joints.

The particular characteristic of the bridge system consists in artificial regulation of its forces to diminish any possible compression forces in the cable truss diagonals under the arbitrary position of live loads. To achieve this, the geometric parameters of the bridge system are so chosen that the dead load causes tension in the diagonals.

The analysis of the many statically indeterminate bridge systems under the action of live loads was performed by computer, considering the nonlinear behavior of the structure. This analysis indicated the substantial influence of geometric nonlinearity of the design forces on the members of the bridge system. Also, considering the space-type character of the bridge system and the influence of the longitudinal forces, the dynamic characteristics of the structure itself were determined.

10.5.10 Material-Handling Cable Truss Bridge over the Volga River, USSR

A typical example of a temporary cable truss bridge is represented by the bridge over the Volga River, having a span of 2867 ft (874.0 m) and erected in 1956 during the construction of the Volgograd Power Station (Fig. 10.34).[31,32,33]

This cable truss bridge, called a ropeway, was erected for the purpose of transferring 18 million tons of granular material across the river, considering the carrying capacity of the ropeway of about 20,000 tons of material per day. The bridge span was 2867 ft (874.0 m), and it consisted of four cable trusses placed abreast at 24.5-ft (7.47 m) centers (Fig. 10.35).

The sag of the cable truss was 262 ft (79.88 m), and the total width of the bridge, 98 ft (29.88 m). The trusses were formed by joining the top and bottom cables by means of inclined wire ropes. In cross section, the load from the cableway was carried by rigid rectangular frames, which were suspended from the cable trusses and spaced at 318.5-ft (97.1 m) centers.

The top chord of each truss consisted of six cables which were $2\frac{5}{16}$ in. in diameter. The bottom chords consisted of two cables, each having a diameter of $1\frac{3}{4}$ in. Single cables, having a diameter of $2\frac{5}{16}$ in., were used for the diagonal

Fig. 10.34   General view of prestressed cable truss bridge over Volga River, USSR.

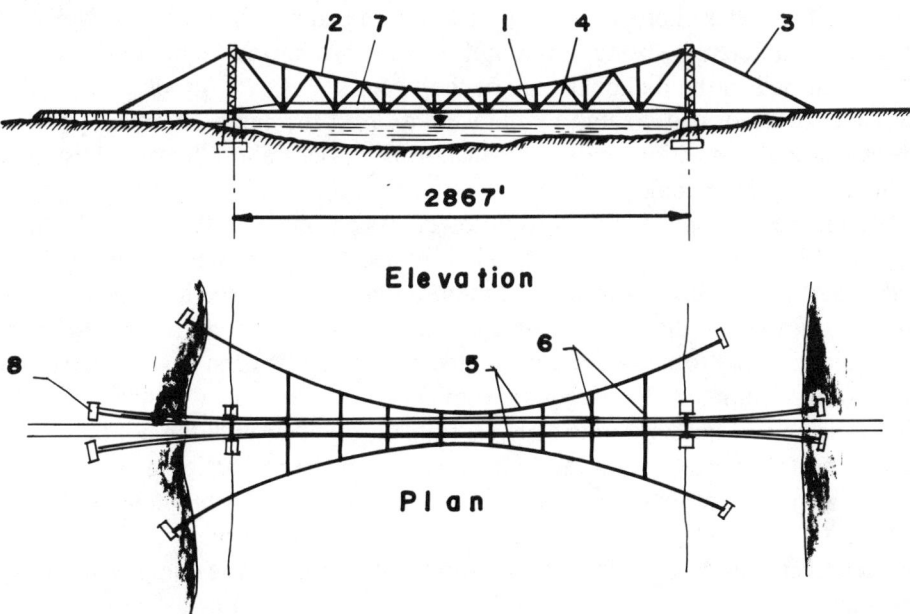

Fig. 10.35   Elevation and plan of cable truss span across the Volga River. 1, Suspended frames—ropeway supports; 2, upper ropes; 3, a pullback of the catenary system; 4, lower ropes; 5, counterwind ropes; 6, horizontal ties; 7, carrying cables of the ropeway; 8, anchorages of the base pullbacks.

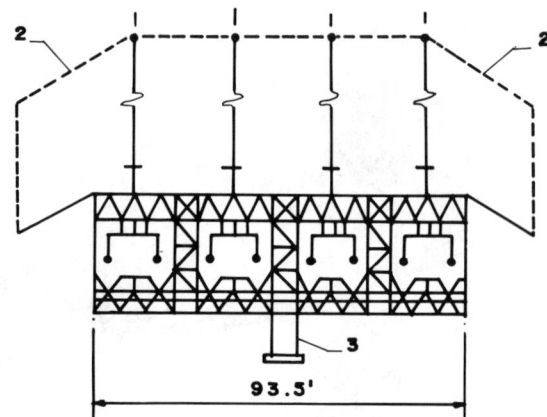

Fig. 10.36 A cross section of the Volga River
bridge. 1, Main cables; 2, ties for connecting the
main and wind cables; 3, pedestrian bridge.

members, and the total dead load was transferred onto the flexible system before
the cable diagonals were connected by joints with the bottom chords. Following
that, the top chord formed a catenary shape and the diagonals were prestressed.
Further, the bottom chords were tensioned and connected to the truss diagonals
at the joints. Following the above operations, the bridge system performed as a
rigid truss.

The cables were tensioned before assembly to 120% of the calculated stress
under working conditions. The rigidity of the whole system is ensured by the
fact that the pulling stresses of the diagonals and cables are always greater than
any compressive stresses likely to occur under live loads. For this purpose, the
cables of the lower chord were anchored onto the shores of the Volga and subjected to a sufficiently strong initial pull by means of special tightening devices.
The initial pull in the diagonals and of the upper set of cables took place under
the effect of dead load created by the weight of all the cables, frames, parts of
ropeways, and protections nets. The anchoring devices allow for an adjustment
in the length of the cables.

The bridge is suspended from two towers, each one 432 ft (131.7 m) high
(Fig. 10.37), and was erected in the following sequence. The towers were erected
with the help of creeping cranes, and then, on the towers, top chord cables were
raised, followed by the erection of a suspended bridge structure. For this purpose, erection platforms were constructed at the top of each tower, on which
joints were assembled. To the joints were connected vertical suspenders. On the
platforms below, bottom joints were assembled, together with the bottom-chord
cables and transverse frames, as well as cables for the protection nets.

Such a method of erection permitted the elimination of cable cranes, the or-
ganization of the whole assembly at one place, and to complete of the erection
in a relatively short time. The construction of the bridge was completed within
one year.

After this bridge served its purpose and the power station at Volgograd was
completed, the bridge was dismantled and its structure was used to build a highway bridge over the Volga River.

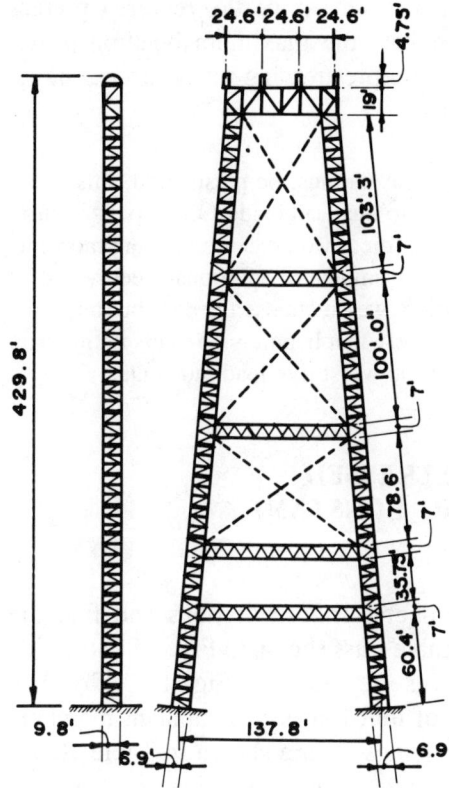

Fig. 10.37 The Volga River bridge tower.

10.6 CABLE-STAYED BRIDGES

Cable prestressing is generally applied during the erection of cable-stayed bridges, mainly by the cantilevering method. Stiffening girders are cantilevered continuously and the cables are adjusted where necessary to satisfy geometric and static conditions. The bridge structure is analyzed both during erection and at the service stage. It is necessary to predict the initial camber during erection and calculate the required pretensioning forces in the cable stays to produce this camber.

The camber required at the different erection stages can be provided by using auxiliary and permanent cable stays. The camber required after erection is completed can also be provided by adjusting the tension force in the cable stays.

During erection, the stay cables are stressed (pulled) from their lower ends. The upper ends are anchored to the tower tops in welded steel-plate assemblies fixed above the tower.

When the construction is advanced and the required span length has been erected, permanent cable stays are provided at the support and all the temporary cables are removed. The required pretensioning force is applied to the permanent cable stays to produce the necessary camber which can be calculated in the same way as is done for temporary stays.

The experimental procedure may be used to determine the required pretensioning force in cable stays in order to reduce the maximum bending girder
moments.[34] Regarding the application of cable prestressing in its service stage,
Leonhardt expressed his opinion as follows:[35]

> In some publications it is said that the stay-cables must be prestressed, this is not
> correct because the cable forces are mainly due to the dead load of the superstructure.
> Prestressing in the sense of increasing the cable force, would cause bending moments
> and normal forces in the main girders and a large amount of additional steel would be
> needed, which would not make sense. However, we do choose the distribution of the
> cable forces over the different cables in such a way that bending moments in the main
> girders under dead load conditions counteract the worst live load moments.

10.7 NUMERICAL EXAMPLE—PRESTRESSED CABLE TRUSS UNDER SYMMETRIC CONSTANT LOADING

To find the total forces in redundant members, choose sections and find the
prestressing of the statically indeterminate cable truss shown in Fig. 10.38(a).[36,37]

The basic system and redundant unknowns are shown in Fig. 10.38(b). The
values of forces in members under action of unit forces in redundant members
P_{1i}, P_{2i} and under the action of external loadings P_{ei} are shown in Table 10.1.

The performance conditions of members under symmetrical loading are

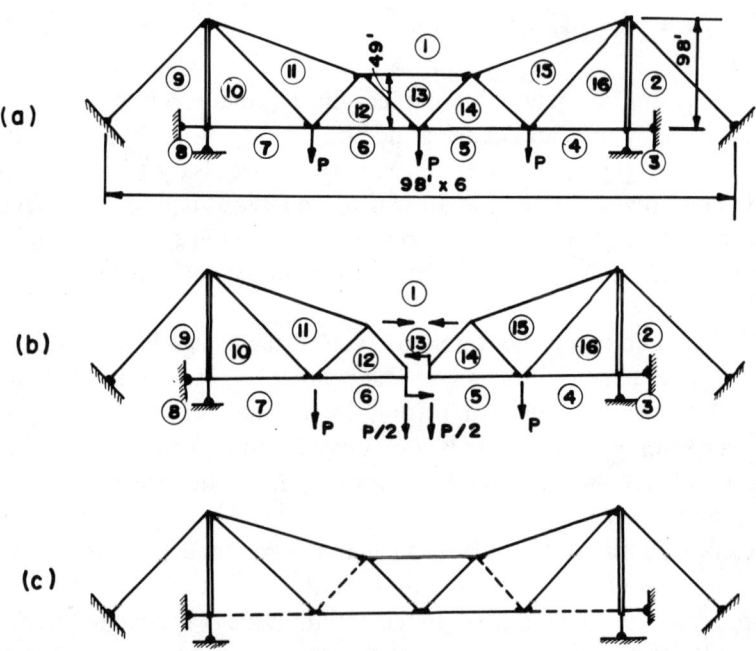

Fig. 10.38   (a) Geometric scheme of the truss. (b) Basic system under symmetrical loading. (c)
Statically determined system under constant loading.

Table 10.1

Member	l (m)	$F_1 = 1$	$F_2 = 1$	p	Forces in Basic System Under		F_p (Design Force)	A_p, cm² (Cross Sect.)	F, t (t)	A (cm²)	P_{ok} (t)	A_{ok} (cm²)
					$\grave{F}_1 = 3.028$ (t)	$F_2 = 4.486$ (t)						
1-11	4.7	0.8	0	2.36	2.4224	0	4.7824	0.48	4.7824	0.48	4.7824	0.48
1-13	3.0	1.0	0	0	3.028	0	3.028	0.305	3.028	0.305	3.692	0.37
1-15	4.7	0.8	0	2.36	2.4224	0	4.7824	0.48	4.7824	0.48	4.7824	0.48
7-10	3.0	0.5	1	−6.0	1.514	4.486	0	0	6.0	0.6	0.6	0.6
6-12	3.0	0	1	−1.5	0	4.486	2.986	0.299	4.486	0.45	4.486	0.45
5-14	3.0	0	1	−1.5	0	4.486	2.986	0.299	4.486	0.45	4.486	0.45
4-16	3.0	0.5	1	−6.0	1.514	4.486	0	0	6.0	0.6	6.0	0.6
10-11	4.25	−0.35	0	5.31	−1.06	0	4.25	0.43	4.25	0.43	4.25	0.43
11-12	2.12	0.35	0	−1.06	1.06	0	0	0	1.06	0.11	1.06	0.11
12-13	2.12	0	0	2.12	0	0	2.12	0.21	2.12	0.21	2.12	0.21
13-14	2.12	0	0	2.12	0	0	2.12	0.21	2.12	0.21	2.12	0.21
14-15	2.12	0.35	0	−1.06	1.06	0	0	0	1.06	0.11	1.06	0.11
15-16	4.25	−0.35	0	5.31	−1.06	0	4.25	0.43	4.25	0.43	4.25	0.43
1-9	4.25	0.707	0	8.5	2.14	0	10.64	1.07	10.64	1.07	10.64	1.07
1-2	4.25	0.707	0	8.5	2.14	0	10.64	1.07	10.64	1.07	10.64	1.07

$(F_3 = 0)$

$y_1 = P_{7\text{-}10} = 0.5F_1 + F_2 - 6 \geq 0$

$y_2 = P_{6\text{-}12} = F_2 - 1.5 \geq 0$

$y_3 = P_{10\text{-}11} = -0.35F_1 + 5.31 \geq 0$

$y_4 = P_{11\text{-}12} = 0.35F_1 - 1.06 \geq 0$

$F_1 \geq 0; F_2 \geq 0$

Let us form an expression for the volume of all steel members of the truss. For this, using equation (7) we calculate the sum of products of forces of all members by length. (the volume is increased R_1 times).

$$P_{1\text{-}11}l_{1\text{-}11} = 3.76F_1 + 11.1 \qquad\qquad P_{1\text{-}13}l_{1\text{-}13} = 3F_1$$

$$P_{1\text{-}15}l_{1\text{-}15} = 3.76F_1 + 11.1 \qquad\qquad P_{7\text{-}10}l_{7\text{-}10} = 1.5F_1 + 3F_2 - 18$$

$$P_{6\text{-}12}l_{6\text{-}12} = 3F_2 - 4.5 \qquad\qquad P_{5\text{-}14}l_{5\text{-}14} = 3F_{1\text{-}45}$$

$$P_{4\text{-}16}l_{4\text{-}16} = 1.5F_1 + 3F_2 - 18 \qquad\qquad P_{10\text{-}1}l_{10\text{-}11} = 1.498F_1 + 226$$

$$P_{13\text{-}14}l_{13\text{-}14} = 4.5 \qquad\qquad P_{12\text{-}13}l_{12\text{-}13} = 4.5$$

$$P_{15\text{-}16}l_{15\text{-}16} = -1.498F_1 + 226 \qquad\qquad P_{14\text{-}15}l_{14\text{-}15} = 0.743F_1 - 2.25$$

$$P_{1\text{-}2}l_{1\text{-}2} = 3F_1 + 36.1 \qquad\qquad P_{1\text{-}9}l_{1\text{-}9} = 3F_1 + 36.1$$

The summation of these inequalities gives

$$\sum_{i-=1}^{m} P_i l_i = 18F_1 + 12F_2 + 99.1$$

By neglecting free terms, the expression for the function is

$$Z = 18F_1 + 12F_2$$

Thus, the problem lies in minimizing the function Z, considering limitations

$$y_1 \geq 0 \quad y_2 \geq 0 \quad y_3 \geq 0 \quad y_4 \geq 0 \quad F_1 \geq 0 \quad F_2 \geq 0$$

which is equivalent to the problem of maximizing the expression $-Z = -18F_1 - 12F_2$

The conditions of the problem in matrix form are:

	$-F_1$	$-F_2$	l
$y_1 =$	-0.5	(-1)	-6
$y_2 =$	0	-1	-1.5
$y_3 =$	0.35	0	5.31
$y_4 =$	0.35	0	-1.06
$Z =$	18	12	0

For solution, the coefficient in the first row of the second column is chosen. After performing one step of the modified Jordan exception, we obtain

	$-F_1$	$-y_1$	l
$y_1 =$	0.5	-1	6
$y_2 =$	0.5	-1	4.5
$y_3 =$	0.35	0	5.31
$y_4 =$	(-0.35)	0	-1.06
$Z =$	12	12	-72

The sign of F_i has a limitation ($F_i \geq 0$); therefore values of F_i are not free variables. For this reason the values of F_i are not excluded from the table. Again, for solution, the negative coefficient of the fourth row is chosen and the following step of a modified Jordan exception is carried:

	$-y_i$	$-y$	l
$F_1 =$	1.43	-1	4.486
$y_2 =$	1.43	-1	2.986
$y_3 =$	1	0	4.25
$F_2 =$	-2.86	0	3.028
$Z =$	34.29	12	-108.336

In this matrix all free terms are positive, and there are no negative coefficients in the Z row. Therefore, at $y_1 = 0$ and $y_4 = 0$, function Z has a minimum value. Thus, the solution is

$$F_1 = 4.486 \text{ t} \qquad F_2 = 3.028 \text{ t}$$

This problem may be solved graphically (because there are only 2 unknowns). In coordinates F_1-F_2, construct a polygon having edges I, II, and III (Fig. 10.39). Equations of straight lines (sides of the polygon) serve the inequalities

$$y_1 \geq 0 \qquad y_2 \geq 0 \qquad y_3 \geq 0 \qquad y_4 \geq 0$$

By equating F_1 and F_2 to zero in certain sequences, the straight lines can be drawn dividing the plane in coordinates F_1-F_2 into two half-planes, satisfying the given inequalities at any value of F_1 and F_2 in one of the half-planes only. The straight line passing through the origin of the coordinates is

$$Z = 18F_1 + 12F_2 = 0 \qquad F_1 = \frac{-12}{18} F_2$$

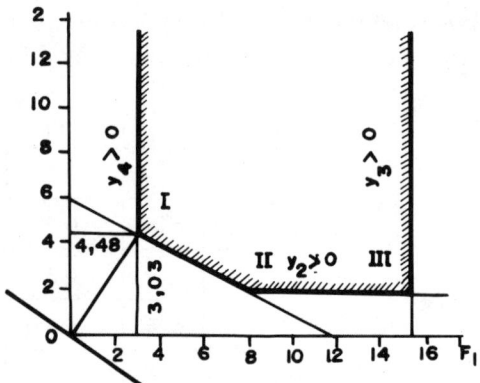

Fig. 10.39 Graphical solution of numerical example.

As shown in Fig. 10.39, the edge closest to the straight line $Z = 0$ will be point I, having coordinates $F_1 = 3.03$ and $F_2 = 4.48$.

After determination of the total forces in redundant members corresponding to the minimal volume of the truss, the sections of the members may be chosen.

For constant loading, the design forces F_p and the corresponding values of cross sections A_p are shown in Table 10.1. In the case of variable loading the design forces and cross-sectional areas can be found from columns F and A of Table 10.1.

Now, let us solve the statically indeterminate system to find the prestressing force F_{sn_1} and F_{sn_2}. Under symmetrical loading these forces are:

$$F_{sn_1} = 0.626 \text{ t} \qquad F_{sn_2} = 3.57 \ t$$

The necessary values of prestressing are

$$F_{pn_1} = F_1 - F_{sn_1} = 3.028 + 0.626 = 3.654 \text{ t}$$

$$F_{pn_2} = F_2 - F_{sn_2} = 4.486 - 3.57 = 0.916$$

Since the force F_{pn_1} is greater than the design force, for which the cross section of member 1-13 was chosen, the section for member 1-13 is modified for this requirement. The other member sections need not be changed. Again, the statically indeterminate system is solved for the modified sections to find the revised prestressing forces:

$$F_{sn_2} = -0.664 \text{ t} \qquad F_{sn_2} = 3.57 \text{ t}$$

Thus, the prestressing forces are

$$F_{pn_1} = 3.028 + 0.664 = 3.692 \text{ t}$$

$$F_{pn_2} = 4.486 - 3.57 = 0.916 \text{ t}$$

The required initial loading considering the performance of member 10-11 is found as follows:

$$F_{10\text{-}11} = (-0.35)3.692 + \frac{5.32}{8}\, p' \geq 0$$

$$p' \geq \frac{1.2922}{1.77} = 0.73 \text{ t}$$

That is, under symmetrical loading the system may perform between force interval $p' = 0.73$ t and $P = 3$ t.

It can be noted from Table 10.1 that different sections are provided for different members, which is practically inconvenient. However, this may be compensated considering the achieved saving in material. For example, in member 10-7, a 47% savings in material is achieved.

Thus, it is necessary to consider the distribution of internal forces corresponding to the minimum volume condition as compared with the conventional approach.

REFERENCES

1. Maney, G. A., "New Type Suspension Bridge Proposed," *Engineering News-Record*, April 24, 1941, pp. 64–65.
2. Modjeski and Masters, staff, "Suspension Bridges and Wind Resistance," *Engineering News-Record*, October 23, 1941, pp. 97–100.
3. Ostashevskii, J. A., "Suspension Bridges Having Inclined Suspenders," *Proceedings LIIKS*, issue VII, Leningrad-Moscow, State Edition of Literature on Buildings, 1940, pp. 116–142 (in Russian).
4. Grove, W. G., "Cable-Stiffened Suspension Bridge Developed Through Engineering Board Research," *Engineering News-Record*, March 7, 1946, pp. 91–94.
5. Grove, W. G., "Probable Future Trends in Suspension Bridge Design," 62nd Annual Meeting of the Connecticut Society of Civil Engineers, Inc., at Hartford, Connecticut, March 20, 1946, pp. 60–83.
6. Shearwood, F. P., "The Stabilization of Suspension Bridges," *The Engineering Journal*, January 1954, pp. 1–6.
7. Ostashewsky, J. A., "Suspension Bridges with Inclined Suspenders," *Trudy MIISK*, vol. 11, Leningrad, 1940 (in Russian).
8. Kireenko, V. I., *Cable-Stayed Bridges*, Edition Budivelnik, Kiev, 1967 (in Russian).
9. Trofimovich, V. V., and Permiakov, V. A., *Design of Prestressed Cable Systems*, Edition Budivelnik, Kiev, 1970, pp. 33–42 (in Russian).
10. Spencer, A. J. M. et al., *Engineering Mathematics*, vol. 2, Van Nostrand Reinhold, New York, 1977, ch. 1, pp. 2–39.
11. Birdsall, B., "Cable-Stiffened Suspension Bridge Updated," *Engineering News-Record*, May 21, 1953, pp. 32–40.
12. Birdsall, B., "A Prophetic Design in Out-of-the Way Place," *Civil Engineering*, September 1954, pp. 40–41.
13. Sollenberger, N. J., Cable-Truss Design Greatly Increases Stiffness, *Civil Engineering*, September 1954, pp. 42–45.
14. Hills, H. W., "Cable Erection Complicated by Floods and Tropical Climate," *Civil Engineering*, September 1954, pp. 48–50.
15. Nixon, J. E., "Construction Methods Adapted to Local Labor," *Civil Engineering*, September 1954, pp. 50–52.
16. Matos, M. E. C., "Experimental Model Tests for Design of a New Type of Suspension

Bridge," IABSE Proceedings Eighth Congress, Final Report, New York, September 9–14, 1968, Zurich, pp. 813–825.

17. "Monthly Review of Engineering Development in the USSR" *Civil Engineering and Public Works Review*, vol. 51, no. 596, February 1956, p. 179.

18. Cain, J. F., "Fifty-Year Development—Construction of Steel Pipeline Suspension Bridges," ASCE Journal of the Construction Division, December, 1975. pp. 733–749.

19. Merkblatt 313, "Advisory Bureau for Steel Use. Pipelines and Energy Bridges," 2d edition, Düsseldorf, 1972, (in German).

20. Lamb, T., and McManus, R. N., "Materials Handling Suspension Bridges," Engineering Institute of Canada, 1962 Annual General Meeting, Paper No. 2, pp. 1–8.

21. Troitsky, M. S., "On the Aerodynamics of Pipeline Suspension Bridges," Proceedings IASS, Calgary, July 3–6, 1972, pp. 235–242.

22. Tesar, A., "Calculation and Construction of Prestressed Pipeline Bridge in CSSR," International Special Convention "Prestressed Steel Constructions" from 10 to 13 September, 1963, Wissenschaftliche Veroffentlichungen aus der Fakultät für Bauwesen der Technishe Universität Dresden, no. 30, 1964.

23. Focardi, F., "Suspension Structure for Carrying Gas Pipeline over the River Po (Italy)," *Acier-Stahl-Steel*, no. 9, 1971, pp. 354–357.

24. Schrefler, B., "Special Problems of Designing Multi-Span Suspension Bridges for Passage of Pipeline," Der Stahlbau, no. 1, 1978, pp. 22–29, (in German).

25. Goschy, B., "The Torsion of Skew-Cable Suspension Bridges," in *Space Structures*, ed. Davies, R. M., Blackwell Scientific Publications, Oxford, 1967, pp. 213–220.

26. Slokim, E. J., *Cable-Truss Crossing at Span of 390 m. Across River Amu-Darya on the Gas Pipeline Buchara-Ural, Metal Structures*, N.S. Streletskii, Moscow, 1966, pp. 244–253 (in Russian).

27. Streletzkii, N. N., "Evaluation and Prospects for Development of Prestressed Steel and Steel-Reinforced Concrete Bridges in the USSR," The 3rd International Conference on Pre-stressed Metal Structures, Leningrad, Reports, vol. II, 1971, pp. 323–333 (in Russian).

28. Anon., "Suspended Gas Pipeline of 660 m. Span (USSR)," *Bulletin IABSE*, no. 32, 1976, pp. 25–27.

29. Meljnikov, N. P., *Metal Constructions—Present State and Development*, Stroiizdat, Moscow, 1983, pp. 377–379 (in Russian).

30. Ibid., pp. 379–381.

31. Anon., "Monthly Review of Engineering Development in the USSR," *Civil Engineering and Public Works Review*, vol. 51, no. 596, February 1956, p. 179.

32. Mansel, S. E., Uskov, A. N., and Jacobson, A. G., "Ropeway over the Volga," *Engineering Digest*, April 1961, pp. 20–23.

33. Meljnikov, op. cit., pp. 367–370.

34. Troitsky, M. S., and Lazar, B., "Model Investigation of Cable-Stayed Bridges—Post-Tensioning of Cables," Report no. 5, Sir George Williams University, Montreal, Canada, 1970.

35. Leonhardt, F., and Zellner, W., "Cable-Stayed Bridges: Report on Latest Developments," Canadian Struct. Engineering Conference—1970, preprint, pp. 1–27.

36. Trofimovich, V. V., and Permiakov, V. A., *Design of Prestressed Cable Systems*, Ed. Budivelnik, Kiev, 1970, (in Russian).

37. Troitsky, M. S., Zielinski, Z. A., Pimprikar, M. S., and Poorooshasb, H. B., "Prestressed Steel Cable Bridge Trusses," Internal Symposium on Structural Steel Design and Construction, Applied Technology Conference, Singapore Structural Steel Society, July 3–4, 1985, pp. 466–487.

Chapter 11
Rehabilitation and Strengthening of Steel Bridges by Prestressing

11.1 INTRODUCTION

The need for rehabilitation of existing bridges is a growing concern in many countries and has been emphasized in various research reports and publications. Many of these bridges, for example, were designed for relatively light loading and have narrow roadways that are inadequate for present traffic; thus they face serious problems such as load and lane limitations and expensive premature bridge replacement. Rehabilitation is also required in cases where the preservation of bridges is necessary as a part of historic and cultural heritage.

Although rehabilitation of a bridge includes all of its components such as substructures, superstructure, and approaches, this study will treat only those concerned with superstructures. Many techniques are interrelated and vary accordingly to the structural components and their field application.

The increasing number of bridges requiring rehabilitation is a common feature and leads to such serious problems as: (1) loss or reduction of structural safety with the resulting necessity for load limitation, and (2) expenditure of large sums for premature bridge replacement which could be avoided if adequate funds were deviated to timely rehabilitation.

Bridge deficiencies evolve from a variety of situations and conditions. Basic design criteria, traffic usage, environmental factors, and other site conditions are all involved to some extent and are responsible for specific deficiencies.

Deficiency causes can be categorized into two broad areas: (1) those which result from the design of the facility and are thus inherent deficiencies, and (2) those which result from the use of the facility and are essentially the result of wear. Deficiencies from either cause may be subdivided into the following four areas: structural, mechanical, geometric, and safety.

1. Structural deficiencies are those which affect the structure's ability to carry imposed loads. These deficiencies are caused most frequently by lack of proper maintenance, poor design details, and "light" original designs.

Many structures that are currently an integral part of highway systems were not designed initially to carry the load being imposed on them by modern traffic. These light designs were based on vehicle weights which are less than those presently being used, and the load frequently rated is only a small percentage of the loads now utilizing the crossings.

2. Mechanical deficiencies are those which prohibit the structure from reacting in a controlled manner to environmental factors. These deficiencies are primarily caused by corrosion of metal elements, the accumulation of debris and silt around bearings and joints, and poor design details.

3. Geometric deficiencies are those that relate to the geometrics of the roadway as it approaches and traverses the bridge. Vertical and horizontal alignment, roadway width, vehicle sight distance, and traffic capacity are included.

4. Safety deficiencies relating to the safety of the motorist include those that affect the safety of the vehicle as it passes over the structure. Many of these are geometric in nature, such as roadway width and clearance.

11.2 Bridge Strengthening—Brief History

Since the early 1950s there have been many reports of bridge strengthening by posttensioning. In 1952, Lee reported the strengthening of British steel highway and railway bridges by posttensioning.[1] Both beam and truss bridges were strengthened. Berridge and Lee described strengthening of a steel truss bridge in 1956,[2] and Knee mentioned strengthening of British steel railway bridges by posttensioning as if it were a fairly common practice.[3]

Sterian described Rumanian practices in strengthening bridges by various methods, including posttensioning by cables or rods, prior to 1969.[4] Although Sterian described several methods of strengthening, including the addition of cover plates, he viewed posttensioning as having the most potential.

A proposal by Kandall in 1969 for strengthening steel structures by posttensioning was unique because he recommended additional material to the compression regions of members.[5] The additional material had to be carefully fitted through or around cross members, making for a relatively complicated strengthening operation.

Vernigana and colleagues described the successful strengthening of a five-span reinforced concrete bridge in Ontario, Canada.[6] The five spans were posttensioned by means of draped cables so as to make the repaired bridge continuous rather than simple span.

Belenya and Gorovskii of the Soviet Union presented a rather complete analysis of steel beams strengthened by posttensioning.[7] According to their analysis, prestressing can add up to 90% capacity to an unprestressed steel beam. They recommended a tie rod length of 0.5 to 0.7 of the span length and recommended considering P-Δ effects only when the depth-to-span ratio is less than $1:20$.

During the past ten years, several Minnesota bridges have been strengthened by posttensioning. A prestressed concrete bridge damaged by vehicle impact was repaired using posttensioning.[8] It appears that at least two Minnesota steel beam bridges have been repaired temporarily using posttensioning.[9] In one case, salvage cable and timber were utilized for repair.[10]

During the 1970s T. Y. Lin International strengthened a multiple-span steel plate girder bridge in Puerto Rico by posttensioning.[11] The posttensioning scheme removed approximately 6 in. of dead-load deflection at mid-span.

In 1983, Lamberson reported numerous examples of strengthening by posttensioning in the United States.[12] The Indiana Department of Highways posttensioned the girders in a reinforced concrete bridge with threadbars harped at mid-span of each girder, using essentially a kingpost truss concept. The Illinois Department of Transportation utilized a similar harped-tendon arrangement to strengthen the transverse steel floor beams in a steel truss bridge. Eleven other examples, including one of the bridges strengthened as a part of the research at Iowa State University, also are described. The threadbars utilized in the examples have been protected by epoxy coatings, grouted pipes, or grouted plastic tubes.

A four-beam, two-lane composite bridge in Pasco County, Florida, was repaired and strengthened in 1984.[13] The posttensioning, designed by the AASHTO Service Load Design Method, was similar to that illustrated in Fig. 11.1 and

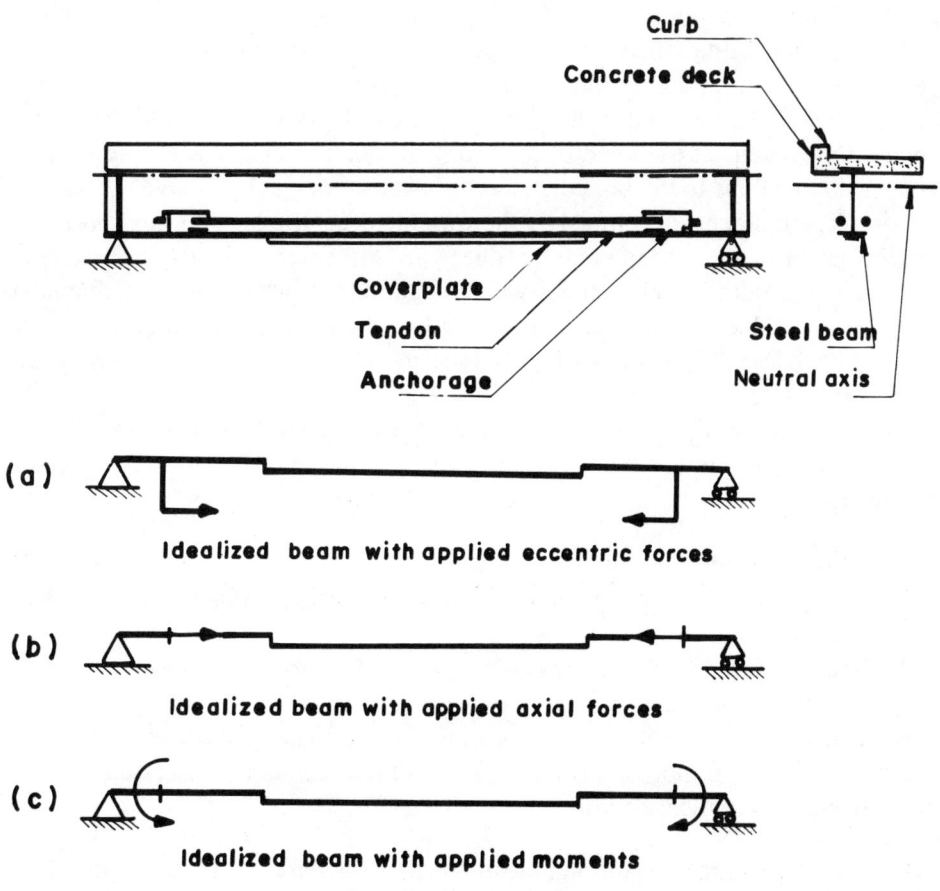

Fig. 11.1   Posttensioned bridge beam.

was applied to all four beams in each of three simple spans.[14] The posttensioning raised the rated capacity of the bridge from an H15-44 to an HS 20-44 at a cost of approximately $20,000.

California has strengthened 7 steel bridges by posttensioning in the period from 1979 to 1984 and is planning to strengthen at least 12 more.[15] Since all the beams are posttensioned equally, lateral distribution of the posttensioning is not a major problem. To date, all of the California bridges have been strengthened using a strand that is enclosed in galvanized pipe and grouted after posttensioning. All prestressing systems and anchorage hardware must be tested and approved by the Caltrans Transportation Laboratory prior to installation.

11.3 REHABILITATION PROCEDURES

The process of rehabilitation of a deficient bridge can vary extensively depending on the degree and severity of the problem needing correction. The work can include a deck replacement and minor repair or can be an involved procedure including strengthening of critical members.[16]

Generally, there are two different approaches to increasing the load-carrying capacity of the existing steel structure. One is reinforcing the existing steel member by adding new ones. The other is prestressing the existing steel member. For bridge rehabilitation the suitable method is prestressing by high-strength steel cables. One of the common methods of prestressing is to place draped tendons below the centroid of the beam or below the beam with holddowns and anchor the tendons at each end of the beam. Another is to attach simple straight posttension tendons to the bolted anchor brackets a few inches above the bottom flanges. The moments developed by the variable eccentric prestressing force will cancel out most of the moments produced by the external loads. Prestressing induces stress in the steel member which is similar to but opposite in character from those produced by the dead load and live loads. These procedures can be developed to carry live loads or to reduce the dead-load stresses, thereby increasing the live-load capacity.

The advantages of prestressing over the reinforcing method of increasing section modulus and higher load-carrying capacity of a steel member include the following:

1. It may be more economical.
2. Normal traffic may be maintained during staging, and if a detour is needed, the period will be short.
3. Jacking of the beams to stress-free the members for the connection of new flange plates is eliminated.
4. In many cases, such as with existing riveted plate girders or trusses, increasing the section modulus is unfeasible and the technique described offers a feasible and economical solution.

Problems in bridge rehabilitation are treated by Park.[17] His book provides common causes of bridge defects and deterioration along with countermeasures for repairs, rehabilitations, retrofits, and replacements. He considers prestressing as the application of a predetermined straight or eccentric force to a steel member so that any external loading will be counterbalanced by this force.

There are three general ways of prestressing beams: The first is using end-anchored high-strength wires; the second is to stress cover plates; and the third is to cast a composite concrete slab to the deflected beam.

The American Society of Civil Engineers prepared a publication related to the rating and strengthening of old bridges built during the early 1900s.[18] Among other concepts of strengthening a truss bridge, they consider prestressing either a single tension member or strengthening the entire bottom chord or some other group of members strengthened by prestressing.

11.4 STRENGTHENING—CONVENTIONAL METHODS

These conventional procedures are well understood and safe but are expensive and difficult to carry out. In each of these cases, the jacking of existing members often presents special problems and may require heavy temporary framing. Furthermore, the alteration interrupts traffic or reduces the number of traffic lanes on a bridge for the duration, which necessitates the construction of temporary structures.

Current practices for increasing the load-carrying capacity of existing steel structures, particularly members of steel bridges, have been discussed by Kandall.[19] Currently structural members are strengthened by bolting or welding steel plates or other structural shapes to the flanges of existing members. These methods increase the section moduli of the strengthened members.

Figures 11.2 and 11.3 show the standard methods of strengthening by adding cover plates to the top and bottom flanges of members carrying a concrete slab. For this strengthening it is necessary to demolish the concrete slab by passing a shallow saw cut on either side of the top flange and then to remove the slab so that the reinforcing steel is left in place. Undamaged loose concrete around the reinforcing steel is removed to provide a proper bond when the slab is repaired.

Demolishing part of the slab while maintaining limited traffic flow on the bridge requires adequate traffic facilities, which may be very costly. Sometimes, where clearance permits, the section modulus may be increased by connecting a certain structural shape to the bottom flange of the member as shown in Fig. 11.4.

In each of the above cases, the beam or girder must be jacked up at a certain location to relieve stress under dead load. If the member is not relieved of stress, the added flange material will be stressed only under live load while the existing

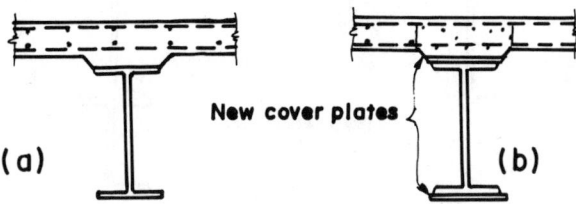

(a)   New cover plates   **(b)**

Fig. 11.2 Standard current practice in reinforcing a rolled section.

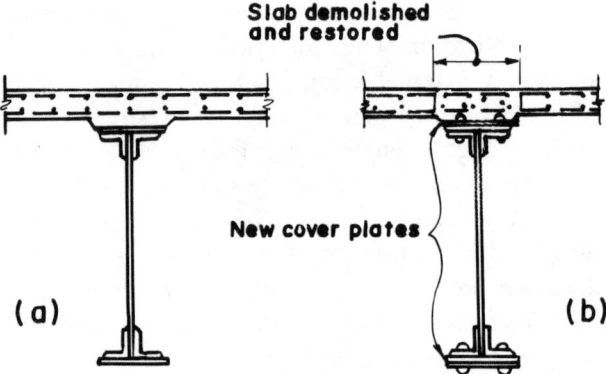

Fig. 11.3 Reinforcing of riveted plate girder.

flange will be stressed under dead and live load. This results in heavy and uneconomical flanges.

The jacking of existing members often presents special problems and may require heavy temporary framing that is costly and complicated.

11.5 STRENGTHENING OF THE PLATE GIRDER

The strengthening of a steel girder by a draped tendon (Fig. 11.5) produces moments and bending stresses that are similar to but opposite in direction from those produced by the design loads, in effect reducing or eliminating the dead-load moment and permitting the structure to take increased dead and live loads without the necessity of strengthening the structural components.

Fins, brackets, or other devices structurally connected to the girder hold the tensioning steel in its predetermined shape. Both the tensioning steel and the compression member can pass through or adjacent to stiffeners of existing girders, or, when other members frame into the beam, pass through holes cut into the webs and brackets of the framing members.

This method is not limited to simple spans but is applicable to continuous members and overhanging beams.

11.6 STRENGTHENING—PRESTRESSING METHODS

Strengthening steel bridges and one of the most effective methods of rehabilitating old ones is prestressing. This method generally may be applied to any steel bridge system.

Fig. 11.4 Reinforcing of steel section by new steel member.

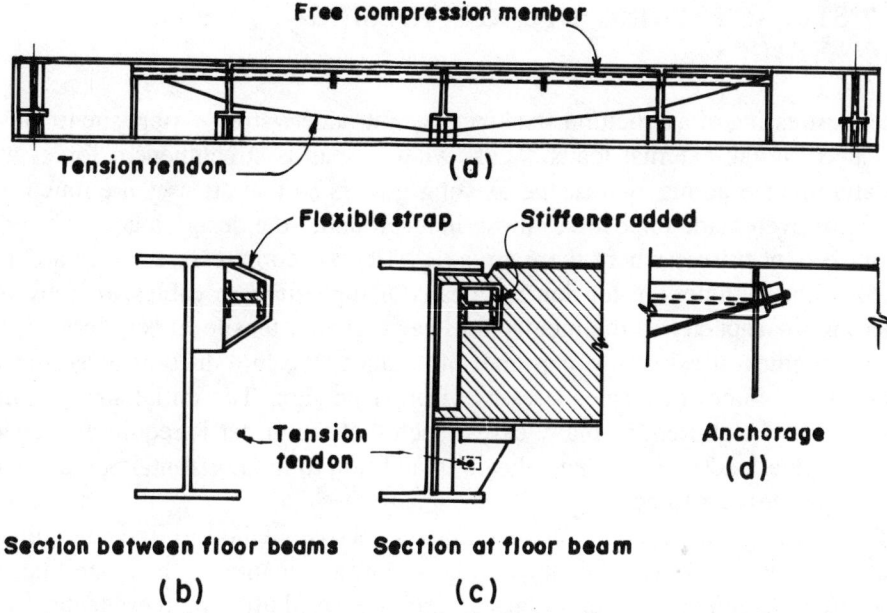

Fig. 11.5   Strengthening of a plate girder.

Older bridges must be either strengthened or replaced in order to carry the heavier loads imposed by today's traffic. A simple tensioning method might prove less time consuming and more economical than replacement by a new structure. In addition, implementing the method will not interfere with traffic, because all work is performed below the roadway deck.

This method of increasing the load-carrying capacity of a structural member involves anchoring a steel tendon to the ends of an independent compression member, draping the tendon to a predetermined polygon, and then tensioning the tendon to the required amount.

Prestressing, which may be performed in different ways, is an effective means of increasing the carrying capacity and strength of existing steel bridges. Several examples of utilization of this technique in Europe, North and South America, and South Africa demonstrate the feasibility of the concept.

In 1968 an ASCE-AASHO Subcommittee prepared a brief review of developments in the use of prestressed steel flexural members and available prestressing methods with comparative economics.[20] During the past two decades considerable fundamental research in the area of prestressed steel construction has been carried out, notably in the USSR. International conferences were organized in 1963, 1966, and 1971, and specifications were developed for designing prestressed steel structures.

To date, a number of steel and composite bridges have been strengthened by using prestressing methods. In the following sections, plate girder and truss-type steel bridges are analyzed for strengthening by available prestressing methods, with possible future trends considered. Further, the rehabilitation of these bridges by the application of prestressing by high-strength cables, or tie rods, is examined in terms of the relative economics and achieved structural capacity.

11.7 STRENGTHENING BY PRESTRESSING— ADVANTAGES

The prestressing of a structural steel member introduces stresses opposite to those produced by the external loads. Thus, when a span is strengthened, forces are usually applied acting against the existing girders so that stresses are much reduced or even made opposite to those induced under the design load.

In cases of trusses where there are axial stresses, compressive forces are applied to the members in tension by means of high-strength cables, thereby increasing the capacity of the members to carry greater tensile forces. In girders, where bending stresses are of prime importance, the introduction of eccentric prestressing forces originates stresses of opposite sign. The initial stress in the tendon may be chosen, in many cases, such that the girder is required to resist only the live-load moments and shears in addition to the horizontal components of the prestressing force.

Prestressing constitutes a convenient and effective method for the strengthening and renovation of existing steel bridges. An examination of the rehabilitation of many steel and composite bridges around the world utilizing these techniques demonstrates that they achieve economy and increase structural capacity while maintaining the original framework for use.

The application of prestress by high-strength cables, or tie rods, offers a wide range of solutions for the strengthening of plate girder, lattice girder, and truss-type bridges. Many proposals have been made in recent years, considering (1) the achieved economy and the fact that often the work is done without interrupting the traffic; (2) the possibility of avoiding temporary falsework; and (3) the amount of high-tensile steel needed is extremely small compared with the tonnage of steel what would otherwise be required to accomplish the same degree of strengthening.

11.8 EXAMPLES OF STRENGTHENED PLATE GIRDER BRIDGES

To date, a number of steel and composite plate girder bridges have been strengthened by using prestressing methods.

11.8.1 Ostrova Bridge; Czechoslovakia

An interesting method was used to strengthen a railway bridge having a 70-ft (20.4-m) span at Ostrova, Czechoslovakia (Fig. 11.6).[21] The carrying capacity of the main girders was increased by placing one truss system outside of each main girder, as shown in Fig. 11.6(c). The prestressing was achieved by placing the tendons outside of each main girder and stretching the tendons against the main girder at its bottom joint. The prestress load was obtained by blocks (Fig. 11.6(b)). To eliminate the substantial deflection, the tendons were connected to the stiffened girders.

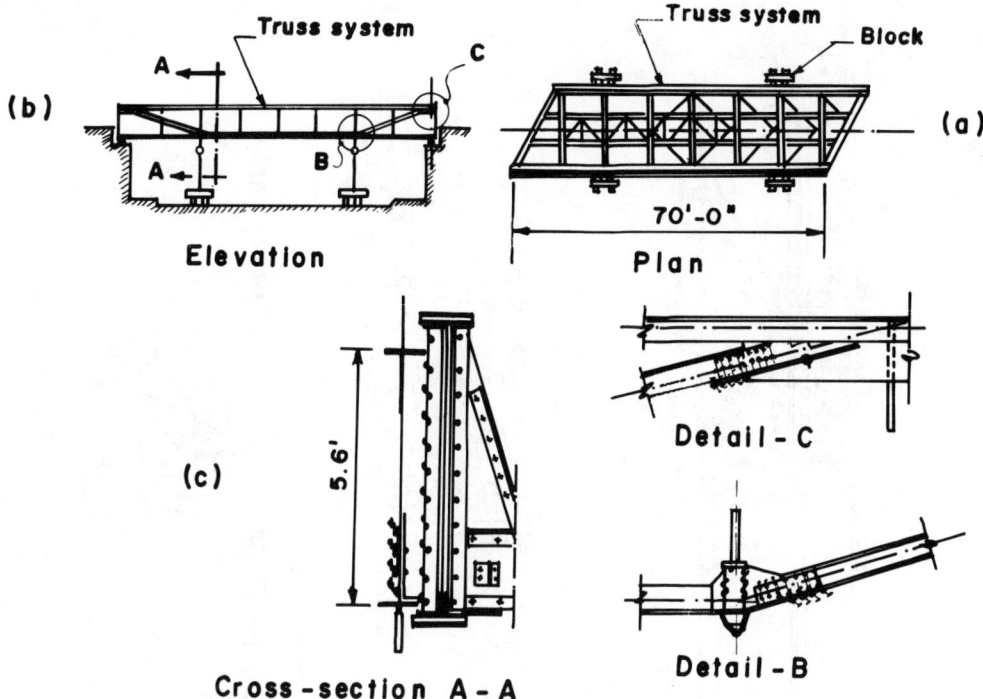

Fig. 11.6   Strengthening by prestressing of a railway bridge at Ostrova, Czechoslovakia.

11.8.2 Restoration of King's Bridge, Melbourne, Australia

The King's Bridge was opened to traffic in 1961, and about one year later the first suspended span collapsed suddenly under a transporter carrying an excavator under a load of 45 tons. Three of the four girders in the span fractured at both ends of the cover plates to the tension flanges. The fracture showed clear evidence of prior cracks in the main plates at the transverse weld at the end of the cover plate. Examination of the fracture showed that the failure was due to a brittle fracture in the steel that may have existed in other girders.[22]

The investigation showed that tension stresses constituted a danger of further failure. The method of restoration was the application of prestressing forces to the girders to reduce the tension stresses and at the same time provide overall protection. The maximum tensile stress in the girder was estimated to be approximately 10,000 psi, and this figure was adopted as the basis of the design of the prestressing system.

Concrete anchor blocks cast between adjacent girder webs were used. The simplest method of attaching the concrete anchor blocks to the girders was by passing transverse prestressing bars through both girder webs and the anchor blocks.

In the suspended spans the lower flanges were in tension and the prestressing cables were placed at about the level of the lower flanges at mid-span (Figs. 11.7 and 11.8). The cables were stressed between concrete anchor blocks at the ends of the girders, and in all of these spans the blocks were fastened to the girders

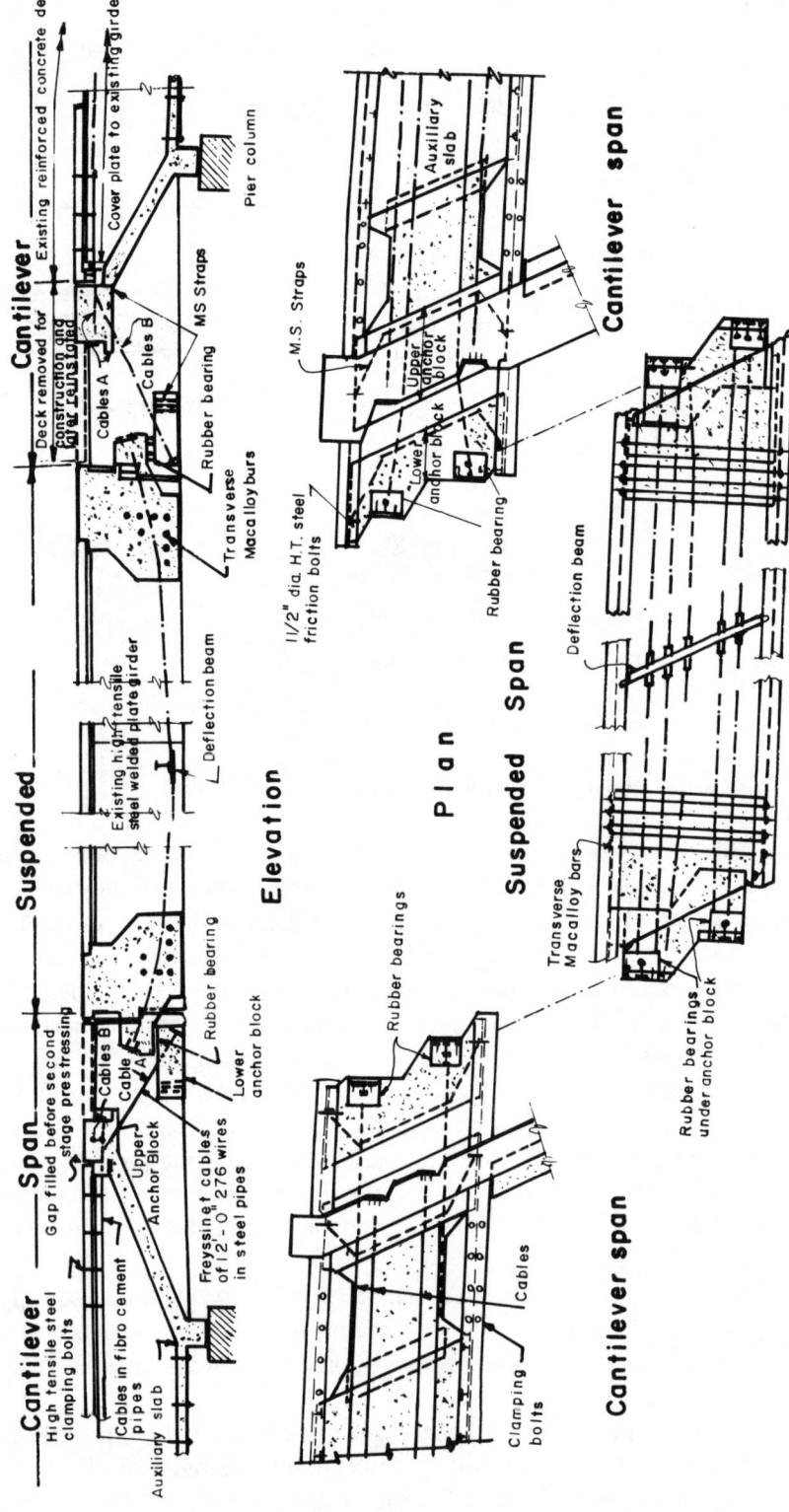

Fig. 11.7   Prestressing of suspended span. Elevation and plan. (Courtesy of the Institution of Engineers, Australia)

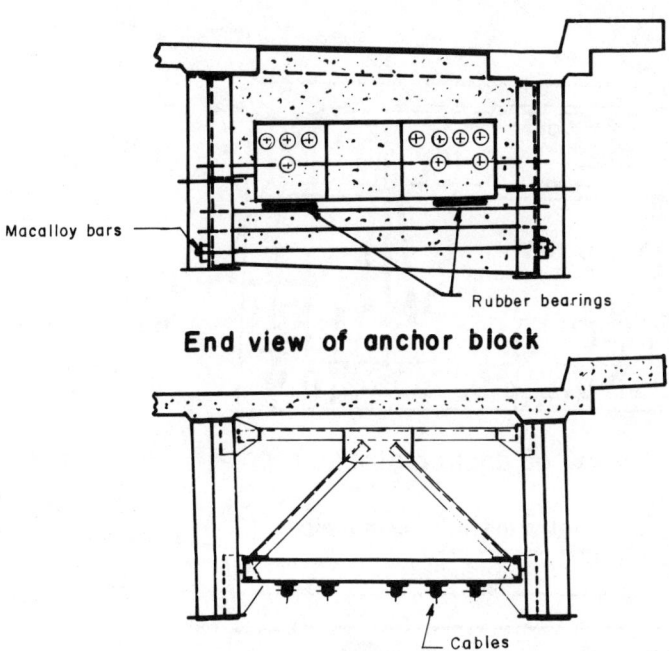

End view of anchor block

Macalloy bars

Rubber bearings

Mid span

Cables

Fig. 11.8 Prestressing of suspended spans. Cross section. (Courtesy of the Institution of Engineers, Australia)

by transverse stressed Macalloy high-tensile steel bars. The transfer of force was by friction between the concrete and steel. Openings were cut on the concrete deck for construction access and the anchor blocks extended the full depth of the girders to make use of the deck to stabilize the block against torsional effects and to apply some prestressing force into the deck.

The prestressing cables were partially draped in the suspended spans and deflected through cast iron deflectors mounted on a new bottom chord, which was welded to the existing steel cross frames.

The application of the prestressing to the cantilever spans was difficult because of the composite deck. The prestressing was applied to both the deck and the girders, except in the cover plate ends (Fig. 11.9). This enabled the full prestressing to be effective in the girders at these critical points.

The tensile stress in the girders at the supports is approximately 10,000 psi, and to safeguard the deck slab the reinforced concrete raking struts were provided from the bottom of the girder at the piers to the upper anchor blocks.

Anchor blocks were cast between the girders just clear of the cover plate ends in the upper flange and also at the ends of the girders just above the lower flanges. The prestressing cables placed under the deck were anchored in one of the upper blocks, passed through the other, and deflected downward through steel tubes into a lower block.

Most of the cantilevers were about 17 ft long, and in these there were 12/12 × 0.276 Freyssinet cables per pair of girders. The prestressing was applied directly to the girders and was sufficient to reduce the tension stress to zero at the cover plate end.

The strength of the shear connectors attaching the deck to the girders was insufficient to transfer the prestressing forces, and additional connectors were

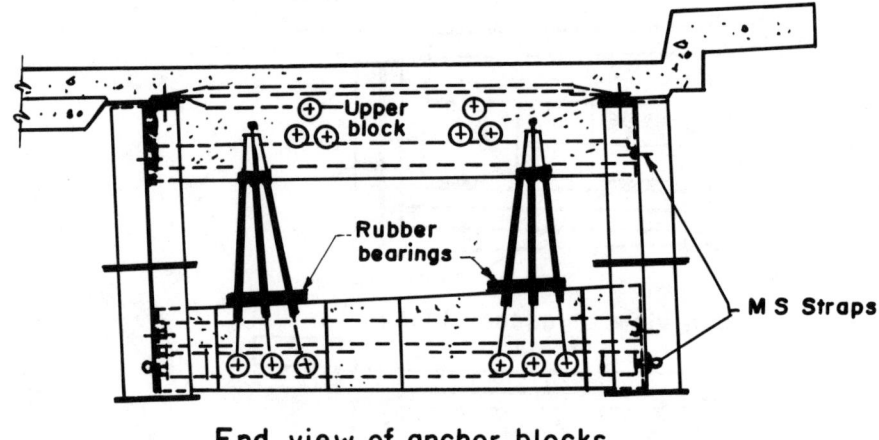

End view of anchor blocks

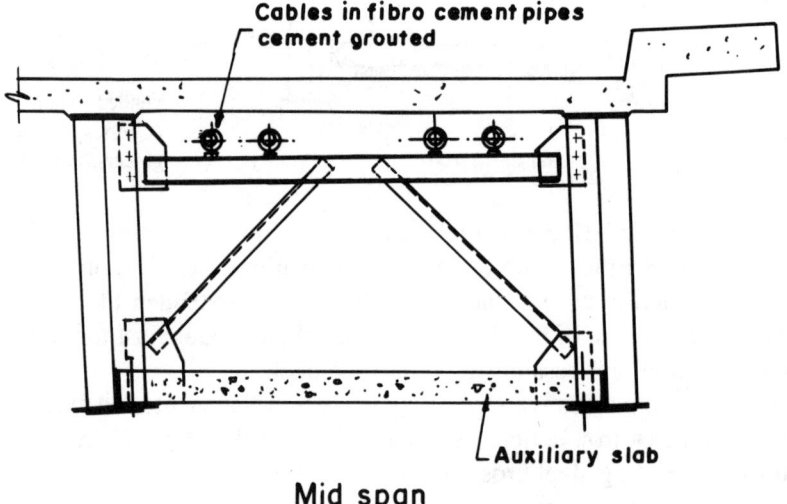

Mid span

Fig. 11.9 Prestressing of cantilever span. Cross section. (Courtesy of the Institution of Engineers, Australia)

provided in the form of high-tensile friction grip bolts with a steel bearing pad on the deck and a nut and washer under the beam flange. The holes were drilled through the deck and flange.

The cables were stressed to provide the required forces, but without exceeding 80% of the ultimate strength of 110 kpi at the jacking end.

All cables are contained in fiber-cement pipes, which were cement-grouted after the completion of stressing.

11.8.3 STRENGTHENING OF HIGHWAY BRIDGES IN IOWA

During the period between 1940 and 1960, a number of single-span steel beam composite concrete deck highway bridges were constructed in Iowa and throughout the United States. Design criteria at that time resulted in most of the exterior beams of these bridges being significantly smaller than the interior beams. As a result of changes in design standards and increases in legal load limits, many of these bridges cannot be rated for legal loads due to the capacity of their exterior beams. Rather than posting or replacing the bridges, a more attractive and economical alternative is to strengthen them. With research grants from the Iowa Department of Transportation and the Iowa Highway Research Board, a method of strengthening these bridges by posttensioning has been developed. This study was divided into three phases. Phase I involved the development of bracket (anchorage) design and the posttension strengthening and testing of a half-scale model bridge in the laboratory.[23]

Phase II, the majority of which was completed during the summer of 1982, involved the strengthening and testing (before and after strengthening) of two actual bridges.[24]

The final phase of the study (phase III) involves the periodic inspection of the two bridges, their retesting during the summer of 1984, and the development of a practical design methodology for the strengthening technique.[25] The major finding of phases I and II was that posttensioning is an economical and viable strengthening method.

The primary emphasis of phase II was on strengthening two full-scale existing bridges, a right-angle highway bridge (Fig. 11.10) and a 45° skewed bridge (Fig. 11.11), referred to as bridges No. 1 and No. 2, respectively.

To determine the feasibility of strengthening composite bridges by posttensioning, the study was undertaken in Iowa on a half-size bridge model (Fig. 11.2).[26] For the model bridge all plan dimensions, along with deck thickness and curb depth, were set at half the prototype dimensions. Model material properties were matched as closely as possible to properties typical for bridges constructed in the 1940s. Section properties resulting from the half-scale dimensions follow principles of similitude. A series of tests was performed on several representative specimens of the concrete and the steel used in the bridge model to determine values of the pertinent material properties.

Numerous test were performed on the bridge. Variables included magnitude and location of posttensioning force, magnitude and location of vertical load, presence or absence of curves, and presence or absence of diaphragms. Each of the 5/8-in.-diameter Dywidag threadbars used for posttensioning the bridge was instrumented with two longitudinal strain gauges connected to cancel the bending strain and sense twice the axial strain, thus improving sensitivity.

Posttensioning force was applied to the various beams, utilizing 60-kips-capacity hollow-core hydraulic jacks. However, the model bridge required 20 kips in each exterior beam to produce the desired results. Tests also involved a combination of vertical load and posttensioning.

To posttension a bridge successfully and to provide a simple method of transmitting the required force from the prestressing tendons to the bridge beam, three different brackets (anchorages) were tested.

This half-scale model bridge provided an excellent means for evaluating the behavior of a composite bridge. On the basis of this research study the following conclusions were made:

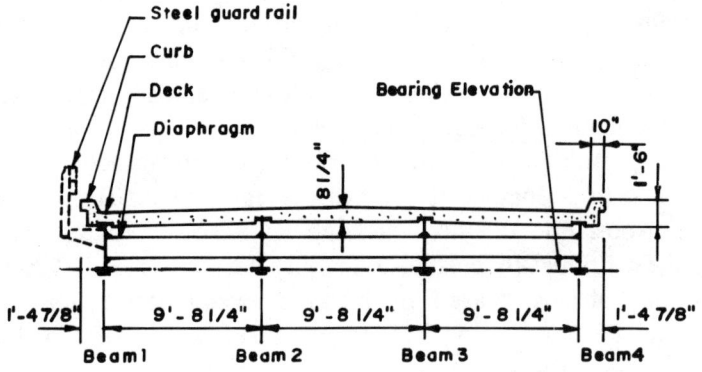

(a) Cross section at midspan

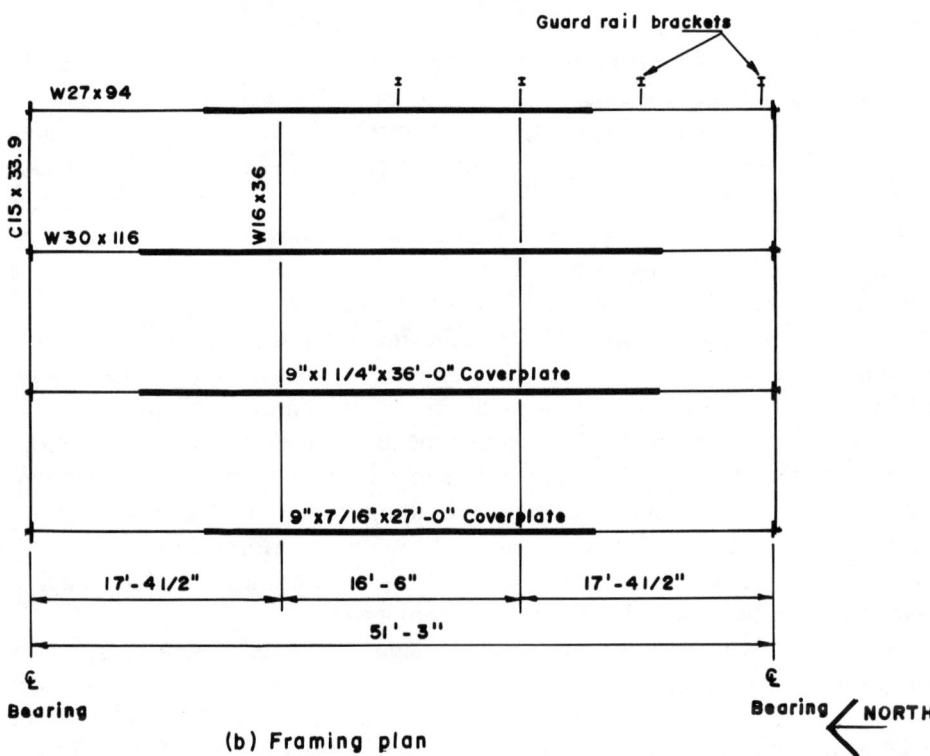

(b) Framing plan

Fig. 11.10 Cross section and plan of bridge no. 1. (Courtesy of Dr. K. F. Dunker, from Ref. 26)

1. Posttensioning can be used to provide strengthening for composite bridges.

2. During the actual posttensioning of a given bridge, posttensioning need not be applied symmetrically to the beams (i.e., external beams stressed simultaneously, etc.).

3. A posttensioning strengthening design requires the checking of flexural stresses at five different locations within a span: the two posttensioned bracket (anchorage) locations, the two cover plate cutoff points, and the mid-span section.

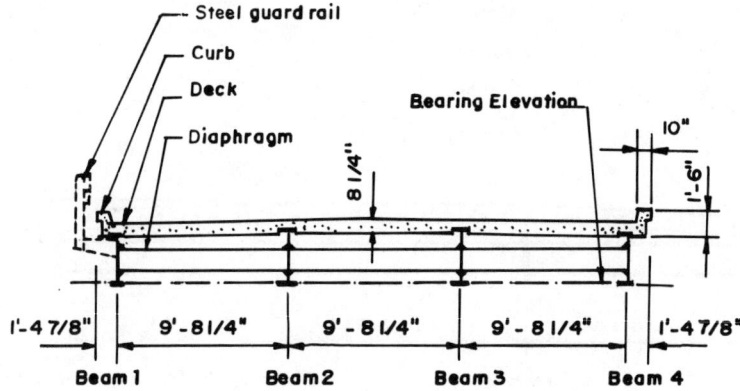

(a) Cross section at midspan.

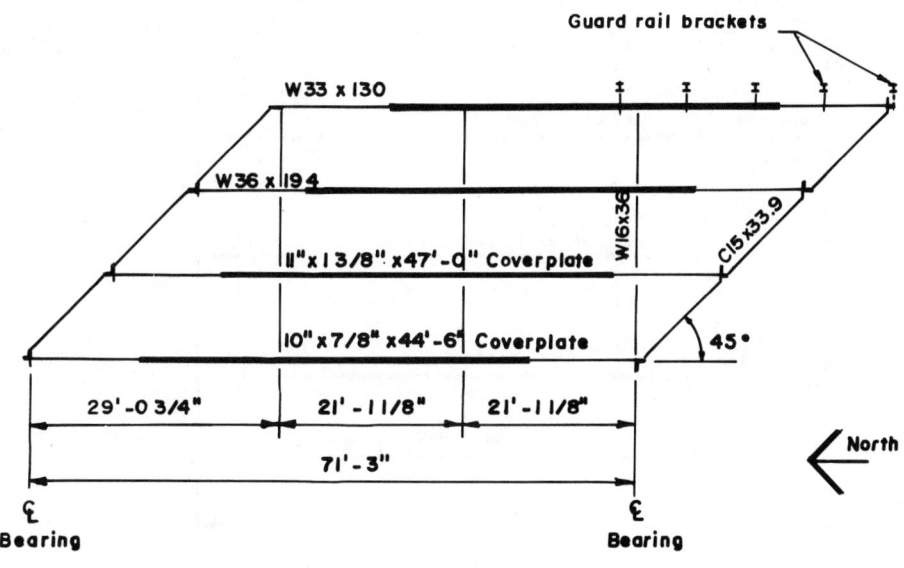

(b) Framing plan

Fig. 11.11   Cross section and plan of bridge no. 2. (Courtesy of Dr. K. F. Dunker, from Ref. 26)

4. In a posttensioning scheme in which only exterior beams are posttensioned, approximately two-thirds of the posttensioning force (for a four-beam bridge) affects the exterior beams. For adequate posttensioning, therefore, more force must be applied in order to compensate for posttensioning distribution to the remainder of the bridge. Obviously the amount of posttensioning forces affecting the exterior beam is a function of beam spacing, beam size, span length, and so on, and thus the 2/3 fraction will be different for various bridges.

5. Posttensioning did not significantly affect the overall load distribution characteristics of the bridge tested.

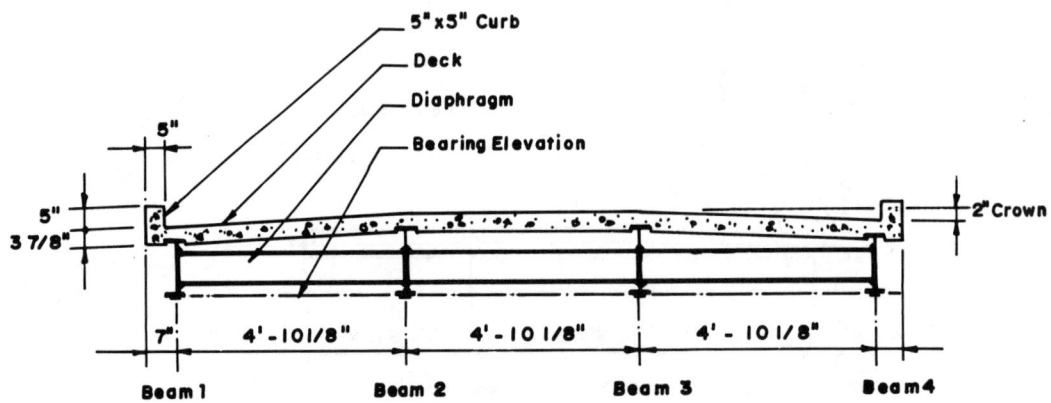

(a) Cross section at midspan

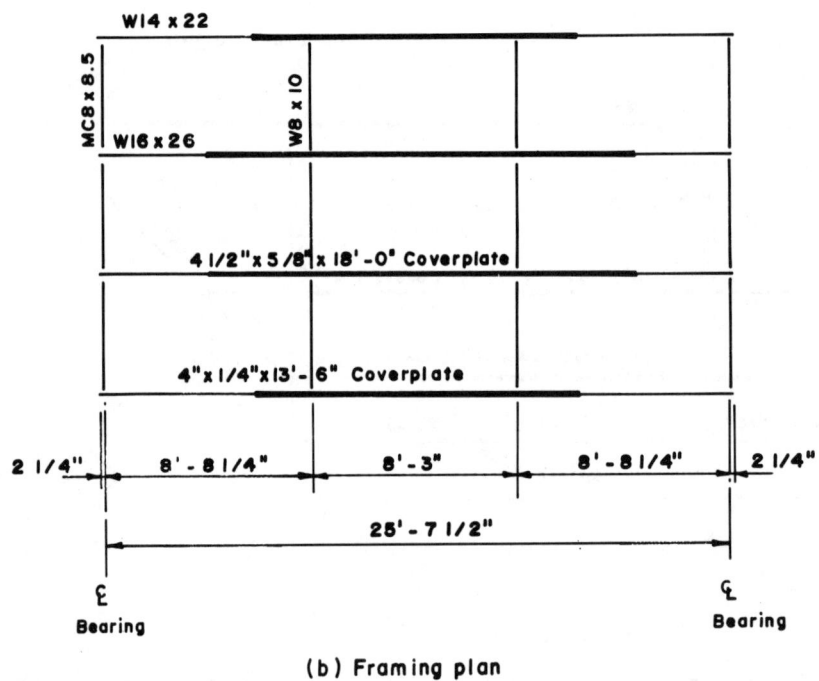

(b) Framing plan

Fig. 11.12   Model bridge. (Courtesy of Dr. K. F. Dunker, from Ref. 26)

6. Orthotropic plate theory may be used to predict approximate distribution of posttensioning axial forces and moments.

7. In the bridge tested, P-Δ effects due to posttensioning were secondary and thus could be neglected.

8. Δ-T effects in posttensioning tendons are small and have a conservative effect. Thus Δ-T effects may be neglected in most designs.

9. Posttensioning forces generally are of sufficient magnitude to cause significant cracking of concrete.

10. The stiffening effect of low curbs has minimal effects on the behavior of

posttensioned bridges. The added stiffness of curbs increases the vertical loading and posttensioning forces carried by exterior beams; however, the net effect may still be a reduction in the maximum stresses in exterior beams.

11. Diaphragms have minimal effects at service loads; however, removal of diaphragms does reduce vertical load transfer and posttensioning transfer among beams.

12. It is extremely important in the design of brackets (anchorages) for transmitting force from the bracket to the beam to consider the strength of the bracket, the strength of the beam to which the bracket is being attached, and the strength of the bolts.

13. A finite-element model was developed which accurately predicted the behavior of a composite bridge under posttensioning and vertical load. The model was verified with test results from a half-scale bridge and the model was more accurate than previous finite-element models which did not account for the flexibility of the webs of the steel beam. Comparison of the finite-element analysis with field data obtained during posttensioning of the Iowa bridges indicated that typical construction details caused considerable restraint. The field-measured strains and deflections generally fell about halfway between the values computed for theoretical simple support and fixed end conditions.

14. For distribution of posttensioning to exterior beams, it is conservative to neglect the effects of field restraints and a skew of 45° or less. Deck crown, shear connector stiffness, and cover plate length have virtually no effect on the posttensioning distribution for typical Iowa bridges. Length of posttensioned region, relative stiffness of exterior beams, and transverse stiffness of deck and diaphragms generally have the largest effects on posttensioning distribution.

15. Posttensioning distribution varies along the span of a bridge; distribution factors at any point on the span can be computed accurately by linear interpolation. Posttensioning of the exterior beams can relieve their tension overstress at mid-span and at cover plate cutoff points. An excess of posttensioning, however, can overstress exterior and interior beams in compression and can overstress curbs and deck in tension.

16. Vertical live-load distribution is virtually the same for a composite bridge with or without posttensioning.

The results of the load tests on the strengthened bridges demonstrate that the posttensioning procedure is an effective method to strengthen the above bridges. A review of the data also indicates that the procedure is suitable for strengthening bridges in which all beams need to have increased flexural strength.

11.8.4 Highway Bridges in California

General Data

In 1975 California implemented substantial increases in live loads for highway bridges.[27] It was necessary to strengthen existing highway bridges designed for lighter loads. To increase the moment capacity in critical areas, longitudinal prestressing tendons were installed. This was done on seven steel girder bridges and was planned for a dozen more. Prestressing has proven to be a quick, economical, esthetically pleasing method of strengthening steel bridges.

Normal prestressing criteria were used for the design of prestressed tendons. Special attention was paid to tendon paths, tendon encasement, and fastening devices, which were designed to match tendon ultimate values. Tendon paths were generally straight, although haunched girders were required at angle points to align forces with girder flanges. Tendon paths cleared girder stiffeners and lateral bracing. Tendons were kept free from corrosion by encasement in galvanized pipes and grouting after tensioning. Anchorages were also sealed.

All strengthening by prestressing in California to date has employed strands. The design process for a simple span was carried out by the following steps:

1. Determine moments for applied dead load plus live load, plus impact ($DL + LL + I$) at center of span.

2. Calculate girder stresses.

3. Determine allowable girder stresses based on as-built material.

4. Calculate stressing forces required to compensate for the difference between the allowable stresses and the $DL + LL + I$ stresses in the tension flange. Assume an eccentricity for the stressing force that will allow adequate space for mounting the stressing anchorage brackets. Check compression of flange steel and concrete stresses.

5. Repeat the process for other critical points within the span at flange reduction locations.

6. Determine location where stressing force may be terminated based on allowable unit stresses. Termination point must be between existing transverse stiffeners if anchorage brackets are mounted on the girder web.

7. Design anchorage brackets. Use sufficient transverse offset to clear girder stiffeners if brackets are mounted on the girder web. Check bracket size to ensure that its dimensions are consistent with the available space and the assumed stressing eccentricity.

8. Check the existing girder web for bearing stresses generated by the anchorage bracket.

Example of Strengthened Bridge—Avenue 328 Overcrossing on Highway 99

This composite steel girder structure, with six simple spans and five lines of girders, was designed for AASHTO HS20-44 live loading. This bridge was found to be deficient in moment capacity for permit loading in the three central 90-ft (27.44 m) spans.

Strengthening was accomplished by adding 120 kips of force to each of the 15 girders by means of two 60-kip tendons, 60 ft (18.3 m) long, symmetrically placed on either side of the web 6 in. above the bottom flange. The attachment was secured to the girder web by $14\frac{7}{8}$ in. high-strength bolts (Fig. 11.13).

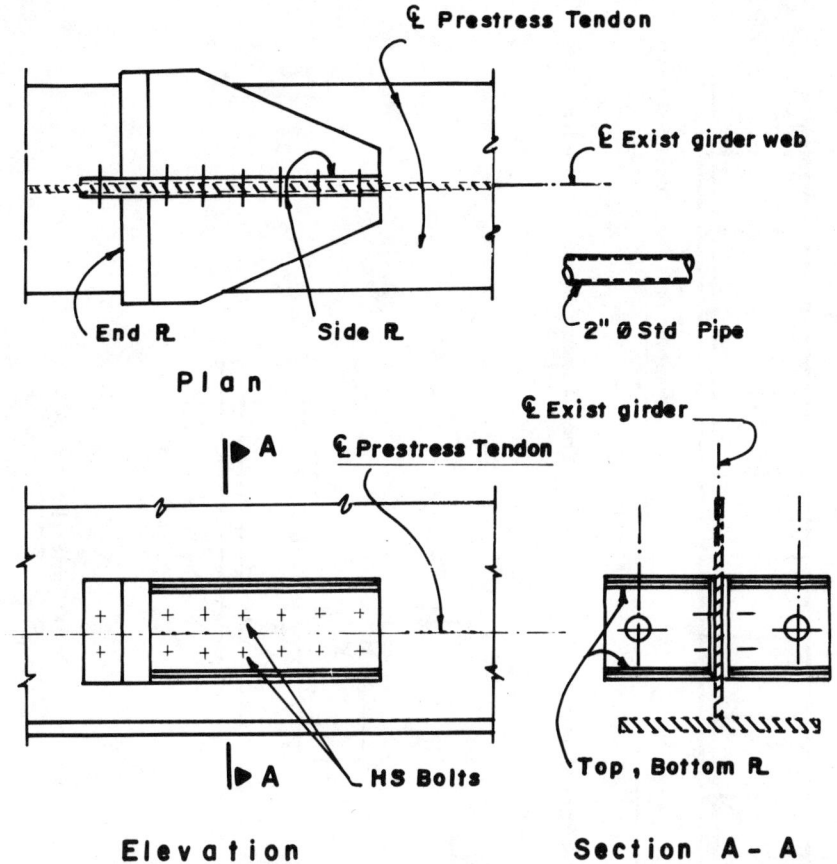

Fig. 11.13   Avenue 328 overcrossing. Prestress tendon anchorage, positive moment zone. (Courtesy of the Transportation Research Board, Washington, DC)

All hardware was galvanized. Creep loss in the prestressing system was assumed to be 5000 psi plus any losses characteristic of the prestressing and anchorage system. The tendons of each girder web were stressed simultaneously and enclosed in 2-in. standard galvanized pipe, which was grouted after stressing. The longitudinal centerline of the tendons was placed outboard of the girder stiffeners, and tendon supports (Fig. 11.14) were placed at 15-ft (4.57 m) intervals.

All structures strengthened by prestressing have performed well. There has been no evidence of loss of prestress due to slippage in the anchor system. Prestressing to upgrade the load-carrying capacity of steel girder structures has become the mainstay of California practice.

11.8.5   Railway Bridge, Rumania

In Rumania a large number of railway bridges as well as some highway bridges have been strengthened by the aid of prestressed rigid high-strength steel tie

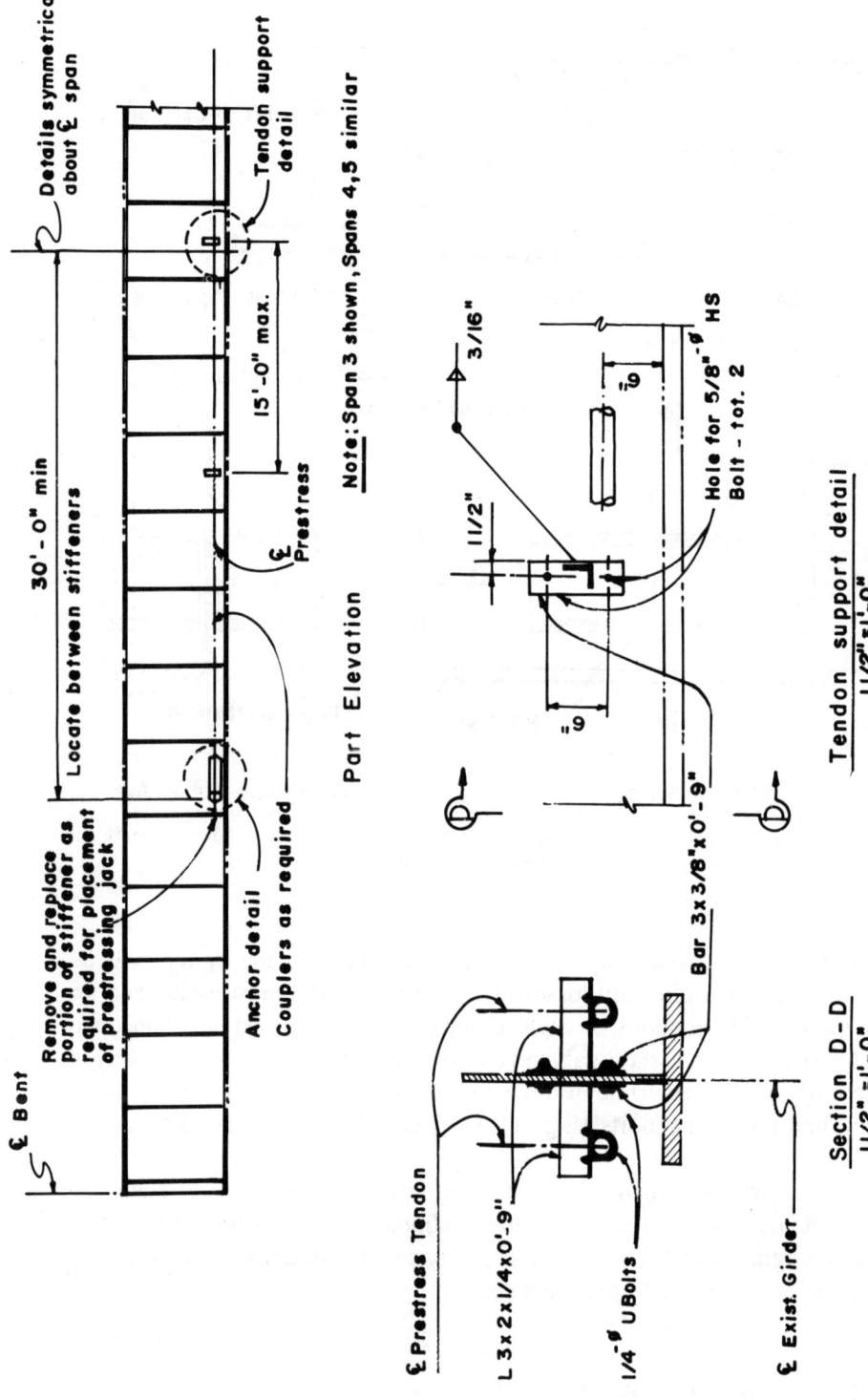

Part Elevation

Note: Span 3 shown, Spans 4,5 similar

Tendon support detail
1 1/2" = 1'-0"

Section D-D
1 1/2" = 1'-0"

Fig. 11.14 Avenue 328 overcrossing. Tendon supports. (Courtesy of the Transportation Research Board, Washington, DC)

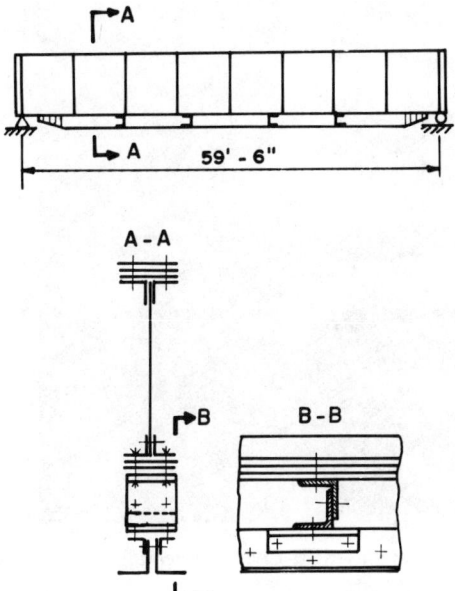

Fig. 11.15   Skew bridge over Sasar River.

members. As a typical example, two railway bridges comprised of deck girders have been strengthened by means of prestressed additional members.[28] The procedure was as follows.

Prestressed tie members were used to strengthen two bridges crossing the Sasar River. The first deck girder bridge has a span of 59 ft 6 in. (18.15 m) and a right-hand skew of 45° (Fig. 11.15). Figure 11.16 shows the straight bridge over river Sasar. Details of strengthening of both bridges consist of rigid tie members each comprising two angle sections 5.9 in. × 5.9 in. × 6.3 in. (150 mm × 150 mm × 16 mm).

11.9 STRENGTHENING OF TRUSS BRIDGES

Prestressing is widely applied for the strengthening of truss bridges to increase their carrying capacity. This method is illustrated by the following examples.

11.9.1 Monmouth Railway Bridge, England

This wrought-iron structure, built in 1876, has a 148-ft 6-in. (45.27 m) span deck-type truss (Fig. 11.17).[29] The lower box sections were added to strengthen

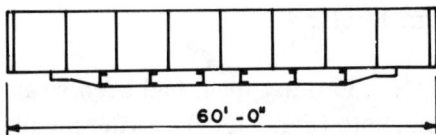

Fig. 11.16   Straight bridge over Sasar River.

Fig. 11.17   Elevation of Monmouth Bridge. (Courtesy of ASCE-*Civil Engineering*)

the original structure some time after it was built (Fig. 11.18). Severe corrosion had taken place inside this box section, resulting in the loss of cross-sectional area.

Prestressing was accomplished in 1957 by the use of four high-tensile-strength stress steel Macalloy bars for each of the two bottom chords, or eight in all. These bars were grouped symmetrically about each chord, as shown in Fig. 11.18. The top two were $\frac{1}{2}$ in. in diameter, and the bottom two 1 in. in diameter. They were tensioned between built-up welded anchorages spaced 128 ft (39.0 m) apart. The eight bars were tensioned simultaneously by hydraulic jacks. The prestressing force increased the camber by $\frac{7}{32}$ in.

The stress in the weakest panel of the bottom chords before and after tensioning is summarized as follows, in pounds per square inch.

	Dead Load	Live Load	Total
Before tensioning	7,056	15,680	22,736
After tensioning	896	15,299	16,195

To reduce any sag in the prestressing bars, permanent flexible supports were provided for them at intervals of 14 ft 8 in. (4.47 m) at centers. To protect the bars against corrosion, they were wrapped with a plastic-coated tape, and their projecting threaded ends at the jacking end were protected with plastic sheaths.

11.9.2 Livery Street Bridge, Birmingham, England

This bridge, built in 1906 and having a 115-ft (35.0 m) span, had been weakened by severe corrosion of the truss bottom chord (Fig. 11.19).[30] After the old

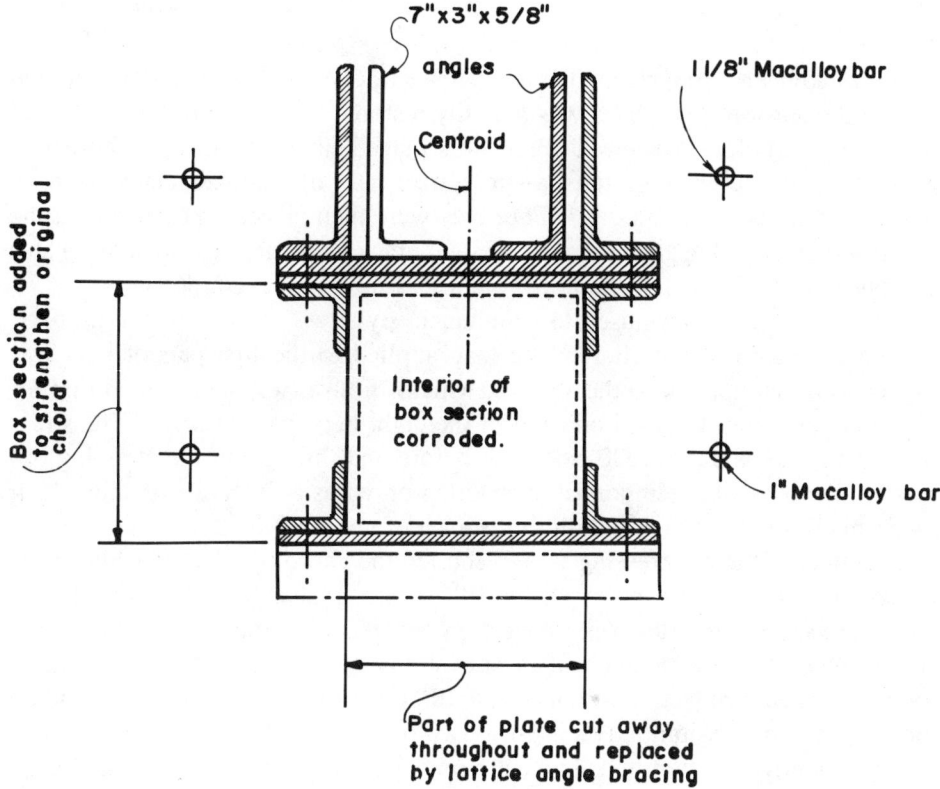

Fig. 11.18 Monmouth Bridge. A cross-sectional view of a bottom chord. (Courtesy of ASCE-*Civil Engineering*)

Fig. 11.19 A view of the Livery Street Bridge, Birmingham, England. (*The Structural Engineer*, August 1964, No. 8, Vol. 42, p. 277. The Institution of Structural Engineers, London).

decking had been removed and the new cross girders fitted into place, a sectional area of the tension truss chord was partially restored by substituting new welded plates. Then, before the new decking was placed, the bottom truss chord was applied through eight high-tensile-stress steel bars of 1-in. diameter, with an ultimate strength of 155,000 psi. Four bars were then placed on each side of the chord and anchored against the welded steel brackets secured to the chord plates and outstanding flanges of the upper angles with high-strength bolts.

Two 45-ton jacks were used to simultaneously stress the corresponding bars, one on each side of the chord. The load applied to the first pair of bars was higher than that applied to the second, which, in turn, was higher than that for the third pair, until the load on each of the eight bars was the same. The stress in the bars was about 82,000 psi. Both before and throughout the stressing operation, the bars were temporarily supported on wires at intervals of about 15 ft (4.57 m) throughout their length, a distance of about 90 ft (27.44 m).

The effect of the prestressing was to shorten the chord by 1/2 in., while at the center, the camber was increased by 7/8 in. Stress changes in the chords and web members of the truss were measured and recorded during prestressing by strain gauges. The recorded changes were similar to those calculated. Maximum totals indicated a reduction in tension of 10,500 psi in the tension chord and an increase in compression of 1,800 psi in two of the web struts.

Subsequently, the high-tensile steel bars were wrapped in canvas and coated with bitumen before being encased in concrete, which filled the whole chord.

11.9.3 Aare River Bridge, Switzerland

This bridge consists of two simple deck truss spans, each having a length of 157.44 ft (48 m). Each truss was strengthened in 1969 by two high-strength steel cables having a polygonal configuration and each having a diameter of 2.56 in. (65 mm) (Fig. 11.20).[31]

In cross section the deck has two lanes each 9 ft (2.74 m) wide. The deck consists of a reinforced concrete slab having a thickness of $4\frac{3}{4}$ in. (Fig. 11.21). Strengthening cables were prestressed up to 60% of their carrying capacity and connected to every fourth vertical member of the truss. At mid-span each truss is supported by the vertical end saddle (Fig. 11.22). The end vertical of each

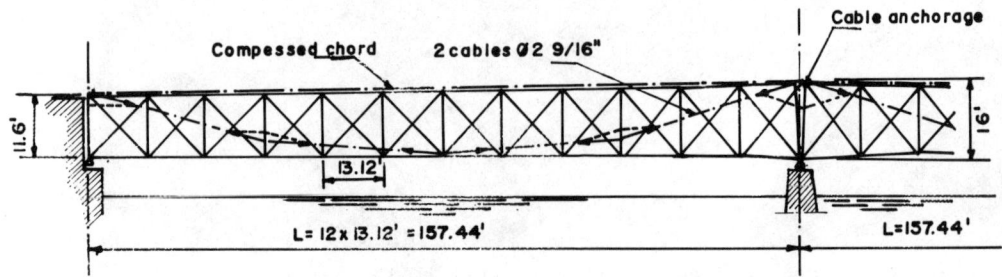

Fig. 11.20   Aare Bridge. Elevation of trusses, strengthened by polygonal cables.

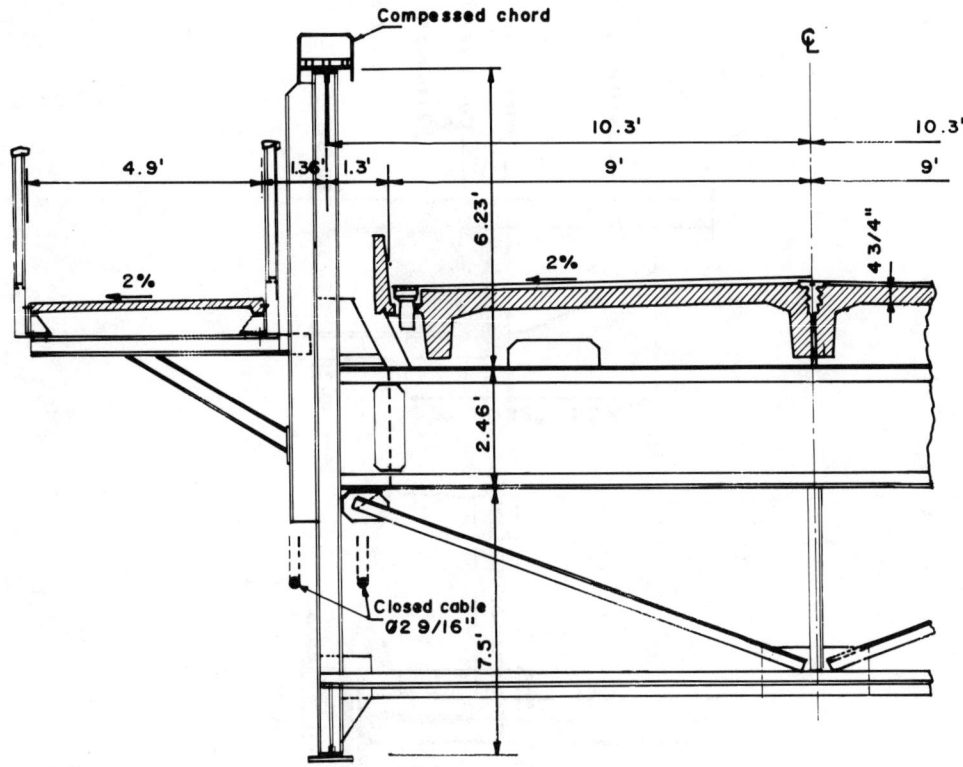

Fig. 11.21   Cross section of the Aare Bridge.

truss above the middle pier consists of a pylon on top of which is installed a cable anchorage (Fig. 11.23).

11.9.4 Bistrita River Bridge, Rumania

This is a railway bridge over Bistrita River having trusses spanning 177 ft (54 m) and supporting the deck at the bottom chord level (Fig. 11.24).[32] Strengthening was performed by means of prestressed tie members installed on the same plane as the bottom chords with the aid of a cross girder and diagonal members at each attachment. The tie member, tensioned by means of hydraulic jacks, relieves the existing chords of some of their dead and live loading. Because of their great length, these tie members were each composed of two parts which were joined after being tensioned. Loading tests with two locomotives were performed after the work had been completed. The measurements obtained in the course of these tests confirmed the values predicted by the calculations.

11.9.5 Yukon River Bridge, Canada

The Yukon River Bridge was damaged by a traffic accident in 1982 and the structure was repaired and strengthened by the application of prestressing tech-

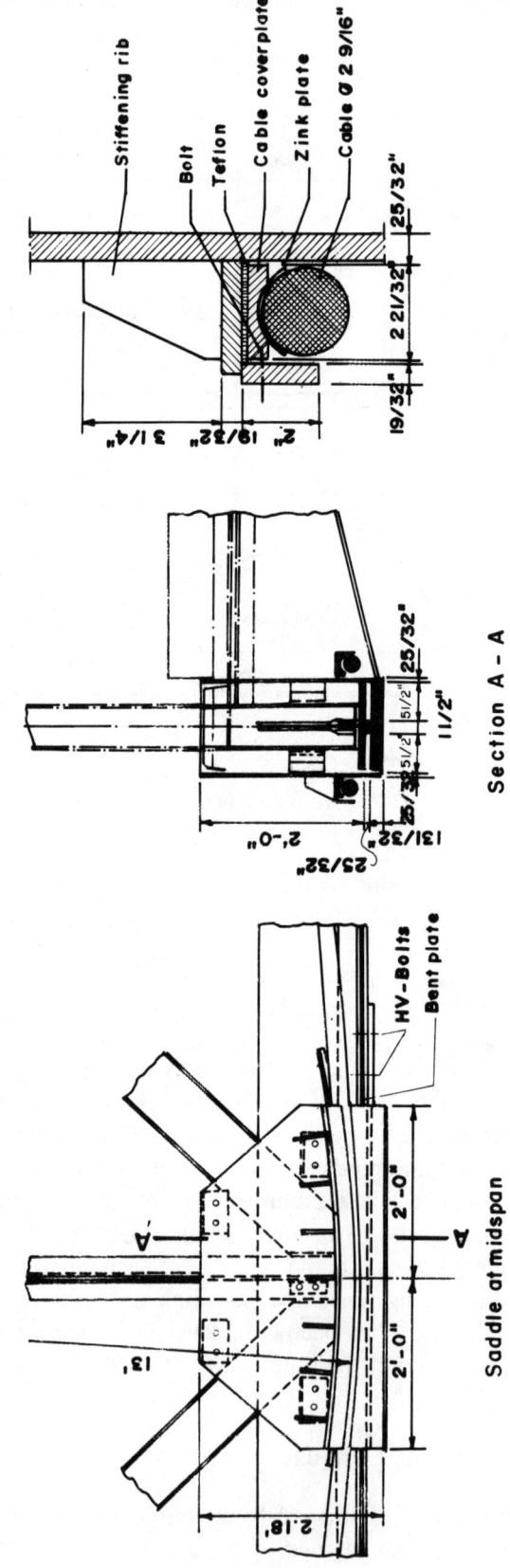

Fig. 11.22 Details of cable support of mid-span.

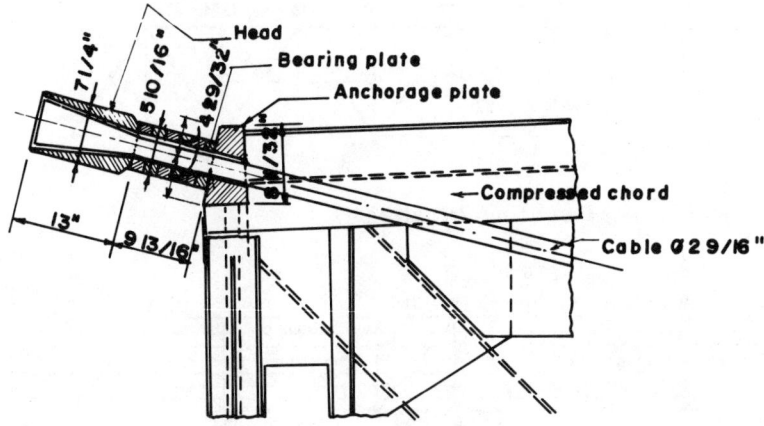

Fig. 11.23   Detail of cable anchorage.

nology.[33] The bridge was damaged by a tractor trailer carrying a large backhoe. The impact of the load broke all the bottom struts of the overhead sway frames and portals of both spans. This in turn precipitated the brittle fracture of all the hangers to which the frames were connected, culminating in the brittle fracture of the bottom chord at panel point L5 of the truss (Fig. 11.25).

The fractured bottom chord of the truss opened a gap of approximately 2 in. and failure appeared to have propagated from a rivet in the hanger gusset plate connection. The repair plan called for the replacement of fractured members, the strengthening of tensile joints along the bottom chord, and most importantly, the restoration of the broken chord. It was decided that the main members would be repaired first. However, to achieve that, the truss should act as an integral structural unit. For this reason all the broken hangers were first repaired. When the bottom chord fractured, the total dead-load tensile force of 950 kips was redistributed and resisted by the longitudinal re-bars, stringers, floor beams, and downstream truss, all of which acted together as an equivalent bottom chord. The objective of the restoration was to release the locked-in stresses in the deck

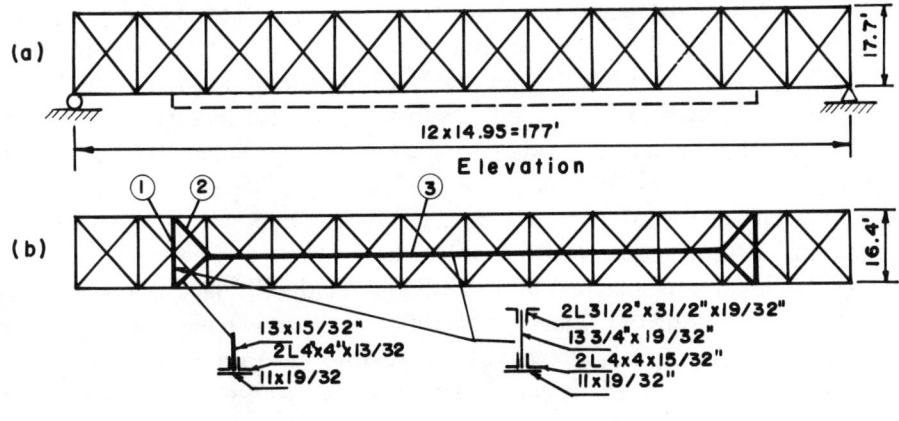

Fig. 11.24   Bistrita Bridge. 1, Cross girder; 2, diagonal member; 3, rigid tie member.

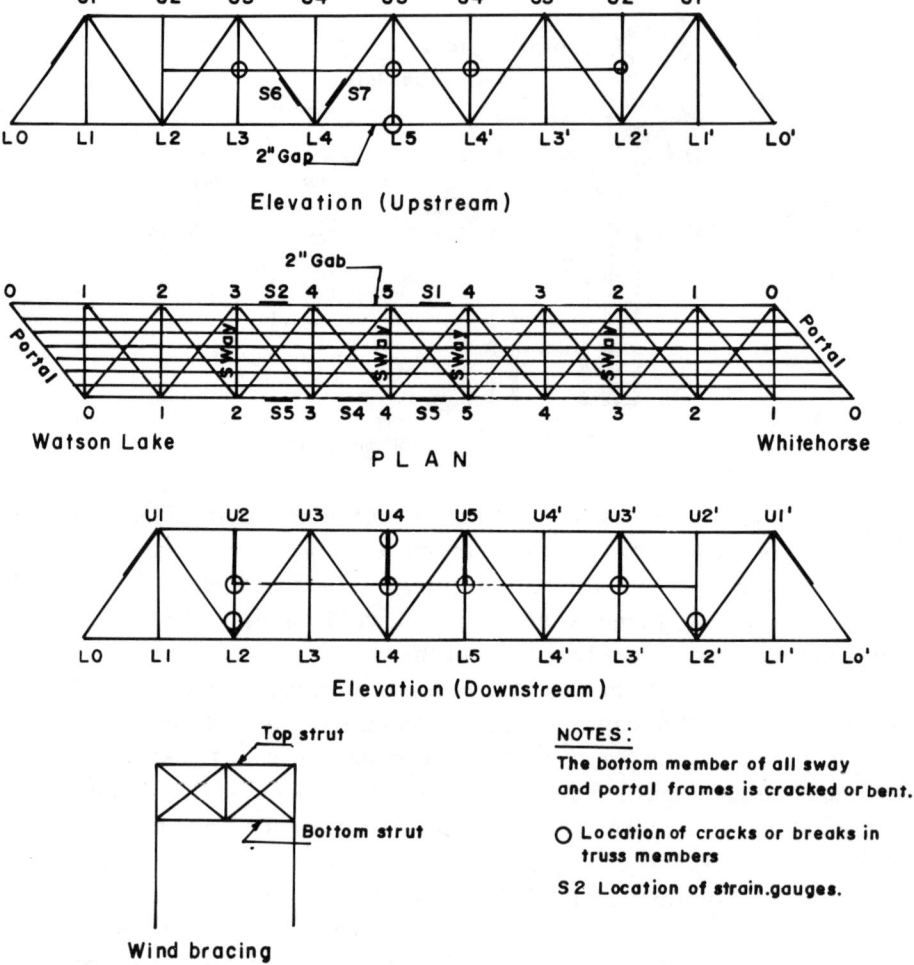

Fig. 11.25 The Yukon River Bridge. Location of damages and strain gauges.

and the downstream truss, and restore the bridge to its original geometry. To achieve this, half of the dead-load tensile force originally carried by the upstream truss, or 480 kips, would have to be reintroduced into the fractured bottom chord, and the 2-in. gap should ideally be closed. The restoration scheme adopted involved the application of hydraulic jacks to introduce the required tensile force in the broken chord member. Anchoring brackets were installed 10 ft (3 m) from both sides of the cracked section.

Four 1- and 1.2-in. (31.8-mm) diameter Dywidag bars with a specified yield strength of 150 ksi were threaded through the brackets and across the cracked section. Each bar was attached to an individual hydraulic jack at one end. To assist in the monitoring of the jacking operation, five strain gauges were installed on critical members directly affected by the jacking. The location of these gauges are indicated in Fig. 11.25. Weldable strain gauges with temperature compensation dummies were used.

Once the bars were in place and the nuts at the ends of the bars were snug tight, the temporary splice was removed. Jacking forces in increments of about

20 kips per bar were introduced. At the end of each load increment, strain gauge readings were taken. The jacking was stopped when the jacking forces reached 147 kips per bar, and the load was then released to 126 kips per bar and rejacked to the 147-kip level.

The operation was repeated several times until no significant changes in the strain gauge readings were detected. The load was then maintained at the level of 126 kips per bar by locking the nuts at both ends of the bars while a new splice was being bolted on. Once the splice had been installed, the jacking load was then released at intervals of 21 kips per bar until the total jacking load was reduced to zero.

A detailed computer structural analysis of the entire bridge was conducted after the restoration to verify the field-measured member forces as well as the assumptions made for the repair. The analysis confirmed the effectiveness of the repair.

The repair actions taken in this project have proven feasible, economical, and expedient. The operation resulted in a substantial saving in both time and expense, since it eliminated the need for the construction of temporary pile supports and the installation of elaborate shoring systems.

11.9.6 Highway Bridge in Illinois

Lamberson described the rehabilitation of an existing structural steel truss bridge using externally applied Dywidag posttensioning tendons.[34] The bridge consists of five spans, with a total length of 2045 ft (623.48 m) and a main span of 560 ft (170.73 m).

The rehabilitation of the structure included redecking of the entire bridge, replacement of damaged structural steel members, the application of posttensioning to the floor beams, and complete repainting of the structure. The external posttensioning system allowed the bridge rating to be increased.

Specifically, 55 transverse floor beams, 30 in. deep and spanning 25 ft (7.62 m), were strengthened utilizing Dywidag bars harped below the floor beams Fig. (11.26). The tendons were connected to structural steel weldments

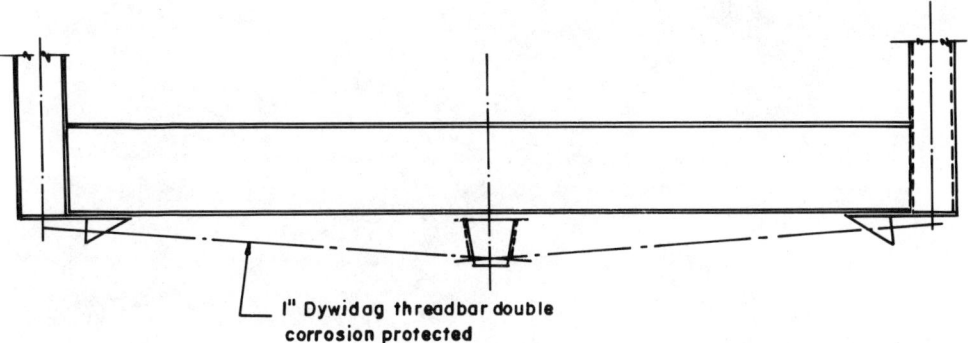

1" Dywidag threadbar double
corrosion protected

Fig. 11.26 Scheme of externally applied posttensioning system, using Dywidaz DCP tendons, to increase bridge rating.

at each main weldment. Tendons were pregrouted in PVC sheathing before installation to simplify the on-site work. The Dywidag double corrosion system, normally used for earth anchor installations, was selected because of its durability and because prestress forces could be monitored or adjusted during the life of the structure.

11.9.7 Chesapeake and Delaware Canal Railroad Bridge

A railroad lift bridge over the Chesapeake and Delaware Canal was damaged when a freighter rammed it at the north end of the left span.[35] The main damage included a torn lower chord, bent wind bracing, and a buckled diagonal. The damaged bottom chord is shown in Fig. 11.27.

The Whiting-Turner Contracting Co. hired consultant Ewell, Bomhardt & Associates of Baltimore to design the yokes and other elements for the repair work. Each yoke consisted of a bracket mounted above and below the existing bottom chord, splices on each side of the damaged area, and prestressing cables between the brackets. The brackets had 1-in.-thick diaphragm plates drilled to the same bolt pattern as the chord splices and stiffened to take the compressive load of plate girder jacking beams, which rested on milled surfaces on the back of the diaphragms at one end of the cross bracing, and the attachment to the bottom chords provided lateral stability. Each yoke assembly had 16 prestressing cables that spanned about 173 ft (52.74 m) between the jacking beams (Fig. 11.28).

To erect each yoke assembly, workers removed some of the bolts from the existing chord splice, leaving enough to carry the dead load without slipping. They mounted the diaphragms with the jacking beams and attached bracing using new bolts. Then they threaded the cable through cutouts in the existing floor

Fig. 11.27 Damaged bottom chord (arrow) had to be replaced along with other members. (Courtesy of the Engineering News Record)

Fig. 11.28 Brackets attached above and below chord on each side of damaged area. (Courtesy of the Engineering News Record)

beams. Cables were tensioned to 1050 lb (the dead-load tension in the bottom chord). The workers replaced the damaged bottom chord section in two halves. They made a new bolted splice in the bottom chord at one end and used the existing splice at the other. After replacing each chord, the workers removed the yokes. Then the splices were restored.

The use of prestressing steel for the yoke resulted in substantial savings. Total weight of the steel, the brackets, and the prestressing cables was 49 tons compared to 101 tons using mill fabrication of specially rolled steel. This reduction in weight resulted in less handling and a reduction in erection cost. Also, the jacking required was one-fourth of a more conventional yoke, which does not use prestressed cables. And because the prestressed cables could span the long distance between the existing field splices, the brackets were removed from the repair area and drilling was reduced. The contractor also used the same type of yoke to replace a damaged section of one of the diagonals.

REFERENCES

1. Lee, D. H., "Prestressed Concrete Bridges and Other Structures," *The Structural Engineer*, vol. 30, no. 12, December 1952, pp. 302–313.
2. Berridge, P. S. A., and Lee, D. H., "Prestressing Restores Weakened Truss Bridge," *Civil Engineering*, vol. 26, no. 9, September 1956, pp. 578–579.
3. Knee, D. W., "The Prestressing of Steel Girders," *The Structural Engineer*, vol. 44, no. 10, October 1966, pp. 351–353.

4. Sterian, D. S., "Introducing Artificial Initial Forces into Steel Bridge Decks," *Acier-Stahl-Steel*, vol. 34, no. 1, January 1969, pp. 31–37.

5. Kandall, C., "Increasing the Load Carrying Capacity of Existing Steel Structures," *Civil Engineering*, vol. 38, no. 9, October 1968, pp. 48–57.

6. Vernigana et al., "Bridge Rehabilitation and Strengthening by Continuous Post-Tensioning," *Prestressing Concrete Institute Journal*, vol. 14, no. 2, April 1969, pp. 88–104.

7. Belenya, E. I., and Gorevskii, D. M., "The Analysis of Steel Beams Strengthened by a Tie Rod," *International Civil Engineering*, vol. II, no. 9, March 1972, pp. 412–419.

8. Palmer, R. A., "Emergency Bridge Repair in Minnesota," *Journal of the Prestressed Concrete Institute*, vol. 19, no. 1, January–February 1974, pp. 73–75.

9. Anon., "How to Keep That Old Bridge Up," *American City and County*, vol. 96, no. 1, January 1981, pp. 29–32.

10. Benthin, K., Letter and Sketch Plans for Strengthening Minnesota Bridge, No. 3699, 1975.

11. Joy, D. D., Senior Vice-President T. Y. Lin International, unpublished letter to C. Pestotnise, Bridge Engineer, Iowa Department of Transportation, October 2, 1980.

12. Lamberson, E. A., *Post-tensioning Concepts for Strengthening and Rehabilitation of Bridges and Special Structures*, DYWIDAG Systems International, USA, Inc., Lincoln Park, New Jersey, 1983.

13. Beck, B. L., Klaiber, F. W., and Sanders, W. W., Jr., *Field Testing of Country Road 54 Bridge over Anclote River, Pasco County, Florida*, Engineering Research Institute, Iowa State University, Ames, Iowa, 1984.

14. American Association of State Highway and Transportation Officials, *Standard Specifications for Highway Bridges*, 13th ed., American Association of State Highway and Transportation Officials, Washington, D.C., 1983.

15. Mancarti, G. D., "Strengthening California's Bridges by Prestressing," *Transportation Research Record* 950, vol. 1, TRB, 1984, pp. 183–187.

16. Troitsky, M. S., Zielinski, Z. A., and Pimprikar, M. S. "Rehabilitation of Steel Bridges by Prestressing Technique," Canadian Society for Civil Engineering, Annual Conference, Saskatoon, Saskatchewan, May 27–31, 1985, pp. 25–43.

17. Park, S. H., *Bridge Rehabilitation and Replacement (Bridge Repair Practice)*, Trenton, N.J., P.O. Box 7474, 1984, pp. 189–191.

18. Bakht, B. et al., *Task Committee on Strength of Old Truss Bridges of the Working Committee on Safety of Bridges Under the Technical Committee on Structural Safety and Reliability, Repair and Strengthening of Old Steel Truss Bridges*, American Society of Civil Engineers, New York, 1979, pp. 1–137.

19. Kandall, op. cit., pp. 48–51.

20. Subcommittee 3 on Prestressed Steel of Joint ASCE-AASHO Committee, "Development and Use of Prestressed Steel Flexural Members," Proceedings ASCE Str. Division, vol. 94, no. ST9, 1968 pp. 2033–2060.

21. Bustin, J., "Reinforcing of Steel Structures by Prestressing," Inz. Stavby, no. 6, 1964, (in Czechoslovakian).

22. Burren, W. H., and Day, H. B., "King's Bridge, Melbourne; Restoration Works," *The Journal of the Institution of Engineers, Australia*, vol. 40, December, 1968, pp. 279–286.

23. Klaber, F. W., Dunker, K. F., and Sanders, W. W., Jr., "Feasibility Study of Strengthening Existing Single Span Steel Beam Concrete Deck Bridges," Department of Civil Engineering, Engineering Research Institute, Iowa State University, Ames, Final Report, June 1981, pp. 1–141.

24. Dunker, K. F., et al., "Strengthening of Existing Single Span Steel Beam and Concrete Deck Bridges," Department of Civil Engineering, Engineering Research Institute, Iowa State University, Ames, Final Report—part II, March 1985, pp. 1–146.

25. Dunker, K. F., Klaiber, F. W., and Sanders, W. W., Jr., "Design Manual for Strengthening Single-Span Composite Bridges by Post-tensioning," Department of Civil Engineering, Engineering Research Institute, Iowa State University, Ames, Final Report—Part III, March 1985, pp. 1–101.

26. Dunker, K. F., "Strengthening of Simple Span Composite Bridges by Post-Tensioning," Ph.D. Dissertation, Iowa State University, Ames, Iowa, 1986, pp. 1–247.

27. Mancarti, G. D., "Strengthening California's Steel Bridges by Prestressing," Transportation Research Record 950, Second Bridge Engineering Conference Vol. 1, Washington, D.C., 1984, pp. 183–187.

28. Sterian, D., "Introducing Artificial Initial Forces into Bridge Decks," *Acier-Stahl-Steel*, no. 1, 1969, pp. 31–37.

29. Berridge, P. S. A., "Prestressing Strengthens a Wrought-Iron Bridge," *Civil Engineering*, August 1957, pp. 38–39.

30. Berridge and Lee, op. cit., pp. 48–49.

31. Muller, Th., "Rebuilding of Highway Bridge Over the Aare River at Aarwangln," Schweizerische Bauzeitung, no. 11, March 1969, pp. 199–203 (in German).

32. Beauchamp, J. C., Chan, M. Y. T., and Pion, R. H., "Repair and Evaluation of a Damaged Truss Bridge—Lewes, Yukon River," *Canadian Journal of Civil Engineering*, vol. 11, no. 3, 1984, pp. 494–504.

33. Ibid.

34. Lamberson, E. A., "Post-tensioning Concepts for Strengthening and Rehabilitation of Bridges," Structures Congress ASCE, Houston, Texas, October 17–19, 1983, Reprint SC-10.

35. "Prestressed Yoke Assembly Cuts Time and Cost on Bridge Repair," *Engineering News-Record*, July 26, 1973, p. 22.

Index

Index

Conversion Factors
From Imperial Units to SI - Units

Units	Imperial Units	SI - Units
LENGTH	1 in	25.40 mm
	1 in	0.0254 m
	1 ft	304.799 mm
	1 ft	0.3048 m
AREA	1 in^2	645.2 mm^2
	1 in^2	6.451cm^2
	1 ft^2	0.0929 m^2
VOLUME	1 in^3	16.387 mm^3
	1 in^3	16.387 cm^3
	1 ft^3	0.0283 m^3
MASS	1 lb	0.454 kg
	1 ton (2000 lb)	907.2 kg

Conversion Factors
From Imperial Units to SI - Units

Units	Imperial Units	SI - Units
MOMENT OF INERTIA	1 in^4	416.200 mm^4
	1 in^4	41.62 cm^4
	1 in^4	0.4162 x 10^{-6}m^4
SECTION MODULUS	1 in^3	16.387 mm^3
FORCE	1 lb	4.448 N
	1 kip (k)	4.448 kN
STRESS	1 psi	6.895 kPa (kN/m^2)
	1 psi	0.070 kg/cm^2
	1 ksi	6.895MN/m^2
	1 psf	47.88 N/m^2
	1 psf	47.88 Pa
	1 ksf	47.88 kN/m^2
	1 ksf	47.88 kPa